Grundlehren der mathematischen Wissenschaften 250

A Series of Comprehensive Studies in Mathematics

Grundlehren der mathematischen Wissenschaften

A Series of Comprehensive Studies in Mathematics

A Selection

continued after Index

V.I. Arnold

Geometrical Methods in the Theory of Ordinary Differential Equations

Second Edition

Translated by Joseph Szücs
English Translation Edited by Mark Levi

With 162 Illustrations

Springer-Verlag
New York Berlin Heidelberg
London Paris Tokyo

V.I. Arnold
Steklov Mathematical Institute
Vavilova 42
117966 Moscow GSP-1
U.S.S.R.

Translator
Joseph Szücs
2303 Hollywood
Galveston, TX 77551
U.S.A.

Editor of the English translation

Mark Levi
Department of Mathematics
Boston University
Boston, MA 02215
U.S.A.

Mathematics Subject Classification (1980): 34CXX, 34D10

Library of Congress Cataloging-in-Publication Data
Arnol'd, V.I. (Vladimir Igorevich)
  Geometrical methods in the theory of ordinary
differential equations.
  (Grundlehren der mathematischen Wissenschaften;
250)
  Translation of: Dopolnitel'nye glavy teorii
obyknovennykh differenfsial'nykh uravneniĭ.
  1. Differential equations. I. Levi, Mark,
1951–   . II. Title. III. Series.
QA372.A6913 1988   515.3′5   87-27502

Original Russian edition: *Dopolnitel'nye glavy teorii obyknovennykh differentsial'nykh uravneniĭ.*
Nauka, Moscow, 1978.

Typeset by Asco Trade Typesetting Ltd., Hong Kong.
Printed and bound by R.R. Donnelley & Sons, Harrisonburg, Virginia.
Printed in the United States of America.

9 8 7 6 5 4 3 2 1

ISBN 0-387-96649-8 Springer-Verlag New York Berlin Heidelberg
ISBN 3-540-96649-8 Springer-Verlag Berlin Heidelberg New York

Preface to the Second Edition

Since 1978, when the first Russian edition of this book appeared, geometrical methods in the theory of ordinary differential equations have become very popular. A lot of computer experiments have been performed and some theorems have been proved. In this edition, this progress is (partially) represented by some additions to the first English text. I mention here some of these recent discoveries.

1. The Feigenbaum universality of period doubling cascades and its extensions—the renormalization group analysis of bifurcations (Percival, Landford, Sinai, ...).
2. The Żołądek solution of the two-parameter bifurcation problem (cases of two imaginary pairs of eigenvalues and of a zero eigenvalue and a pair).
3. The Iljashenko proof of the "Dulac theorem" on the finiteness of the number of limit cycles of polynomial planar vector fields.
4. The Ecalle and Voronin theory of holomorphic invariants for formally equivalent dynamical systems at resonances.
5. The Varchenko and Hovanski theorems on the finiteness of the number of limit cycles generated by a polynomial perturbation of a polynomial Hamiltonian system (the Dulac form of the weakened version of Hilbert's sixteenth problem).
6. The Petrov estimates of the number of zeros of the elliptic integrals responsible for the birth of limit cycles for polynomial perturbations of the Hamiltonian system $\ddot{x} = x^2 - 1$ (solution of the weakened sixteenth Hilbert problem for cubic Hamiltonians).
7. The Bachtin theorems on averaging in systems with several frequencies.
8. The Davydov theory of normal forms for singularities of implicit differential equations and relaxation oscillations.
9. The Neistadt and Cary–Escande–Tennyson theory of adiabatic invariant's change under separatrix crossings (explaining, according to Wisdom, the Kirkwood gaps in the distribution of asteroids).
10. The Neistadt theory of dynamical bifurcations.

The problem of bifurcations at $1:4$ resonance seems to be still unsolved, but I present the conjectural answer supported by both computer experiments and asymptotic analysis.

I mention here some other important recent results:

(1) the bifurcation theory of fundamental systems of solutions of linear equations (related to the Schubert stratification of the Grassmannians and to the Weierstrass points on algebraic curves) by M. Kazarian;
(2) the theory of normal forms of vector fields with Jordan linear part (related to sl(2)-modules) by Bogaevski, Povzner, and Givental;
(3) the bifurcation theory of cycles in reversible systems (related to the metamorphoses of the umbrella's sections) by M. Sevrjuk;
(4) the theory of nonoscillatory linear equations (related to the geometry of the Schubert stratification of the flag manifolds);
(5) the classification of the local topological bifurcations in generic gradient systems depending on three parameters (related to the Thom conjecture on catastrophes) by B. Hessin;
(6) the theory of versal deformations for the vector fields on a line (related to the differential forms of complex degree) by V. Kostov;
(7) the tunelling asymptotics in systems with many competing attractors (related to the Fokker–Planck equation and to the Witten inequalities) by V. Fok;
(8) the bifurcation theory for planar homogeneous vector fields (related to the higher dimensional umbrellas) by B. Hessin.

These subjects need too much space to be discussed here.

The reader may find more details and an extensive bibliography, for many of the subjects discussed in this textbook, in *The Encyclopaedia of Mathematical Sciences*, Volumes 1, 3, and 5, also published by Springer-Verlag.

Moscow   V. ARNOLD
September 10, 1987

Preface to the First Edition

Newton's fundamental discovery, the one which he considered necessary to keep secret and published only in the form of an anagram, consists of the following: *Data aequatione quotcunque fluentes quantitae involvente fluxiones invenire et vice versa.* In contemporary mathematical language, this means: "It is useful to solve differential equations".

At present, the theory of differential equations represents a vast conglomerate of a great many ideas and methods of different nature, very useful for many applications and constantly stimulating theoretical investigations in all areas of mathematics.

Many of the routes connecting abstract mathematical theories to applications in the natural sciences lead through differential equations. Many topics of the theory of differential equations grew so much that they became disciplines in themselves; problems from the theory of differential equations had great significance in the origins of such disciplines as linear algebra, the theory of Lie groups, functional analysis, quantum mechanics, etc. Consequently, differential equations lie at the basis of scientific mathematical philosophy (Weltanschauung).

In the selection of material for this book, the author intended to expound basic ideas and methods applicable to the study of differential equations. Special efforts were made to keep the basic ideas (which are, as a rule, simple and intuitive) free from technical details. The most fundamental and simple questions are considered in the greatest detail, whereas the exposition of the more special and difficult parts of the theory has been given the character of a survey.

The book begins with the study of some special differential equations integrable by quadrature. Attention is paid mainly to connections with general mathematical ideas, methods, and concepts (resolution of singularities, Lie groups, and Newton diagrams) on the one hand, and to applications to the natural sciences on the other, rather than to the formal cookbook aspect of the elementary theory of integration.

The theory of partial differential equations of the first order is considered by means of the natural contact structure in the manifold of 1-jets of functions. The necessary elements of the geometry of contact structures are developed incidentally, making the entire theory independent of other sources.

A significant portion of the book is concerned with methods which are usually called *qualitative*. The recent development of the qualitative theory of differential equations, originated by Poincaré, led to the realization that similar to the fact that the explicit integration of differential equations is generally impossible, the qualitative study of general differential equations with a multidimensional phase space turns out to be impossible. The book discusses the analysis of differential equations from the point of view of structural stability, that is, the stability of the qualitative picture with respect to a small change in the differential equations. The basic results obtained after the first publications of Andronov and Pontrjagin in this area are expounded: the elements of the theory of structurally stable Anosov systems, all trajectories of which are exponentially unstable, and Smale's theorem on the nondensity of structurally stable systems. We also discuss the significance of these mathematical discoveries to applications. (We speak of the description of stable chaotic regimes of motion like turbulence.)

The most powerful and frequently applicable methods of study of differential equations are the various asymptotic methods. We develop the basic ideas of the averaging method going back to the work of the founders of celestial mechanics and widely usable in all those areas of application, where a slow evolution has to be separated from fast oscillations (Bogoljubov, Mitropol'skiĭ, and others).

In spite of the abundant research in averaging, in the problem of evolution even for the simplest multifrequency systems, everything is not entirely clear. We give a survey of the work concerning passage through resonances and capture to resonance in an attempt to illuminate the problem.

The basis of the averaging method is the idea of annihilating perturbations by means of an appropriate choice of the coordinate system. This very idea lies at the basis of the theory of Poincaré normal forms. The method of normal forms is the fundamental method of the local theory of differential equations, which describes the behavior of phase curves in the neighborhood of a singular point or a closed phase curve. In this book, we describe the main results of the method of Poincaré normal forms, including a proof of Siegel's fundamental theorem on the linearization of a holomorphic mapping.

Important applications of the method of Poincaré normal forms come across not only in the study of a single differential equation, but also in bifurcation theory, where the subject of research is a family of equations depending on parameters.

Bifurcation theory studies the qualitative change under the variation of the parameters on which the system depends. For general values of the parameters, we usually have to deal with generic systems (all singular points are simple, etc.). However, if a system depends on parameters, then for some values of the parameters we cannot avoid degeneracies (for example, the fusion of two singular points of a vector field).

In a one-parameter system, we generically encounter only simple degeneracies (those which we cannot get rid of by a small perturbation of the family). Consequently, there arises a hierarchy of degeneracies according to the codimensions of the corresponding surfaces in the function space of all systems under study: in one-parameter generic families, only degeneracies corresponding to surfaces of codimension 1 occur, and so on.

Recent progress in bifurcation theory is connected with the application of ideas and methods of the general theory of singularities of differentiable mappings due to Whitney.

This book concludes with a chapter on bifurcation theory, in which the methods developed in the preceding chapters are applied, and main results obtained in this field, beginning with the fundamental work of Poincaré and Andronov, are described.

In discussing all of these subjects, the author attempts to avoid the axiomatic-deductive style, with its unmotivated definitions concealing the fundamental ideas and methods; similar to parables, they are explained only to disciples in private.

The axiomization and algebraization of mathematics, after more than 50 years, has led to the illegibility of such a large number of mathematical texts that the threat of complete loss of contact with physics and the natural sciences has been realized. The author attempts to write in such a way that this book can be read by not only mathematicians, but also all users of the theory of differential equations.

We only assume a little general mathematical knowledge on the part of the reader, let us say roughly the first two courses of a university program; for example, familiarity with the textbook V. I. Arnold, *Ordinary Differential Equations*, Moscow, Nauka, 1974 [in English, Cambridge, MA, MIT Press, 1973, 1978]* is sufficient (but not necessary).

The exposition is developed in such a way that the reader can omit passages that turn out to be difficult for him, without much harm to the understanding of what follows: as much as possible, we avoid references from one chapter to another, and even from one paragraph to another.

The content of this book constitutes the material of a series of mandatory and special courses delivered by the author at the Department of Mechanics and Mathematics of Moscow State University, 1970–1976, to students of mathematics in grades II–III, and to mathematicians working in applications.

The author expresses his gratitude to students O. E. Hadin, A. K. Koval'dzhi, E. M. Kaganova, and to Professor Ju. S. Il'jašenko, whose notes were very useful in the preparation of this book. The notes of a special course composed by Il'jašenko and the notes of the lectures given in the experimental group have been in the department library for a number of

---

*In the exposition of some special questions, we have also used or recalled elementary information on differential forms, Lie groups, and functions of a complex variable. This information is not necessary for the understanding of most of the book.

years. The author is grateful to the many readers and students of these courses for a series of valuable remarks used in the preparation of the book. The author is grateful to referees D. V. Anosov and V. A. Pliss for a careful and helpful review of the manuscript.

June, 1977   V. ARNOLD

Contents

Notation

$\mathbb{R}$	the set of real numbers
$\mathbb{C}$	the set of complex numbers
$\mathbb{Z}$	the set of integers
$\mathbb{R}^n$	the n-dimensional real linear space
$\exists$	there exists
$\forall$	for every
$a \in A$	the element a of the set A
$A \subset B$	the subset A of the set B
$A \cap B$	intersection of the sets A and B
$A \cup B$	union of the sets A and B
$A \backslash B$	difference of the sets A and B (the part of A outside B)
$A \times B$	direct product of the sets A and B (the set of pairs (a, b), $a \in A, b \in B$)
$A \oplus B$	direct sum of linear spaces
$f : A \to B$	a mapping f of A into B
$x \mapsto y$ or $y = f(x)$	the mapping f maps the element x onto the element y
Im f or $f(A)$	image under the mapping f (but Im z is the imaginary part of z)
$f^{-1}(y)$	complete inverse image of the point y under the mapping f (the set of all x for which $f(x) = y$)
Ker f	kernel of the linear operator f (the complete inverse image of zero)
$\dot{f}$	rate of change of the function f (derivative with respect to time t)
$f', f_*, df/dx,$ Df/Dx	derivative of the mapping f
$T_x M$	the tangent space of the manifold M at the point x
$A \Rightarrow B$	assertion A implies B
$A \Leftrightarrow B$	assertions A and B are equivalent
$\omega_1 \wedge \omega_2$	exterior product of the differential forms ω_1 and ω_2
$f \circ g$	composition of mappings $[(f \circ g)(x) = f(g(x))]$
$L_v f$	derivative of the function f in the direction of the vector field v

Chapter 1

Special Equations

In the study of differential equations, methods from all fields of mathematics are used. In this chapter, we discuss selected special equations and types of equation. Special attention is paid, on the one hand, to the significance in applications of these equations and, on the other hand, to the connection between research methods and various general mathematical problems (resolution of singularities, Newton diagrams, Lie groups of symmetries, etc.). This chapter concludes with the elementary theory of the one-dimensional stationary Schrödinger equation and the geometric theory of a nonlinear equation of the second order.

§ 1. Differential Equations Invariant under Groups of Symmetries

In this section, general arguments are discussed on which the methods of integration of differential equations in explicit form are based. As an example, we discuss the theory of similarity, i.e., the theory of homogeneous and quasihomogeneous equations.

A. Groups of Symmetries of Differential Equations

Let us consider a vector field v in a phase space U.

Definition. A diffeomorphism $g: U \to U$ is called a *symmetry* of v if it transforms v into itself:

$$v(gx) = g_{*x}v(x).$$

Then v is said to be *invariant* under g.

EXAMPLES

1. A vector field with components independent of x on the (x, y)-plane is invariant under translations along the x-axis (Fig. 1).

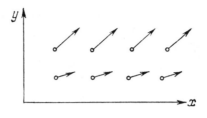

Figure 1.

2. The vector field $x\partial_x + y\partial_y$ on the Euclidean (x, y)-plane is invariant under the dilations $g(x, y) = (\lambda x, \lambda y)$ and rotations.

All the symmetries of a given field form a group.

Exercise. Determine the symmetry group of the field $x\partial_x + y\partial_y$ in the coordinate plane (x, y).

Let us consider a direction field in the extended phase space.

Definition. A diffeomorphism of the extended phase space is called a *symmetry of the direction field* if it maps the field into itself. The direction field is then said to be *invariant* with respect to this diffeomorphism.

EXAMPLES

1. The direction field of the equation $\dot{x} = v(x)$ is invariant under translations along the t-axis (Fig. 2a).

2. The direction field of the equation $\dot{x} = v(t)$ is invariant under translations along the x-axis (Fig. 2b).

Definition. The differential equation $\dot{x} = v(x)$ (respectively, $\dot{x} = v(x, t)$) is said to be *invariant* under the diffeomorphism g of the phase space (respectively, of the extended phase space) if the vector field v (respectively, the direction field v) is invariant under this diffeomorphism g. The diffeomorphism g is then called a *symmetry* of the given equation.

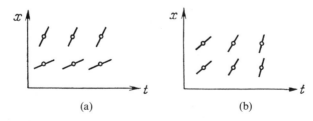

(a)   (b)

Figure 2.

Theorem. *Any symmetry of a given equation transforms phase (integral) curves of the equation into phase (integral) curves of the same equation.*

◀ Let $x = \varphi(t)$ be a solution of the equation $\dot{x} = v(x)$ and let g be a symmetry. Then $x = g(\varphi(t))$ is also a solution. Consequently, symmetries transform phase curves into phase curves. The proof is analogous for integral curves. ▶

EXAMPLE

The family of integral curves of the equation $\dot{x} = v(t)$ is transformed into itself by translations along the x-axis, and that of the equation $\dot{x} = v(x)$ by translations along the t-axis.

The following examples are frequently encountered in applications under the names "theory of similarity", "comparison of dimensions", or "scaling".

B. Homogeneous Equations

Definition. The direction field in the plane without the origin 0 is said to be *homogeneous* if it is invariant under all the dilations

$$g^{\lambda}(x, y) = (e^{\lambda}x, e^{\lambda}y), \qquad \lambda \in \mathbb{R}.$$

The differential equation $dy/dx = v(x, y)$ is said to be *homogeneous* if its direction field is homogeneous (Fig. 3).
In other words, the directions of the field have to be parallel to each other at all points of the same ray starting from the origin of the coordinate system:

$$v(e^{\lambda}x, e^{\lambda}y) \equiv v(x, y).$$

EXAMPLE

The function f is said to be homogeneous of degree d if $f(e^{\lambda}x, e^{\lambda}y) \equiv e^{\lambda d}f(x, y)$. Any form (homogeneous polynomial) of degree d can serve as an example. Let P and Q be two forms of degree d depending on the variables x and y. The differential equation

$$\dot{x} = P, \qquad \dot{y} = Q$$

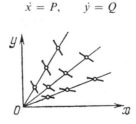

Figure 3.

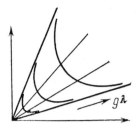

Figure 4.

is given by the vector field (P, Q) in the plane. The corresponding direction field in the domain $P \neq 0$ is the direction field of the homogeneous equation

$$\frac{dy}{dx} = \frac{Q}{P} \quad \left(\text{e.g.,} \ \frac{dy}{dx} = \frac{ax + by}{cx + dy}, \frac{dy}{dx} = \frac{x^2 - y^2}{x^2 + y^2}, \text{etc.} \right)$$

Remark. The domain (of definition) of a homogeneous field does not necessarily have to be the whole punctured plane. Homogeneous fields may be defined in any homogeneous (i.e., invariant under dilations) domains, e.g., in an angular sector with vertex 0, etc.

Theorem. *Every integral curve of an arbitrary homogeneous equation is transformed by any dilation g^λ into an integral curve of the same equation.*

Consequently, given a homogeneous equation, it is sufficient to study only one integral curve in each sector of the plane.

The proof may be obtained by an immediate application of the theorem in § 1A.

Exercise. Let P and Q be forms of degree d. Prove that the phase curves of the system $\dot{x} = P, \dot{y} = Q$ are obtained from each other by dilations (Fig. 4).

If one of these curves is closed and has period T, then the dilation g^λ transforms it into another closed phase curve having period $T/e^{\lambda(d-1)}$.

C. Quasi-Homogeneous Equations and the "Comparison of Dimensions"

Let us fix the real numbers α and β and consider the family of transformations

$$g^s(x, y) = (e^{\alpha s}x, e^{\beta s}y) \tag{1}$$

which dilate in the x and y directions by different amounts.

Note that Eq. (1) defines a one-parameter group of linear transformations in the plane (Fig. 5).

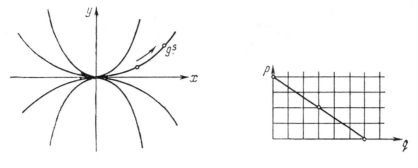

Figure 5.   Figure 6.

Definition. The function f is said to be *quasi-homogeneous of degree d* if

$$f(g^s(x, y)) \equiv e^{ds} f(x, y).$$

EXAMPLE

If $\alpha = \beta = 1$, then we obtain the ordinary homogeneous functions of degree d.

Quasi-homogeneous degrees add under the multiplication of functions. They are also called *weights*. For example, x has weight α, y has weight β, $x^2 y$ has weight $2\alpha + \beta$, and so on. All quasi-homogeneous monomials of a fixed degree can be easily seen in the following *Newton diagram* (Fig. 6). We identify the monomial $x^p y^q$ with the point (p, q) of the integer lattice. Exponents of all possible monomials of degree d are the integer lattice points on the segment given by the equation $d = \alpha p + \beta q$ in the (p, q)-plane.

Exercise. Choose weights in such a way that the function $x^2 + xy^3$ is quasi-homogeneous.

Definition. The differential equation $dy/dx = v(x, y)$ is said to be *quasi-homogeneous (with weights α and β)* if its direction field v is invariant under the transformations in Eq. (1).

From the general theorem of § 1A on symmetries, we can derive the following.

Theorem. *The integral curves of a quasi-homogeneous equation are obtained from each other by using the transformations of Eq. (1).*

Exercise. Prove that the function $v(x, y)$ is the right-hand side of a quasi-homogeneous differential equation (with weights (α, β)) if and only if it is quasi-homogeneous of degree $d = \beta - \alpha$.

Remark. The above definitions and theorems can be easily extended to the case of more than two variables and to differential equations of order greater than 1. In particular, it is easy to prove the following.

Theorem. *Let γ: $y = y(x)$ be a curve in the (x, y)- plane and let $d^k y/dx^k = F$ at the point (x_0, y_0). Then for the curve $g^s \gamma$ we have*

$$\frac{d^k y}{dx^k} = e^{(\beta - k\alpha)s} F$$

at the corresponding point.

In other words, $d^k y/dx^k$ is transformed, as is y/x^k, by the transformations of Eq. (1), which in turn explains the convenience of the notation $d^k y/dx^k$.

Exercise. Prove that if a particle in a homogeneous force field of degree d moves along the trajectory Γ in time T, then the same particle moves along the dilated trajectory $\lambda\Gamma$ in time

$$T' = \lambda^{(1-d)/2} T.$$

Solution. The Newtonian equation $d^2 x/dt^2 = F(x)$, where F is homogeneous of degree d, is transformed into itself by appropriate transformations of the form of Eq. (1). Namely, it is sufficient to choose weights α (for x) and β (for t) such that $\alpha - 2\beta = \alpha d$. Then $\beta = ((1 - d)/2)\alpha$. Consequently, $T' = \lambda^{(1-d)/2} T$ corresponds to the dilation $x' = \lambda x$.

Exercise. Prove Kepler's third law: The squares of the times needed for similar trajectories in the gravitational field are proportional to the cubes of the linear measurements of these trajectories.

Solution. From the solution of the preceding exercise with $d = -2$ (the law of universal gravity), we obtain $T' = \lambda^{3/2} T$.

Exercise. Determine how the period of oscillation depends on the amplitude in the case of a restoring force proportional to the elongation (linear oscillator) and to the cube of the elongation (weak force).

Answer. In the case of a harmonic oscillator, the period does not depend on the amplitude; in the case of a weak oscillator, it is inversely proportional to the amplitude.

Exercise. It is known that a top with a vertical axis has a critical angular velocity: if the angular velocity is greater than the critical velocity, then the top stands up firmly vertically, and if it is less, it falls.

How does the critical angular velocity change if we take the top to the moon, where the gravitational force is six times less than on the earth?

Answer. It decreases by a factor of $\sqrt{6}$.

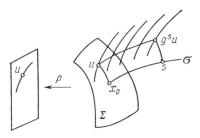

Figure 7.

D. Applications of One-Parameter Groups of Symmetries to Lowering the Order

Theorem. *If a one-parameter group of symmetries of a direction field in $\mathbb{R}^n$ is known, then the problem of integration of the corresponding differential equation reduces to the problem of integrating a differential equation in $\mathbb{R}^{n-1}$.*

In particular, if a one-parameter group of symmetries of a direction field on the plane is known, then the corresponding equation $dy/dx = f(x, y)$ can be integrated explicitly.

◀ Let $\{g^s\}$ be the group of symmetries under consideration. Let us consider the orbits $\{g^s x\}$ of the flow $\{g^s\}$. We can define (at least locally) an $(n-1)$-dimensional *orbit space* (the quotient space with respect to the action of g^s) and a mapping p of the initial space onto the quotient space (p maps the orbits of the flow $\{g^s\}$ into points). It turns out that the initial direction field is mapped by p into a new direction field on the $(n-1)$-dimensional orbit space; one only has to integrate it. ▶

More precisely, consider a point $x_0 \in \mathbb{R}^n$ and assume that the orbit of $\{g^s\}$ going through x_0 is a curve σ. Through x_0 let us draw an $(n-1)$-dimensional local transversal Σ to σ. In the neighborhood of x_0, introduce the local coordinate system (s, u), where the point $g^s u$ of the initial space corresponds to the pair $s \in \mathbb{R}$, $u \in \Sigma$. Then in the neighborhood of x_0, the projection p onto the orbit space and the action of the group g^s of symmetries are given by the formulas

$$p(s, v) = u, \qquad g^{s_1}(s_2, u) = (s_1 + s_2, u)$$

(the points on the surface Σ parameterize the local orbits.)

We note that if the group g^s is given explicitly, then the coordinates (s, u) can be found explicitly. We write the initial differential equation in these coordinates. If our direction field is not tangent to Σ at x_0 (which can always be achieved by the choice of Σ), then in the neighborhood of this point our equation takes the form

$$\frac{du}{ds} = v(s, u).$$

Moreover, $\{g^s\}$ consists of translations along the s-axis; therefore, v *does not depend on* s. The vector field $v(u)$ on Σ defines a direction field on this $(n-1)$-dimensional surface. If we know its integral curves, we can find the solutions to the equation $du/ds = v(u)$ (by quadrature) and, consequently, the integral curves of the initial equation.

In the special case $n = 2$, the passage to the coordinates (s, u) immediately leads to the integrable equation $du/ds = v(u)$.

Remark. In practice, it is often more convenient to use an appropriate function z of s instead of s itself. In such a coordinate system an equation having the group g^s of symmetries will read

$$\frac{du}{dz} = v(u)f(z)$$

(in case $n = 2$, we obtain an equation with separated variables).

EXAMPLE

A homogeneous equation reduces to an equation with separated variables in polar coordinates and also in the coordinates $u = y/x$, $z = x$ (Fig. 8a).

Here $\{g^s\}$ is the one-parameter group of dilations by e^s, Σ is the circle $x^2 + y^2 = 1$ in the case of polar coordinates and the straight line $x = 1$, $z = e^s$ in the case of the other coordinate system.

Exercise. In what coordinates can the quasi-homogeneous equation

$$\frac{dy}{dx} = v(x, y)$$

be integrated explicitly, where the weight of x is α and that of y is β (so that v is a quasi-homogeneous function of degree $\beta - \alpha$).

Solution. We may take $u = y^\alpha/x^\beta$, $z = x$ (in a domain where $x \neq 0$) (cf., Fig. 8b.)

Exercise. Give the equation with separated variables explicitly to which the equation of the preceding exercise reduces in the coordinates (u, z).

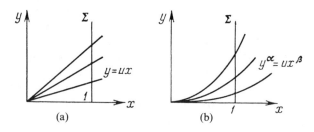

Figure 8.

Solution. $y^\alpha = ux^\beta$; therefore, $\alpha y^{\alpha-1} dy = x^\beta du + \beta u x^{\beta-1} dx$. If $dy = v\, dx$, then $\alpha y^{\alpha-1} v\, dx = x^\beta du + \beta u x^{\beta-1} dx$, i.e.,

$$\frac{du}{dx} = \frac{\alpha y^{\alpha-1} v - \beta u x^{\beta-1}}{x^\beta}.$$

But $v(x, y) = x^{(\beta/\alpha)-1} w(u)$; therefore,

$$\frac{du}{dx} = \frac{\alpha u^{(\alpha-1/\alpha)} w(u) - \beta u}{x}.$$

§ 2. Resolution of Singularities of Differential Equations

Here we briefly describe an important general mathematical technique called resolution of singularities or the σ-process.

A. σ-Process

In the neighborhood of a nonsingular point, all vector fields are simple and have identical structures.

For the study of fine details of all mathematical objects near a singular point a special apparatus is devised, having fine resolution similar to a microscope. It is called the resolution of singularities. From the analytical point of view, we speak of the choice of a coordinate system near the singular point in which, to a small displacement near the singularity, there corresponds a great change in coordinates.

The polar coordinate system has this property; however, the passage to polar coordinates requires transcendental (trigonometric) functions; therefore, algebraically, it is often more convenient to use another procedure: the so-called σ-process or resolution of singularities.

We begin with an auxiliary construction. Let $p : \mathbb{R}^2\backslash O \to \mathbb{R}P^1$ be the standard fibration defining the projective line. (The projective line is a manifold whose points are the straight lines in the plane going through the origin of the coordinate system. To a point in the plane, the mapping p assigns the straight line connecting the point with the origin.)

Let us consider the graph Γ of the mapping p. This graph represents a smooth surface in the Cartesian product $(\mathbb{R}^2\backslash O) \times \mathbb{R}P^1$ (Fig. 9). Embedding the punctured plane in the plane, we may consider the graph as a smooth surface Γ in the Cartesian product $\mathbb{R}^2 \times \mathbb{R}P^1$. The natural projection $\pi_1 : \mathbb{R}^2 \times \mathbb{R}P^1 \to \mathbb{R}^2$ maps Γ onto the punctured plane $\mathbb{R}^2\backslash O$ diffeomorphically. (In order to picture this more clearly, it is useful to note that Γ locally has the form of a spiral ladder; globally, the projective line is diffeormorphic to the circle, and the product $\mathbb{R}^2 \times \mathbb{R}P^1$ to the interior of the torus.)

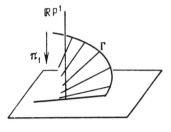

Figure 9.   Figure 10.

Theorem. *The closure of the graph Γ of the mapping p in $\mathbb{R}^2 \times \mathbb{R}P^1$ is a smooth surface $\Gamma_1 = \Gamma \cup (O \times \mathbb{R}P^1)$. The surface Γ_1 is diffeomorphic to the Möbius strip (Fig. 10).*

◀ Let (x, y) be the coordinates in the plane, and $u = y/x$ the affine local coordinate in $\mathbb{R}P^1$. Then (x, y, u) is a local coordinate system in $\mathbb{R}^2 \times \mathbb{R}P^1$. In this coordinate system, Γ is given by $y = ux$, $x \neq 0$, and Γ_1 by the local equation $y = ux$. This surface is smooth; it is obtained by adding to the part of Γ covered by our coordinate system the part of the projective line $O \times \mathbb{R}P^1$ falling there.

The proof of the smoothness of Γ_1 can be completed by considering a second coordinate system (x, y, v), where $x = vy$.

The projection $\pi_2 \colon \mathbb{R}^2 \times \mathbb{R}P^1 \to \mathbb{R}P^1$ foliates Γ_1 into straight lines. When the circle $\mathbb{R}P^1$ is traversed once, the corresponding straight line in $\mathbb{R}^2$ turns by the angle π. It follows from this that Γ_1 is a Möbius strip. ▶

Definition. The passage from $\mathbb{R}^2$ to Γ_1 is called a *σ-process with center O* or the *blowing up* of the point O into the straight line $O \times \mathbb{R}P^1$. The mapping $\pi_1 \colon \Gamma_1 \to \mathbb{R}^2$ is called an *antisigma process* or *collapsing* the circle $O \times \mathbb{R}P^1$ into the point O.

The mapping $\pi_1 \colon \Gamma_1 \to \mathbb{R}^2$ restricted to Γ is a diffeomorphism onto the punctured plane. Therefore, all geometric objects in the plane having a singularity at O are carried over to Γ_1. At the same time, singularities may become simpler or may be "resolved".

EXAMPLE

Consider three straight lines passing through the point O. On Γ_1, this corresponds to three straight lines intersecting $\mathbb{R}P^1$ at distinct points (Fig. 11).

Exercise. Consider two curves which are tangent to each other of order n (for example, $y = 0$ and $y = x^2$, $n = 2$). Prove that on Γ_1 two corresponding curves have tangency of order $n - 1$ at the corresponding point O_1 (Fig. 12).

If after the σ-process singularities do not reduce to transversal intersections, then

Figure 11.   Figure 12.

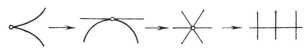

Figure 13.

another σ-process can be performed at the remaining singular points, etc. until all singularities reduce to transversal intersections. It can be proved that every algebraic curve can be resolved (reduced to transversal intersections) in a finite number of steps.

Exercise. Resolve the singularities of the curve $x^2 = y^3$.

Solution. Cf., Fig. 13.

B. Resolution Formulas

In practice, a σ-process means the passage from the coordinates (x, y) to the coordinates $(x, u = y/x)$, where $x \neq 0$, and to the coordinates $(\sigma = x/y, y)$, where $y \neq 0$ (Fig. 14). Let us see what this does to the differential equation given by a vector field in the plane (x, y). We shall assume that the point O is a singular point of our vector field.

Theorem. *A smooth vector field w with singular point at O after the σ-process turns into a vector field on Γ extendible to a smooth field on Γ_1.*

◀ Let w be the field given by the system $\dot{x} = P(x, y)$, $\dot{y} = Q(x, y)$. Using the coordinates $(x, u = y/x)$, we find

$$\dot{x} = P(x, ux), \qquad \dot{u} = \left(\frac{Q(x, ux) - uP(x, ux)}{x} \right).$$

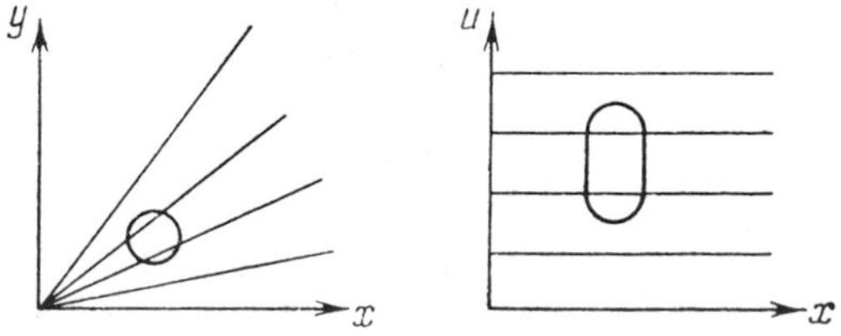

Figure 14.

The right-hand sides are smooth, since $P(0, 0) = Q(0, 0) = 0$. In the other coordinate system ($v = x/y, y$), we also obtain a smooth field. ▶

Remark. It can happen that the vector field obtained in the σ-process vanishes on the whole straight line pasted in the σ-process. Then, one can divide the field by x in the domain of the first coordinate system, and by y in the domain of the second. Division does not change the direction of the vectors of the field. One obtains a direction field on Γ_1 with singular points lying on the pasted line but not filling it out completely. In the neighborhood of every singular point, the direction field is given by a smooth vector field.

To every "entering direction" of phase curves of the initial field at O there corresponds a singular point of such a field on Γ_1 lying on the straight line $\mathbb{RP}^1$ pasted in the σ-process.

If these singular points O_i are not sufficiently simple, then σ-processes can be performed at them. Continuing in this way, we finally arrive at the case where at least one of the eigenvalues of the linearization of the field is different from 0 at every singular point.

In many cases, the first σ-process enables us to analyze the behavior of phase or integral curves near a singular point. For example, the integral curves of a homogeneous equation turn into the integral curves of an equation with separated variables after our change of variables $(x, y) \mapsto (x, u = y/x)$.

C. Example: The Study of a Pendulum with Friction

We illustrate the method by the trivial example of a linear equation. The equation of a pendulum with a friction coefficient k has the form $\ddot{x} + k\dot{x} + x = 0$. In the phase plane, this equation is equivalent to the system

$$\dot{x} = y, \qquad \dot{y} = -ky - x.$$

The homogeneous equation reads

$$\frac{dy}{dx} = -k - \frac{x}{y}.$$

According to the general theory, the variables should separate after the σ-process, i.e., in the coordinate system ($x, u = y/x$). Indeed, $du/dx = -(u^2 + ku + 1)/ux$. Introducing $\log|x| = z$, we obtain

$$\frac{du}{dz} = -k - \left(u + \frac{1}{u}\right).$$

We study the integral curves of this equation for different values of the coefficient $k > 0$. The graph of the function $f = u + 1/u$ is a hyperbola (Fig. 15). Consequently, the graph of the function $-k - f(u)$ has the form shown in Fig. 16. Correspondingly,

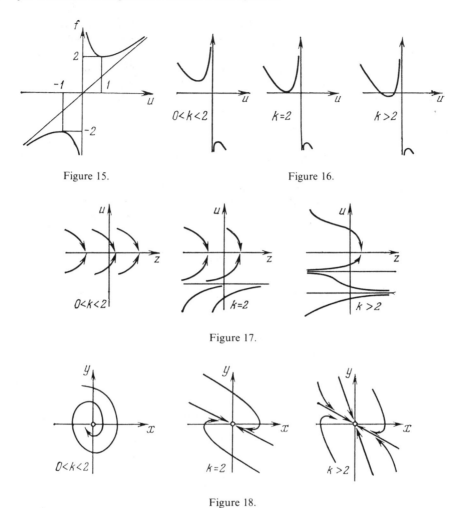

Figure 15.   Figure 16.

Figure 17.

Figure 18.

the integral curves of the equation $du/dz = -k - f(u)$ have the form shown in Fig. 17. Returning to the phase plane (x, y), we obtain Fig. 18.

Hence, for small values of the coefficient of friction ($0 < k < 2$), the pendulum performs an infinite number of oscillations, and for $k \geqslant 2$ the direction of the motion of the pendulum changes no more than once.

Exercise. Determine the phase curves of the equations $\dot{z} = az^n$ and $\dot{z} = a\bar{z}^n$, $z \in \mathbb{C}$.

D. Example: The Period of Small Oscillations

Theorem. *Assume that all phase curves passing through points close to the equilibrium point O are closed. Then the period of oscillation in the neighborhood of O tends to the period of oscillation in the linearized system as the amplitude of the oscillation tends to 0.*

◀ In the σ-process, the closed phase curves going around 0 turn into curves on the Möbius strip closing after two revolutions; that the amplitude of the oscillation converges to 0 corresponds to the fact that the phase curve on the Möbius strip converges to the projective line pasted in the σ-process (the middle line of the Möbius strip).

Since a solution depends continuously on the initial condition, the limit of the period of oscillation, as the amplitude tends to 0, is equal to the doubled period of revolution on the pasted straight line $\mathbb{R}P^1$ in the system obtained in the σ-process. The velocities of motion on the pasted straight line for the field under consideration and for its linearization are identical (cf., the equation for u in § 1B). It is easy to verify that all phase curves of the linearized equation are closed. These closed curves in the linear system are traversed in the same time, since the linear vector field is transformed into itself by dilations of the phase plane. Consequently, the limit of the period of oscillation in the initial system is equal to the limit of the period of oscillation in the linearized system, and, consequently, to the period of oscillation in the linearized system. ▶

Remark. This limit is called the *period of small oscillations.*

Exercise. Calculate the period of small oscillations of the pendulum $\ddot{x} = -\sin x$ near the equilibrium point $x = 0$.

§ 3. Implicit Equations

In this section we consider the basic notions of the theory of differential equations not solved for the derivatives. We will do this using the general theory of singularities of smooth mappings and the geometry of spaces of jets.

A. Basic Definitions

We consider the equation

$$F(x, y, p) = 0, \tag{1}$$

where $p = dy/dx$.

EXAMPLES

(1) $p^2 = x$; (2) $p^2 = y$; (3) $y = px + p^2$. The three-dimensional space with coordinates (x, y, p) is called the *space of 1-jets of the functions* $y(x)$. (Two smooth functions y_1 and y_2 have the same k-jet at the point x_0 if $|y_1(x) - y_2(x)| = o(|x - x_0|^k)$; therefore, the 1-jet of a function is defined by the choice of the point x, the value y at this point, and the value p of the derivative.)

In the space of jets Eq. (1) describes a surface. It turns out that a direction field arises on this surface. Its construction is as follows. Consider a point in the space of jets. The components of the vector ξ applied at this point will be

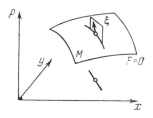

Figure 19.

denoted by $dx(\xi)$, $dy(\xi)$, $dp(\xi)$. Hence, dx, dy, and dp are not mysterious, infinitely small quantities, but completely defined linear functions of ξ.

At the point (x, y, p) of the space of jets, consider the plane consisting of the vectors ξ for which $dy = p\,dx$. In other words, the vector ξ at the point (x, y, p) lies in the indicated plane (Fig. 19) if its projection onto the Euclidean (x, y)-plane forms an angle with the x-axis whose tangent is equal to p. The plane thus constructed is called a *contact plane*. Hence, a contact plane passes through every point of the space of 1-jets; all of these planes form the *contact field of planes* (or, in other words, a *contact structure*) in the space of 1-jets.

Exercise.* Are there any surfaces in the space of 1-jets whose tangent at every point is the contact plane drawn at that point?

Answer. No.

Assume that the surface given by Eq. (1) in the space of 1-jets is smooth. [This is not a strong restriction, since for a generic smooth (infinitely differentiable) function F, the value 0 is not critical and the set of zeros is smooth. If this is not so for the function under consideration, then for almost every arbitrarily small perturbation of F the set of zeros becomes smooth: for example, it is sufficient to add a small constant to F (cf., Sard's theorem, § 10E)].

We consider a point on the smooth surface M given by Eq. (1) and assume that at this point the tangent of the surface does not coincide with the contact plane. Then these two planes intersect in a straight line. Moreover, the tangent and contact planes intersect in straight lines at all nearby points of M, so a direction field arises on M in the neighborhood of the point being considered.

The *integral curves* of Eq. (1) are, by definition, the integral curves of the direction field thus obtained on M. To solve Eq. (1) it is necessary to find these curves. The connection between the integral curves on M and the graphs of solutions of Eq. (1) on the (x, y)-plane is discussed below. We emphasize that the integral curves on M are defined in terms of contact planes and not solutions of Eq. (1).

B. Regular Points and the Discriminant Curve

The direction of the p-axis in the space of jets will be called *vertical direction*. Let M be the smooth surface given by Eq. (1) in the space of jets. Consider the projection

$$\pi \colon M \to \mathbb{R}^2, \qquad \pi(x, y, p) = (x, y)$$

in the vertical direction.

Definition. A point of the surface M is said to be *regular* if it is not a critical point of the mapping π.

In other words, a point of M is regular if the tangent plane at this point is not vertical, or, equivalently, if the projection onto the (x, y)-plane is a diffeomorphism in the neighborhood of the point.

The set of critical values of the mapping π (i.e, the projection of the set of critical points) is called the *discriminant curve* of Eq. (1).

EXAMPLE

The discriminant curve of the equation $p^2 = x$ is the y-axis, and that of the equation $p^2 = y$ is the x-axis (Fig. 20).

Consider a regular point of M. By the implicit function theorem, in the neighborhood of this point M is the graph of a smooth function $p = v(x, y)$.

Theorem. *Projection onto the (x, y)-plane turns the integral curves of Eq. (1) on M in the neighborhood of a regular point exactly into the integral curves of the equation*

$$\frac{dy}{dx} = v(x, y) \tag{2}$$

in the neighborhood of the projection of this point (Fig. 21).

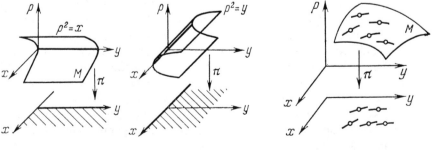

Figure 20.   Figure 21.

◀ By definition, the projection of a contact plane onto the (x, y)-plane is a straight line in the direction field of Eq. (2). Therefore, under the local diffeomorphism π, the direction field of Eq. (1) in the neighborhood of the regular point on M under consideration turns into the direction field of Eq. (2); consequently, the integral curves turn into each other, as well. ▶

Remark. Globally, the projections of the integral curves of Eq. (1) onto the (x, y)-plane are not, in general, integral curves of some direction field. The projections of integral curves of Eq. (1) onto the (x, y)-plane have cusps on the discriminant curve in the general case; however, for some cases of Eq. (1), these projections remain smooth at points of the discriminant curve.

C. Examples

1. $p^2 = x$ (Fig. 22). The surface M is a parabolic cylinder. The discriminant curve is the y-axis. In order to find the integral curves, it is convenient to choose the coordinates p and y and not the coordinates x and y on M (the former coordinate system is global).

   We write down the conditions for the components dx, dy, and dp of the vector at the point (x, y, p) of the surface M and belonging to our direction field:

$$\begin{cases} p^2 = x & \text{(the condition of belonging to } M); \\ 2p\,dp = dx & \text{(the condition of being tangent to } M); \\ dy = p\,dx & \text{(the condition of belonging to the contact plane)}. \end{cases}$$

   Consequently, in coordinates (p, y), the integral curves are determined from the equation $dy = 2p^2\,dp$.

   Hence, the integral curves on M are given by the relations $y + C = \frac{2}{3}p^3$, $x = p^2$. Their projections onto the (x, y)-plane are semicubic parabolas.

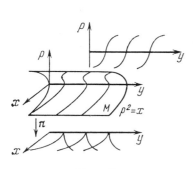

Figure 22.

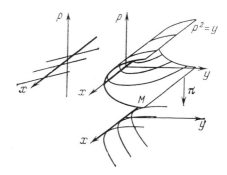

Figure 23.

2. $p^2 = y$ (Fig. 23). Arguing as in the preceding example, we obtain

$$p^2 = y,$$
$$2p\,dp = dy,$$
$$dy = p\,dx.$$

Using the coordinates x and p on M, we obtain $p(dx - 2\,dp) = 0$, whence either

$$p = 0, \quad y = 0 \qquad \text{or} \qquad x = 2p + C, \quad y = p^2.$$

The projections of these curves onto the (x, y)-plane are parabolas tangent to the discriminant curve $y = 0$.

3. (Clairaut's Equation). $y = px + f(p)$ (Fig. 24). The surface M is ruled (its intersections with the planes $p = $ const are straight lines). It is convenient to choose x and p as coordinates on M. We determine the integral curves from the relations

$$y = px + f(p),$$
$$dy = p\,dx + x\,dp + f'\,dp,$$
$$dy = p\,dx.$$

Then $(x + f')\,dp = 0$. The points where $x + f' = 0$ are critical, and the remaining points are regular. On the (x, p)-plane the integral curves are the straight lines $p = $ const $= C$; generally these straight lines intersect the line $x + f' = 0$ of critical points.

The projections of the integral curves onto the (x, y)-plane are the straight lines $y = Cx + f(C)$ tangent to the discriminant curve. (Strictly speaking, the points of intersection with the critical line do not belong to the integral curves on M, since for the equation under consideration, the direction field is not defined at these points: the contact plane is tangent to M.)

The discriminant curve can be determined from the conditions

$$y = px + f(p), \qquad x + f' = 0.$$

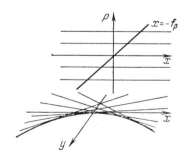

Figure 24.

For example, if $f(p) = -p^2/2$, then the discriminant curve is the parabola $y = x^2/2$, and the projections of the integral curves are its tangents.

Clairaut's equation is connected with important general mathematical notions: the Legendre transformation and projective duality.

D. Legendre Transformation

Let a function f of the variable x be given. The *Legendre transformation** of this function is, by definition, a new function g of a new variable p defined in the following way. Consider the graph of f on the (x, y)-plane. Draw the line $y = px$ with slope p. Determine the point where the graph is the farthest from this straight line in the direction of the ordinate axis. Consider the difference of the ordinates of points of the straight line and the graph. This difference is the value of g at the point p (Fig. 25):

$$g(p) = \sup_{x}(p(x) - f(x)).$$

EXAMPLES

1. Let $f(x) = x^2/2$. Calculating the Legendre transform, we obtain $g(p) = p^2/2$.

2. Let $f(x) = x^\alpha/\alpha$. Then $g(p) = p^\beta/\beta$, where $\alpha^{-1} + \beta^{-1} = 1$. (Here x, p, α, and β are nonnegative.)

Let f be strictly convex ($f'' > 0$) and its derivative define a diffeomorphism of the real line onto itself. Then g is also strictly convex and the supremum is attained at the uniquely determined point where $f'(x) = p$. Points x and p are said to *correspond* to each other under the Legendre transformation.

---

*In the literature several different objects, also associated with the names Minkowski and Young, are called Legendre transforms as well. Nevertheless, we shall not try to be completely pedantic concerning terminology.

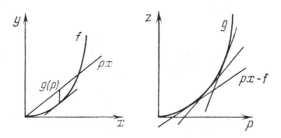

Figure 25.

Theorem. *We have the inequality*

$$f(x) + g(p) \geqslant px.$$

If f is strictly convex and f′ is a diffeomorphism onto, then equality is attained if and only if the points x and p correspond to each other.

◀ The function $px - f(x)$ does not exceed its supremum $g(p)$. ▶

EXAMPLE

For any nonnegative x and p, we have the inequality $px \leqslant x^{\alpha}/\alpha + p^{\beta}/\beta$.

In the corollaries below, we assume that f and g are strictly convex and their derivatives are diffeomorphisms of the real line onto itself.

Corollary. *The Legendre transformation is involutive: the Legendre transform of g(p) is f(x) (with appropriate notation of the coordinate).*

◀ This follows from the fact that the inequality in the preceding theorem is symmetric with respect to f and g. ▶

Corollary. *The passage from the strictly convex function g in Clairaut's equation y = px − g(p) to the function f giving the envelope of solutions by the formula y = f(x) is the Legendre transformation.*

◀ The graph of f is the envelope of its tangents. ▶

Remark. The Legendre transformation for functions of n variables can be defined analogously and has the same properties. If x is a point in $\mathbb{R}^n$, then p is a point in the dual linear space (the space $\mathbb{R}^{n*}$ of linear functions on $\mathbb{R}^n$).

E. Projective Duality

The Legendre transformation is a particular case of a general construction in projective geometry. Let us consider the n-dimensional projective space $\mathbb{RP}^n$.

A point of the projective space can be given by a nonzero vector x of the linear space $\mathbb{R}^{n+1}$ determined up to multiplication by a nonzero number. This definition can briefly be written as

$$\mathbb{RP}^n = (\mathbb{R}^{n+1}\backslash O)/(\mathbb{R}\backslash O).$$

A hyperplane in the projective space consists of those points of the projective space for which the corresponding points of the linear space belong to a hyperplane passing through the origin.

Let us consider the set of all hyperplanes in n-dimensional projective space. This set itself is an n-dimensional projective space in a natural way.

Indeed, a hyperplane in the projective space is given by a homogeneous equation

$$(a, x) = 0, \qquad x \in \mathbb{R}^{n+1}, \qquad a \in \mathbb{R}^{n+1}*\backslash O,$$

where $\mathbb{R}^{n+1}*$ is the space of linear functions on $\mathbb{R}^{n+1}$ (this space is linear of dimension $n + 1$ and called the *conjugate* or the dual space of the initial space $\mathbb{R}^{n+1}$).

Hence, to a hyperplane in the projective space, there corresponds a non-zero vector in $\mathbb{R}^{n+1}*$, determined up to multiplication by a nonzero number.

Consequently, the set of all hyperplanes in $\mathbb{R}P^n$ has the natural structure of a projective space of dimension n:

$$\mathbb{R}P^{n}* = (\mathbb{R}^{n+1}*\backslash O)/(\mathbb{R}\backslash O).$$

The projective space of hyperplanes in the projective space $\mathbb{R}P^n$ is called the dual space of $\mathbb{R}P^n$ and denoted by $\mathbb{R}P^{n}*$. For example, the space of all straight lines on the projective plane is itself a projective plane, dual to the initial plane.

We note that duality is a "mutual" notion, i.e., $\mathbb{R}P^{n}** = \mathbb{R}P^n$. This follows from the symmetry of a and x in the equation $(a, x) = 0$ of a hyperplane.

EXAMPLES

All straight lines passing through one point of the projective plane form a straight line in the dual plane, as is easily seen. All straight lines passing through a given point of the projective plane inside an angle with vertex at this point form an interval in the dual plane.

All tangents of a nondegenerate second-degree curve on the projective plane form a nondegenerate second-degree curve on the dual space. In general, all tangents of any smooth curve form a (not necessarily smooth) curve in the dual plane. This curve is said to be dual to the initial curve.

Theorem. *The graphs of a strictly convex function and its Legendre transform are projectively dual to each other.*

◀ Consider all straight lines on the affine (x, y)-plane not parallel to the y-axis. These straight lines themselves form a plane: a straight line can be given by the equation $y = px - z$, and we can consider (p, z) as affine coordinates in the new plane. In this case, the Legendre transformation

Figure 26.

reduces to the passage from the graph of a function f to the family of tangents of this graph; when the point on the (x, y)-plane runs over the graph of f, the tangent of the graph of f runs over a curve in the (p, z)-plane which is the graph of the Legendre transform, $z = g(p)$. ▶

Therefore, the Legendre transformation is the passage from a curve to the projectively dual curve given in affine coordinates.

EXAMPLE

Let the graph of f be a convex polygon. A supporting line is, by definition, a straight line which is below the graph and which has a point in common with the graph.*
Consider all supporting lines of the convex polygon.

It is easy to see that they form a convex polygon in the dual plane. Indeed, the supporting lines form an angle at every vertex of the initial polygon and, consequently, form a segment or closed interval in the dual plane. Similarly, the vertices in the dual plane are obtained from the sides of the initial polygon.

Projective duality allows us to consider more general cases than that of the Legendre transformation.

Exercise. Determine the curve projectively dual to the curve in Fig. 26.

Hint. To the double tangents of the initial curve, there correspond points of self-intersection of the dual curve. Consequently, the dual curve has four points of self-intersection.

To points of inflection of the initial curve there correspond cusps of the dual curve. Indeed, if $f = x^3$, then the tangent of the graph at the point $x = t$ is given by the coordinates $p = 3t^2$, $z = 2t^3$. These relations define a curve with cusp in the (p, z)-plane. Hence, the dual curve has eight cusps, two between any pair of consecutive self-intersection.

Moreover, consider the initial curve as a pair of intersecting ellipses somewhat smoothed near the points of intersection (the part of each ellipse inside the other is eliminated).

---

*In general, a *supporting hyperplane* of a convex body is a plane which has a common point with the body and is such that the body lies in one of the half-spaces into which the plane divides the space.

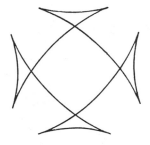

Figure 27.

The dual curve is also connected with a pair of ellipses. To the points of intersection of the initial ellipses there correspond double tangents of the dual ellipses. It is easy to see how to construct the dual curve from the pair of ellipses with their double tangents (Fig. 27).

F. The Legendre Transformation and Dual Norms

Definition. A *norm* in $\mathbb{R}^n$ is, by definition, a real nonnegative convex positively homogeneous even function of degree 1. It is equal to zero only at the origin:

$$f \geq 0, f(x) = 0 \Leftrightarrow x = 0, f(\lambda x) = |\lambda|(x), f(x + y) \leq f(x) + f(y).$$

Let f be a norm in $\mathbb{R}^n$. The function f is determined by the set where it is equal to 1. This set is a convex hypersurface centrally symmetric with respect to zero. Conversely, every convex compact body in $\mathbb{R}^n$ centrally symmetric with respect to zero and containing zero inside it defines a unique norm which is equal to 1 on its boundary. This hypersurface $f = 1$ is called the *unit sphere* of the norm f.

Exercise 1. Determine the unit spheres of the following norms in $\mathbb{R}^3$:

$$\text{(a) } f = \sqrt{(x, x)}; \qquad \text{(b) } f = \max|x_i|; \qquad \text{(c) } f = \sum|x_i|.$$

Let us consider the dual space $\mathbb{R}^{n*}$ of $\mathbb{R}^n$.

Definition. The *dual* norm in $\mathbb{R}^{n*}$ is defined as

$$g(p) = \max_{f(x) \leq 1} |(p, x)|.$$

It is easy to see that g is indeed a norm.

The relation of duality is mutual, since the defining inequality can be written in the symmetric form $|(p, x)| \leq f(x)g(p)$.

For every point p of the dual space, we consider the hyperplane $p = 1$ in the initial space.

Theorem. *The unit sphere of the dual norm is the set of supporting hyperplanes of the unit sphere of the initial norm.*

◀ The condition $g(p) = 1$ means that (p, x) has maximum equal to 1 on the unit sphere, i.e., the plane $p = 1$ is a plane of support for the initial sphere. ▶

The set of all hyperplanes of support of a given convex hypersurface is called the *dual convex hypersurface*. Hence, the unit spheres of dual norms are dual.

Exercise 2. Determine the hypersurfaces in $\mathbb{R}^3$ dual to the (a) sphere, (b) tetrahedron, (c) cube, (d) octahedron.

Exercise 3. Determine the norms dual to the norms in Exercise 1.

From what was said above it is clear that the passage from a convex hypersurface to its dual is locally given by the Legendre transformation.

G. The Problem of Envelopes of Families of Plane Curves

Two smooth functions

$$x = x(s, t), \qquad y = y(s, t)$$

of two variables define a family of curves in the plane parametrized by the parameter t (indicating the point on the curve) and indexed by the parameter s (indicating the index of the curve).

Exercise. Draw the families of curves defined by the following functions:

(a) $x = (s + t^2), y = t$;
(b) $x = s + st + t^3, y = t^2$, s and t small;
(c) $x = (s + t^2)^2, y = t$.

Answer. Cf., Fig. 28.

It can be shown that for a generic family the envelope is a curve whose only singularities are cusps (as in the case of a semicubic parabola) and points of self-intersection; in the neighborhood of every point where the envelope is smooth, the family reduces to one of the normal forms (a), (b), or (c) by means of a smooth transformation $X(x, y)$, $Y(x, y)$ of the coordinates and a smooth transformation $S(s), T(s, t)$ of the parameters. This is not true in the holomorphic case. See J. P. Dufour, *Familles de courbes planes différentiables*, Topology **22**, 4 (1983), 449–474; and V. I. Arnold, *Wave fronts evolution and the equivariant Morse lemma*, Comm. Pure Appl. Math. **29**, 6 (1976), 557–582.

        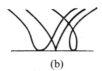

       (a)   (b)   (c)

Figure 28.

In the neighborhood of a generic point of the discriminant curve, an implicit equation can generically be reduced to the normal form

$$\left(\frac{dy}{dx}\right)^2 = x$$

by means of a smooth transformation $X(x, y)$, $Y(x, y)$ of the coordinates (cf., § 4).

The projections of the integral curves onto the (X, Y)-plane are semicubic parabolas in these coordinates. Hence, the discriminant curve is the envelope of the projections of the integral curves only for exceptional equations (for example, Clairaut's equation). In particular, for a small generic perturbation of Clairaut's equation, the discriminant curve turns from the envelope into the locus of cusps of the projections of the integral curves.

§ 4. Normal Form of an Implicit Differential Equation in the Neighborhood of a Regular Singular Point

Here we study singularities of families of integral curves of a generic differential equation implicit in the derivative.

A. Singular Points

Consider the equation

$$F(x, y, p) = 0 \tag{1}$$

where $p = dy/dx$, and F is a smooth function in some domain.

Let Eq. (1) define a smooth surface in the three-dimensional space of jets with the coordinates (x, y, p). By the implicit function theorem, for this it is sufficient to assume that the total differential of F does not vanish at the points where $F = 0$.

Consider the projection of the surface $F = 0$ onto the (x, y)-coordinate plane parallel to the p-direction.

Definition. A point of the surface $F = 0$ is said to be singular for Eq. (1) if the projection $(x, y, p) \mapsto (x, y)$ of the surface to the plane is not a local diffeomorphism in the neighborhood of this point.

By the implicit function theorem, the singular points are those points of the surface $F = 0$ at which $\partial F/\partial p = 0$.

B. Criminant

Consider the set of all singular points of Eq. (1). In the three-dimensional space of jets, this set is given by the two equations $F = 0$ and $\partial F/\partial p = 0$. Therefore, "generally speaking", the singular points form a curve.

Definition. The set of singular points of the equation $F = 0$ in the three-dimensional (x, y, p)-space of jets is called the *criminant* of the equation.

By the implicit function theorem, the criminant is a smooth curve in the three-dimensional space of jets in the neighborhood of each of its points where the rank of the derivative of the mapping $(x, y, p) \mapsto (F, \partial F / \partial p)$ of the three-dimensional space to the plane is maximal (equal to 2).

C. Discriminant Curve

Definition. The projection of the criminant to the (x, y)-plane parallel to the p-direction is called the *discriminant curve**.

By the implicit function theorem, the neighborhood of a point of the criminant is projected diffeomorphically to the (x, y)-plane parallel to the p-direction if the criminant is not parallel to the p-direction.

Remark. Under the above conditions, the discriminant curve can well have singularities.

These singularities arise because, in general, several points of the criminant curve can be mapped onto the same point of the discriminant curve by the projection. Generically, these singularities will be points of self-intersection of the discriminant curve. For the generic equation in the neighborhood of such a point, the discriminant curve consists of two branches intersecting at a nonzero angle.

On the other hand, to the points where the tangent to the criminant is parallel to the p-direction, generically there correspond cusps on the discriminant curve.

All more complicated singularities of the discriminant curve besides points of self-intersection and cusps can be eliminated by a small perturbation of the equation. These two types of singularities are preserved, only a little displaced, for small deformations of the equation.

D. Points of Tangency of the Criminant with the Contact Plane

At every point (x, y, p) of the space of jets, there is a *contact plane $dy = p\,dx$*. In particular, such a plane exists at the points x of the criminant. The tangent of the criminant at a given point may lie in the contact plane or it may intersect it.

Definition. A point of the criminant is called a *point of tangency with the contact plane* if the tangent of the criminant at this point lies in the contact plane.

We note that points of tangency of the criminant with the p-direction are points of tangency with the contact plane. Indeed, the contact plane contains the p-direction at every point.

E. Regular Singular Points

Definition. A singular point of Eq. (1) is said to be *regular* if at this point the condition of smoothness of the criminant[†]

---

*This definition is a reformulation of the definition of a discriminant curve in § 3.
[†] The rank of a mapping is the rank of its derivative.

$$\text{rank}((x, y, p) \mapsto (F, F_p)) = 2$$

is satisfied and the criminant is not tangent to the contact plane.

EXAMPLE

Consider the equation $p^2 = x$. The criminant is given by the equations $p = 0$ and $x = 0$. This is the y-axis. The condition of smoothness is satisfied. The tangent vector $(0, 1, 0)$ of the criminant does not lie in the contact plane $dy = 0\,dx$. Consequently, every singular point of the equation $p^2 = x$ is regular.

Remark. For a generic equation, almost all singular points are regular: the nonregular points lie on the criminant discretely. If this is not so for a given equation, then it can always be achieved by a small perturbation. This and the previous "genericity arguments" can be verified by Sard's theorem in § 10.

F. Theorem on the Normal Form*

Theorem. *Let* (x_0, y_0, p_0) *be a regular singular point of the equation* $F(x, y, p) = 0$. *There exists a diffeomorphism of the neighborhood of the point* (x_0, y_0) *in the* (x, y)-*plane onto the neighborhood of the point* $(0, 0)$ *in the* (X, Y)-*plane reducing the equation* $F = 0$ *to the form* $P^2 = X$ *(where* $P = dY/dX$*).*

Explanation. The equation $F = 0$ defines a surface in the three-dimensional space of linear elements in the (x, y)-plane. A diffeomorphism of the plane transforms every linear element into a new linear element. It is claimed that the part of the surface $F = 0$ near a regular singular point can be transformed into a part of the surface $P^2 = X$ near the point $(X = 0, Y = 0, P = 0)$.

Corollary. *In the neighborhood of a regular singular point, the family of integral curves of Eq. (1) is diffeomorphic to the family of semicubic parabolas* $y = x^{3/2} + C$.

◀ The diffeomorphism mentioned in the theorem maps integral curves of Eq. (1) in the (x, y)-plane to integral curves of the equation $P^2 = X$ in the (X, Y)-plane. These latter integral curves are semicubic parabolas with cusp on the discriminant curve: $dY/dX = \sqrt{X}$, $Y = \frac{2}{3}X^{3/2} + C$. ▶

G. Proof of the Theorem on the Normal Form

◀ *1. Reduction to the Case Where the Criminant Is the* y-*Axis.* Let (x_0, y_0, p_0) be a regular singular point of the equation $F(x, y, p) = 0$. Then the discriminant curve is smooth in the neighborhood of (x_0, y_0). Consider the projections of the contact planes onto the (x, y)-plane at the points of the criminant. In the neighborhood of (x_0, y_0), we obtain a family of straight lines not tangent to the discriminant curve.

    Now choose a local coordinate system near (x_0, y_0) in the (x, y)-plane so that (1)

---

*M. Cibrario, *Sulla reduzione a forma canonica delle equazioni lineari alle derivate parziali di secondo ordine di tipo misto*; Accademia di scienze e lettere, Instituto Lombardo, Rendiconti **65** (1932), 889–906.

the equation of the discriminant curve is $x = 0$; (2) the straight lines $y = $ const intersect the discriminant curve in the directions just constructed.

As before, we shall denote the coordinates by (x, y) and the derivative dy/dx by p. The singular point has the coordinates $(0, 0, 0)$.

2. Analysis of the Regularity Conditions. In our coordinate system, the criminant is the y-axis; we have $x = 0$, $p = 0$ ($y = $ const). From this it follows for our equation that $F(0, y, 0) = 0$, $F_p(0, y, 0) = 0$ in the coordinates introduced above. The regularity condition of the criminant has the form

$$\det \left| \frac{D(F, F_p)}{D(x, p)} \right| \neq 0, \quad \text{i.e.,} \quad \det \begin{vmatrix} F_x & F_{xp} \\ F_p & F_{pp} \end{vmatrix} \neq 0$$

(since $F_y = 0$, $F_{yp} = 0$ at the points of the criminant). Moreover, $F_p = 0$ at the points of the criminant. Consequently, the regularity condition of the criminant can be written in the form

$$F_x(0, y, 0) \neq 0, \qquad F_{pp}(0, y, 0) \neq 0.$$

The condition of nontangency to the contact plane is satisfied automatically.

Expand F into a Taylor series in p with the remainder of order 2:

$$F(x, y, p) = A(x, y) + pB(x, y) + p^2 C(x, y, p).$$

It follows from the above relations that $A(0, y) = 0$, $B(0, y) = 0$. Therefore, we may write $A(x, y) = x\alpha(x, y)$, $B(x, y) = x\beta(x, y)$, where α and β are smooth functions.

The regularity conditions of the criminant curve have the form $A_x(0, y) \neq 0$, $C(0, y, 0) \neq 0$. In what follows we may even assume that $C > 0$, $A_x < 0$ (if this is not so, we change the signs of F and/or x). Thus $\alpha(0, 0) < 0$, $C(0, 0, 0) > 0$.

3. A Quadratic Equation. Consider the relation $F = 0$ as a quadratic equation in p with coefficients C, B, and A. We obtain

$$p = \frac{-B \pm \sqrt{B^2 - 4AC}}{2C} = \frac{-x\beta \pm \sqrt{x\gamma}}{2C},$$

where $\gamma = -4\alpha C + x\beta^2$ is a function of (x, y, p), and $\gamma(0, 0, 0) = -4\alpha(0, 0, 0) \times C(0, 0, 0) > 0$.

Finally, let $x = \xi^2$. Keeping only the sign "$+$" in "$\pm$", we obtain

$$p = \frac{-\xi^2 \beta(\xi^2, y) + \xi\sqrt{\gamma(\xi^2, y, p)}}{2C(\xi^2, y, p)}.$$

We apply the implicit function theorem to this equation in $p(\xi, y)$. We obtain a solution $p = \xi\omega(\xi, y)$, where ω is a smooth function and $\omega(0, 0) \neq 0$.

4. A Differential Equation for $y(\xi)$. We note that $p = dy/dx = dy/2\xi \, d\xi$. Therefore, we obtain the following differential equation for $y(\xi)$:

$$\frac{dy}{d\xi} = 2\xi^2 \omega(\xi, y), \qquad \omega(0, 0) \neq 0. \tag{2}$$

In the (ξ, y)-plane the integral curves intersect the axis $\xi = 0$ and have points of tangency of order 2 with the straight lines $y = \text{const}$. Therefore, the equation has a first integral of the form $I(\xi, \eta) = y - \xi^3 K(\xi, y)$, where K is a smooth function, and $K(0, 0) \neq 0$. (I is the coordinate of the intersection point of the integral curve through (ξ, y) with the axis $\xi = 0$; $K \neq 0$ since $\omega \neq 0$.)

5. *Construction of Normalizing Coordinates.* We decompose K into even and odd parts with respect to ξ:

$$K(\xi, y) = L(\xi^2, y) + \xi M(\xi^2, y).$$

Here L and M are smooth functions of x and y, and $L(0, 0) \neq 0$. With this notation we have $I(\xi, y) = y - \xi^4 M(\xi^2, y) - \xi^3 L(\xi^2, y)$. We introduce new variables Y and Ξ by the formulas

$$\Xi = \xi\sqrt[3]{L(\xi^2, y)}, \qquad Y = y - \xi^4 M(\xi^2, y).$$

Then $I = Y - \Xi^3$.
   Consider also $X = \Xi^2$. Then

$$X = x\sqrt[3]{L^2(x, y)}, \qquad Y = y - x^2 M(x, y).$$

These formulas give a local diffeomorphism of the plane in the neighborhood of $(0, 0)$, since $L(0, 0) \neq 0$. The first integral takes the form

$$I = Y - X^{3/2}.$$

Now $(dY/dX)^2 = \frac{9}{4}X$. The normal form can be obtained by stretching one of the coordinate axes. ▶

H. Remarks

The basic step in the above proof is the substitution $x = \xi^2$, i.e., the passage to a two-sheeted covering of the (x, y)-plane with branching along the discriminant curve. From topological arguments (although, in a complex domain), it is clear *a priori* that the two-valuedness of $p(x, y)$ disappears on this two-sheeted covering, and the equation splits into two. To trouble with the quadratic equation is necessary only for the proof of this fact in a real domain. It remains to reduce Eq. (2) obtained on the covering to its normal form by a diffeomorphism of the covering plane into itself. This can easily be achieved by decomposing the first integral into even and odd parts with respect to ξ.

Remark. Our proof used the representation of an even function by a function of the square of its argument. For analytic functions (or formal series) such a representation is obvious. In the case of a smooth function, it needs a proof.
   Indeed, an infinitely differentiable even function can be considered as a function of the square of the argument with values on the positive semiaxis. It is infinitely differentiable at all points of this semiaxis, including zero. We have to represent it as

the restriction to the positive semiaxis of a function infinitely differentiable on the entire axis.

Such a representation is possible since a smooth extension to the negative semiaxis exists. This follows from the theorem (of Borel) on the existence of an infinitely differentiable function on the real line with an arbitrary Taylor series at zero. We do not discuss the (simple) proof of this theorem.

Besides the regular singular points, a generic equation may have some isolated points where the discriminant curve is smooth, but the contact plane is tangent to the surface defined by the equation. At a neighborhood of such a point, a generic equation is reducible to the normal form

$$y = (p + kx)^2$$

by a diffeomorphism of the (x, y)-plane; for the proof, see A. A. Davydov, *The normal form of the differential equation implicit in the derivative*, Funct. Anal. Appl. **19**, 2 (1985), 1–10. The topological case was settled in A. V. Phakadse, A. A. Shestakov, *On the classification of singular points of a first-order differential equation implicit in the derivative*, Mat. Sbornik **49**, 1 (1959), 3–12.

The direction field on the surface defined by the equation has, at a generic point of tangency of a contact plane with the surface, a singularity of the same type as a direction field of a plane generic vector field at its singular points—that is, a saddle node or a focus. Hence these singular points of the implicit differential equations are called *folded saddle* (*nodes, foci*): they are obtained from ordinary saddles (nodes, foci) by a folding mapping.

It is interesting to note that the foldings generate no new moduli: the parameter k in the normal form is defined by the ratio of the eigenvalues of the linearization of the vector field, whose phase portrait generates the folded singularity under the folding mapping (Fig. 29).

In contrast with these singularities, those corresponding to the cusps of the discriminant curve have functions-moduli not only with respect to diffeomorphisms, but even with respect to homeomorphisms of the plane (x, y) (see the Davydov paper quoted above). These cusps are the projections of Whitney pleats on the (x, y)-plane.

The projections of the integral curves on this plane may be described as the family of projections of the swallowtail surface's generic plane sections under a generic (rank

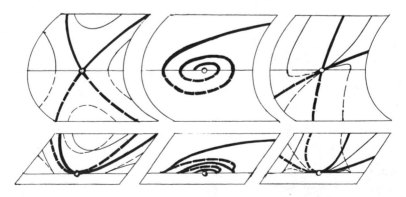

Figure 29.

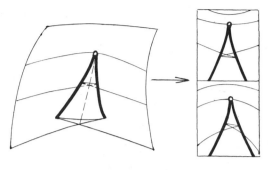

Figure 30.

2) mapping of the ambient space on the plane. The swallowtail surface is given by

$$\{(a, b, c): t^4 + at^2 + bt + c \text{ has a multiple root}\}.$$

The sections may be chosen as $a = $ const (Fig. 30).

The proof of the above statement on the integral curves is published in J. W. Bruce, *A note on the first-order differential equations of degree greater than one and wavefront evolution*, Bull. London Math. Soc. **16** (1984), 139–144 (see also V. I. Arnold, *Wavefronts evolution and the equivariant Morse lemma*, Comm. Pure Appl. Math. **29**, 6 (1976), 557–582).

§ 5. The Stationary Schrödinger Equation

In this section we develop the mathematical rudiments of elementary quantum mechanics. We do not discuss the physical motivation of the concepts introduced here, but use physical terminology for the description of properties of solutions of the equation.

A. Definitions and Notation

In physics the equation

$$\frac{d^2\Psi}{dx^2} + (E - U(x))\Psi = 0 \tag{1}$$

is called the *stationary Schrödinger equation*.

The independent variable x is called the *Cartesian coordinate of the particle*. The unknown, in general, complex-valued function Ψ is called the *wave function of the particle*, and the solutions of the Schrödinger equation are called the *states of the particle*. The spectral parameter E is called the *energy of the particle*, and the given function U the *potential* or *potential energy of the particle*. *Quantum mechanics* deals mainly with the study of the properties of Eq. (1) and equations and systems of partial differential equations generalizing it.

EXAMPLE

Let $U = 0$. Then the particle is said to be *free*. The Schrödinger equation for a free particle with energy $E = k^2$ has the form

$$\Psi_{xx} + k^2\Psi = 0. \tag{2}$$

This equation has the two linearly independent solutions

$$\Psi_+ = e^{ikx}, \qquad \Psi_- = e^{-ikx}.$$

These two solutions describe the *particle moving to the right* (with momentum $k > 0$) and the *particle moving to the left*, respectively. Hence, the space of solutions of a free particle with energy E is a two-dimensional complex space.

Physicists call the square of the absolute value of the wave function the *probability density* of the event that the particle is at a given place. Hence, a free particle with impulse k "can be found at any point with the same probability." (This terminology can be used regardless of the meaning of these words and regardless of how all this is connected with probability theory.)

B. Potential Barriers

We assume that the potential has compact support (is different from zero only in some domain). If $U \geqslant 0$, then we say that we are given a *potential barrier*, if $U \leqslant 0$, then we say that we are given a *potential well*. The domain where the potential is different from zero is called the support of the potential (Fig. 31).

We assume that the energy $E = k^2$ of the particle is *positive*. Left of the support, Eq. (1) coincides with Eq. (2) of a free particle. Consequently, Schrödinger's equation has two solutions which coincide with e^{ikx} and e^{-ikx} *left of the support*. These two solutions are called the *particle incoming from the left* and the *particle outgoing to the left*, respectively. We note that

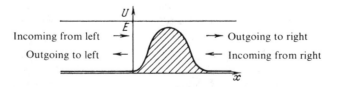

Figure 31.

these solutions are defined for all x, but coincide with e^{ikx} and e^{-ikx} only left of the support.

Exactly in the same way, there exist two solutions which coincide with e^{ikx} and e^{-ikx} *right of the support*. These solutions are called the *particle outgoing to the right* and the *particle incoming from the right*, respectively.

Exercise. Can a particle arriving from the left be entirely reflected to the left (i.e., can a wave function be zero right of the barrier and not zero left of it)? Can the particle depart to the right entirely?

Solution. No, yes.

C. Monodromy Operator

Definition. The *monodromy operator* of the Schrödinger equation (1) with a potential of compact support is a linear operator mapping the state space of a free particle with energy $E = k^2$ into itself. It is defined in the following way.

To a solution of Eq. (2) of a free particle we assign a solution of the Schrödinger equation coinciding with it to the left of the support, and to this solution, in turn, we assign its value to the right of the support.

It turns out that the monodromy operator has the remarkable $(1, 1)$-unitarity property. In order to formulate it, we introduce the following.

Notation. Denote by $\mathbb{R}^2$ the *space of real solutions of the Schrödinger equation* (1). The *space of states of the particle* (i.e., the space of complex solutions of the equation) is the complexification of $\mathbb{R}^2$; we denote it by $\mathbb{C}^2 = {}^{\mathbb{C}}\mathbb{R}^2$. All four states of the particle arriving and departing to the right and left belong to this space.

The *space of real solutions of eq. (2) of a free particle* is denoted by $\mathbb{R}^2_0$ (since $U = 0$ for a free particle). In this space, we have the following natural basis:

$$e_1 = \cos kx, \qquad e_2 = \sin kx.$$

We denote the *space of states of a free particle* by $\mathbb{C}^2_0$; this is the complexification of $\mathbb{R}^2_0$. The states of the particles moving to the right and to the left form a natural basis. We denote them by

$$f_1 = e^{ikx}, \qquad f_2 = e^{-ikx}.$$

Note that e_1 and e_2 determine a basis in the state space, too. These two bases are connected via the relations

$$f_1 = e_1 + ie_2, \qquad f_2 = e_1 - ie_2.$$

Definition. The group $SU(1, 1)$ of $(1, 1)$-*unitary unimodular matrices* consists of all complex 2×2 matrices with determinant 1 preserving the Hermitian form $|z_1|^2 - |z_2|^2$. In other words, these are all matrices $\begin{pmatrix} a & b \\ c & d \end{pmatrix}$ for which $|a|^2 - |b|^2 = |c|^2 - |d|^2 = 1$, $a\bar{c} - b\bar{d} = 0$, $ad - bc = 1$.

Theorem. *The matrix of the monodromy operator in the basis* (f_1, f_2) *belongs to the group* $SU(1, 1)$.

The reason why the monodromy operator belongs to $SU(1, 1)$ is that the phase flow of Eq. (1) preserves area. For the proof, recall some information on the group $SU(1, 1)$.

D. An Algebraic Digression: The Group $SU(1, 1)$

Consider the real linear space $\mathbb{R}^2$ and its complexification $\mathbb{C}^2$. In $\mathbb{R}^2$, choose an area element and denote by $[\xi, \eta]$ the oriented area of the parallelogram generated by the vectors ξ and η. The skew-symmetric inner product $[\ ,\ ]$ is called a *symplectic structure*. If a basis (e_1, e_2) is fixed in $\mathbb{R}^2$ for which $[e_1, e_1] = 1$, then $[\xi, \eta]$ is equal to the determinant consisting of the components of the vectors ξ and η in the basis e_1, e_2.

The complexification of the bilinear form $[\ ,\ ]$ defines a symplectic structure in $\mathbb{C}^2$, which we shall denote by the same square brackets.

We note that the form $[\ ,\ ]$ is nondegenerate: if $[\xi, \eta] = 0$ for all ξ, then $\eta = 0$.

Consider the *Hermitian form* $\langle \xi, \eta \rangle = (i/2)[\xi, \bar{\eta}]$ in $\mathbb{C}^2$. This is indeed a Hermitian form: $\langle \lambda\xi, \eta \rangle = \lambda\langle \xi, \eta \rangle$, $\langle \xi, \eta \rangle = \overline{\langle \eta, \xi \rangle}$. For the following, it is useful to calculate the Hermitian products of the vectors $f_1 = e_1 + ie_2$ and $f_2 = e_1 - ie_2$. It is easy to prove the following lemma.

Lemma. *The following relations hold:*

$$\langle f_1, f_1 \rangle = 1, \qquad \langle f_2, f_2 \rangle = -1, \qquad \langle f_1, f_2 \rangle = 0.$$

◀ For example,

$$\langle f_1, f_1 \rangle = \frac{i}{2}[f_1, \bar{f_1}] = \frac{i}{2}[f_1, f_2] = \frac{i}{2}\begin{vmatrix} 1 & i \\ 1 & -i \end{vmatrix} = 1. \ ▶$$

Hence, the *Hermitian* form $\langle \ , \ \rangle$ is of "type $(1, 1)$" (one positive and one negative square in the canonical form $\langle z, z \rangle = |z_1|^2 - |z_2|^2$).

Now we consider the linear transformations of the plane $\mathbb{C}^2$ preserving the Hermitian, symplectic, and real structures.

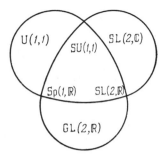

Figure 32.

Definition. The group of linear transformations of the plane $\mathbb{C}^2$ preserving the Hermitian form $\langle\ ,\ \rangle$ is called the $(1, 1)$-*unitary group* and denoted by $U(1, 1)$.

The group of linear transformations of the plane $\mathbb{C}^2$ preserving the symplectic structure $[\ ,\ ]$ is called the *special* (or *unimodular*) *linear group of second order* and denoted by $SL(2, \mathbb{C})$.

The group of all real linear transformations of the plane $\mathbb{C}^2$ (i.e., the group whose elements are complexifications of linear transformations in $\mathbb{R}^2$) is called the *real linear group of the second order* and denoted by $GL(2, \mathbb{R})$.

Hence, we have defined three subgroups of the group $GL(2, \mathbb{C})$ of linear transformations of $\mathbb{C}^2$: the $(1, 1)$-unitary group $U(1, 1)$, the unimodular group $SL(2, \mathbb{C})$, and the real group $GL(2, \mathbb{R})$.

The Hermitian form $\langle\ ,\ \rangle$ defining the unitary group, the symplectic structure $[\ ,\ ]$ defining the unimodular group, and the complex conjugation defining the real group are connected by the relation $\langle a, b\rangle = (i/2)[a, \bar{b}]$.

Theorem. *The intersection of any two of these groups coincides with the intersection of all three groups (Fig. 32).*

This intersection is called the *special* $(1, 1)$-*unitary group** and is denoted by $SU(1, 1)$. [It is also called *real unimodular group* and is denoted by $SL(2, \mathbb{R})$ or the *real symplectic group of the second order* and denoted by $Sp(1, \mathbb{R})$.]

◀ If the transformation A is real and unimodular, then $[A\xi, A\eta] = [\xi, \eta]$ and $A\bar{\xi} = \overline{A\xi}$. Therefore, $\langle A\xi, A\eta\rangle = i/2[A\xi, \overline{A\eta}] = i/2[A\xi, A\bar{\eta}] = i/2[\xi, \eta] = \langle\xi, \eta\rangle$.

If A is real and $(1, 1)$-unitary, then $\langle A\xi, A\eta\rangle = \langle\xi, \eta\rangle$ and $A\bar{\xi} = \overline{A\xi}$. Therefore, $[A\xi, A\eta] = -2i\langle A\xi, \overline{A\eta}\rangle = -2i\langle A\xi, A\bar{\eta}\rangle = -2i\langle\xi, \bar{\eta}\rangle = [\xi, \eta]$.

If A is $(1, 1)$-unitary and unimodular, then $[A\xi, A\eta] = [\xi, \eta]$ and $[A\xi, \overline{A\eta}] = [\xi, \bar{\eta}]$. Therefore, $[A\xi, \overline{A\eta}] = [A\xi, A\bar{\eta}]$ for all ξ and η. Con-

---

*We emphasize that we are dealing with operators and not matrices. The matrices of these operators belong to the matrix group $SU(1, 1)$ for the special choice of basis indicated above.

sequently, $[\xi, \overline{A\eta} - A\overline{\eta}] = 0$ for all ξ, and hence $\overline{A\eta} = A\overline{\eta}$ for all η, i.e., A is real. ▶

Corollary. *If the matrix of an operator in the real basis (e_1, e_2) is real unimodular, then the matrix of this operator in the complex conjugate basis $(f_1 = e_1 + ie_2, f_2 = e_1 - ie_2)$ is special $(1, 1)$-unitary, and conversely.*

◀ The Hermitian scalar square of the vector $z_1 f_1 + z_2 f_2$ can be expressed by the coordinates (z_1, z_2) in the basis (f_1, f_2) by the formula $\langle z, z \rangle = |z_1|^2 - |z_2|^2$ (cf., the lemma above). Therefore, the following statements are equivalent:

(i) the matrix of A in the basis (e_1, e_2) is real and unimodular;
(ii) $A \in GL(2, \mathbb{R}) \cap SL(2, \mathbb{C})$.
(iii) $A \in SU(1, 1)$; and
(iv) the matrix of A in the basis (f_1, f_2) is $(1, 1)$-unitary and unimodular. ▶

E. A Geometric Digression: SU(1, 1) and the Lobachevsky Geometry

The matrix groups $SL(2, \mathbb{R})$ and $SU(1, 1)$ are connected with the Lobachevsky geometry in the following way (Fig. 33).

A real unimodular matrix of the second order defines a linear-fractional transformation $z \to (az + b)/(cz + d)$ mapping the upper half-plane onto itself. This transformation is a motion of the Lobachevsky plane represented in the form of the upper half-plane. All motions of the Lobachevsky plane can be obtained in this way. The group of motions of the Lobachevsky plane is isomorphic to $SL(2, \mathbb{R})/\pm E$.

A 2×2 unimodular $(1, 1)$-unitary matrix defines a linear-fractional transformation of the unit disk onto itself. Indeed, the cone $|z_1|^2 < |z_2|^2$ is mapped onto itself by any $(1, 1)$-unitary unimodular transformation. Under the natural mapping

$$\mathbb{C}^2 \setminus O \to \mathbb{C}P^1, \qquad (z_1, z_2) \mapsto w = z_1/z_2,$$

this cone turns into the unit disk $|w| < 1$ and the linear transformations of $\mathbb{C}^2$ into linear-fractional transformations of $\mathbb{C}P^1$ (Fig. 34).

The linear-fractional transformations of the unit disk onto itself expressed by matrices belonging to $SU(1, 1)$ are the motions of the Lobachevsky plane

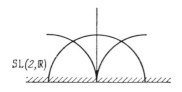

Figure 33.

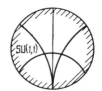

Figure 34.

represented as the interior of the unit disk. All motions of the Lobachevsky plane can be obtained in this way. The group of motions of the Lobachevsky plane is isomorphic to $SU(1, 1)/\pm E$.

The matrix groups $SL(2, \mathbb{R})$ and $SU(1, 1)$ are isomorphic: they are obtained from the same group of operators. The matrices of these operators in the real basis (e_1, e_2) belong to $SL(2, \mathbb{R})$, and in the complex conjugate basis, (f_1, f_2) belongs to $SU(1, 1)$. The passage from $SL(2, \mathbb{R})$ to $SU(1, 1)$ corresponds to the passage from the real basis to the complex conjugate basis, and from the model of the Lobachevsky plane in the upper half-plane to its model in the unit disk.

Problem. Prove that $SL(2, \mathbb{R})$ is homeomorphic to the solid torus $S^1 \times D^2$ (the interior of a doughnut).

F. Properties of the Real Monodromy Operator

We return to the monodromy operator of the Schrödinger equation (1). Besides the spaces $\mathbb{R}^2$ and $\mathbb{R}_0^2$ of solutions of Eq. (1) and Eq. (2) of a free particle, we consider the *phase plane* $\mathbb{R}_p^2$. The points of the phase plane are pairs (Ψ, Ψ_x) of real numbers.

Fix $x \in \mathbb{R}$. Consider the linear operator

$$B^x : \mathbb{R}^2 \to \mathbb{R}_p^2, \ \Psi \mapsto (\Psi(x), \Psi_x(x)),$$

assigning the initial condition at the point x to each (real) solution of Eq. (1). This operator is an isomorphism. The isomorphism $g_{x_1}^{x_2} = B^{x_2}(B^{x_1})^{-1}$ is called the *phase transformation* from x_1 to x_2.

For Eq. (2) of a free particle, the operator (solution $\mapsto$ phase point)

$$B_0^x : \mathbb{R}_0^2 \to \mathbb{R}_p^2$$

is defined in a similar way.

With this notation, the real monodromy operator M is defined by the following commutative diagram of isomorphisms:

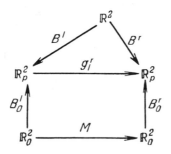

Here l indicates a point left of the support, and r a point to the right of it. The operator M does not depend on the choice of these points.

Theorem. *The determinant of the monodromy operator of the Schrödinger equation is equal to* 1.

◀ In the space $\mathbb{R}_0^2$ of real states of a free particle, we choose the basis $e_1 = \cos kx$, $e_2 = \sin kx$. In the real phase space $\mathbb{R}_p^2$, the coordinates Ψ, Ψ_x are chosen, thereby determining a basis as well. In this basis, the matrix of B_0^x has the form

$$(B_0^x) = \begin{pmatrix} \cos kx & \sin kx \\ -k \sin kx & k \cos kx \end{pmatrix}.$$

Consequently, $\det B_0^x = k$ does not depend on x. In particular, the determinants of the left and right vertical isomorphisms in the diagram are the same. Hence, $\det M = g_l^r$ (the diagram is commutative). By Liouville's theorem, the phase flow is area preserving (the term Ψ_x does not enter into the Schrödinger equation). Therefore, $\det g_l^r = 1$. Consequently, $\det M = 1$. ▶

G. Properties of the Complex Monodromy Operator

◀ Proof of the theorem (from § 5B) on the (1, 1)-unitarity.

The complex monodromy operator is the complexification of the real monodromy operator.

The matrix of the monodromy operator in the real basis (e_1, e_2) is in $\mathrm{SL}(2, \mathbb{R})$ (cf., § 5F). Consequently, the matrix of this operator in the complex conjugate basis $(f_1 = e_1 + ie_2, f_2 = e_1 - ie_2)$ is in $\mathrm{SU}(1, 1)$ (cf., § 5D, Corollary) ▶

Problem. Prove that the Schrödinger equation (1) does not have a nonzero solution coinciding with ae^{ikx} to the left of the support and with be^{-ikx} to the right of it (no particle can arrive without departing).

Solution. The monodromy operator preserves the (1, 1)-Hermitian square $|z_1|^2 - |z_2|^2$. On the other hand, $\langle ae^{ikx}, ae^{ikx} \rangle = |a|^2$, $\langle be^{-ikx}, be^{-ikx} \rangle = -|b|^2$ (cf., the Lemma in § 5D). Consequently, $|a|^2 = -|b|^2$, i.e., $a = b = 0$.

H. Transmission and Reflection Coefficients

Definition. We say that a particle arriving from the left with impulse $k > 0$ passes the barrier with *transmission coefficient* $|A|^2$ and *reflection coefficient* $|B|^2$ if the Schrödinger equation (1) in which $E = k^2$ has a solution equal to

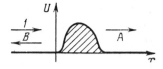

Figure 35.

$$e^{ikx} + Be^{-ikx} \qquad \text{to the left of the barrier;}$$

$$Ae^{ikx} \qquad \text{to the right of the barrier (Fig. 35).}$$

Lemma. *The solution* Ψ *and the complex constants A and B satisfying the above conditions exist and are unique for every k > 0.*

◀ We consider the particle departing to the right (the solution equal to e^{ikx} to the right of the barrier). Left of the barrier this solution, as any other, is a linear combination of e^{ikx} and e^{-ikx}. The coefficient of e^{ikx} is different from zero, since the monodromy operator is (1, 1)-unitary (cf. the Problem in § 5G).

Dividing by this nonzero coefficient, we obtain the required solution. Hence, coefficients A and B are defined uniquely. ▶

Problem. Prove that the transmission coefficient is always different from zero.

Solution. If $\Psi \equiv 0$ to the right of the barrier, then $\Psi \equiv 0$ to the left of it, as well.

Theorem. *The sum of the transmission and reflection coefficients is equal to* 1.

◀ **Lemma.** *In the basis* $(f_1 = e^{ikx}, f_2 = e^{-ikx})$, *the matrix of the monodromy operator can be expressed in terms of the complex coefficients A and B by the formula*

$$(M) = \begin{pmatrix} 1/\bar{A} & -\bar{B}/\bar{A} \\ -B/A & 1/A \end{pmatrix}.$$

◀ By the definition of A and B, the monodromy operator acts according to the formula $f_1 + Bf_2 \mapsto Af_1$. Since the monodromy operator is real, we can also determine the image of the complex conjugate vector. Taking the relation $\bar{f}_1 = \bar{f}_2$ into account, we obtain $f_2 + \bar{B}f_1 \mapsto \bar{A}f_2$. Dividing by $A \neq 0$ and $\bar{A} \neq 0$, we obtain the matrix

$$(M^{-1}) = \begin{pmatrix} 1/A & \bar{B}/\bar{A} \\ B/A & 1/\bar{A} \end{pmatrix}$$

of the inverse of the monodromy operator.

To invert a unimodular matrix of the second order, it is sufficient to interchange the diagonal elements and change the sign of the nondiagonal ones:

$$\begin{pmatrix} a & b \\ c & d \end{pmatrix} \begin{pmatrix} d & -b \\ -c & a \end{pmatrix} = \begin{pmatrix} ad - bc & 0 \\ 0 & ad - bc \end{pmatrix}. \blacktriangleright$$

Since $M \in SU(1, 1)$, we obtain $1/|A|^2 - |B|^2/|A|^2 = 1$. $\blacktriangleright$

Problem. Calculate the transmission and reflection coefficients for the potential equal to a constant U_0 for $0 \leqslant x \leqslant a$ and zero at the remaining points (Fig. 36).

Solution

$$|A|^2 = \left(1 + \frac{U_0^2 \sin^2 ak_1}{4E(E - U_0)}\right)^{-1},$$

where $E = k^2$, $E - U_0 = k_1^2$. (Passing over the barrier, the particle slows down; therefore, the probability density of finding it within the barriers is larger than finding it outside.) For large E, the coefficient of reflection converges to 0,

$$|B|^2 \sim \frac{U_0^2}{4E(E - U_0)} \sin^2 ak_1.$$

If the energy of the particle is less than the height of the barrier, then the transmission coefficient is exponentially small:

$$|A|^2 = \frac{4k^2 \varkappa^2}{(k^2 + \varkappa^2)\mathrm{sh}^2 a\varkappa + 4k^2\varkappa^2},$$

where $E = k^2$, $U_0 - E = \varkappa^2$ (Fig. 37). Although the transmission coefficient through a high and wide barrier is small, it is always different from zero ("tunneling effect": a quantum particle "passes under the barrier" which is not surmountable for a classical particle).

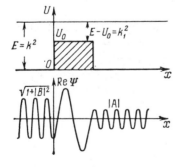

Figure 36.

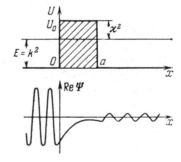

Figure 37.

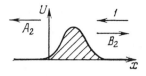

Figure 38.

I. Scattering Matrix

Along with the passage through a barrier from left to right we may consider passage from right to left. The corresponding solution Ψ_2 is equal to

$$e^{-ikx} + B_2 e^{ikx} \qquad \text{to the right of the barrier;}$$

$$A_2 e^{-ikx} \qquad \text{to the left of the barrier (Fig. 38).}$$

Definition. The matrix

$$S = \begin{pmatrix} A & B \\ B_2 & A_2 \end{pmatrix}$$

is called the *scattering matrix* (or *S-matrix*).

Remark. From the viewpoint of the stationary Schrödinger equation, it is difficult to understand which operator corresponds to this matrix and why this matrix has the remarkable properties we are going to prove. The explanation lies in the fact that S "transforms incoming particles into outgoing ones". These words can be given an exact meaning if we consider the nonstationary equation (which we shall not do).

Theorem. *The scattering matrix is unitary, and the transmission coefficients from left to right and from right to left are identical: $A = A_2$.*

◀ The monodromy operator acts in the following way:

$$A_2 f_2 \mapsto f_2 + B_2 f_2, \qquad \bar{A}_2 f_1 \mapsto f_1 + \bar{B}_2 f_2.$$

Consequently,

$$(M) = \begin{pmatrix} 1/\bar{A}_2 & B_2/A_2 \\ \bar{B}_2/\bar{A}_2 & 1/A_2 \end{pmatrix}.$$

Comparing this with the matrix calculated in the Lemma of § 5H, we obtain

$$A_2 = A, \qquad B_2 = -\bar{B}A/\bar{A}.$$

Since $|A|^2 + |B|^2 = |A_2|^2 + |B_2|^2 = 1$ and $A\bar{B}_2 + B\bar{A}_2 = 0$, the matrix S is unitary. ▶

Remark. We have considered the Schrödinger equation (1) with the real spectral parameter $E = k^2$. It proves to be very useful to consider complex vaules of k as well. The scattering matrix remains unitary and symmetric for complex k. Besides, S is "real", $S(-k) = \overline{S(k)}$ and "analytic", $A(k)$ is the boundary value of a function, analytic in the upper half-plane $\text{Im}\,k > 0$, and having a finite number of poles on the imaginary axis.

Since the transmission and reflection coefficients can be measured, there arises the so-called *inverse problem of scattering theory* of determining the potential U from the scattering matrix $S(k)$.

The potential U is given by a real function on the real line or two real functions on the half-line. The coefficients A and B are two complex functions on the semi-axis $k > 0$, i.e., four real functions on this semi-axis. The unitarity condition $|A|^2 + |B|^2 = 1$ decreases the number of real functions on the semi-axis from four to three.

Since $3 > 2$, it can be expected that not every pair A, B satisfying the condition $|A|^2 + |B|^2 = 1$ corresponds to a potential; in order to reconstruct the potential from A and B, another condition has to be imposed on these coefficients. Analyticity turns out to be such a condition.

It is surprising that these heuristic arguments based on the calculation of the number of arbitrary functions can be turned into theorems which can be formulated and proved in an exact manner (however, these theorems go beyond the scope of this course).

J. Bound States

Now we consider the potential in the form of a finite well ($U(x) \leq 0$, $U(\infty) = 0$). We say that a particle is *in the well* if $\Psi \to 0$ as $x \to \pm\infty$ (Fig. 39). It is clear that the particle can be in the well only if its energy E is *negative*. Left or right of the well, the solution is a linear combination of the functions $e^{\varkappa x}$, $e^{-\varkappa x}$, where $\varkappa^2 = -E$, $\varkappa > 0$. A particle is in the well if the coefficient of the exponential function increasing to the left vanishes to

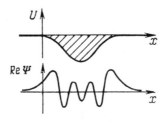

Figure 39.

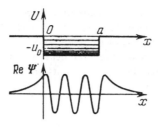

Figure 40.

the left of the well and the coefficient of the exponential function increasing to the right vanishes to the right of the well. A solution with these properties does not exist for every negative value of the energy.

It turns out that if the well is sufficiently deep and wide, then there exist a finite number of negative values of E for which the particle can be stationary in the well; the deeper and wider the well, the larger the number of these values.

These values of E are called *stationary levels*, and the wave functions Ψ *decaying as $x \to \pm \infty$* are called *bound states* (in case the well is not compact, $\int |\Psi|^2 \, dx < \infty$ is required).

Problem. Determine the stationary energy levels in the rectangular well of depth U_0 located between $x = 0$ and $x = a$ (Fig. 40).

Answer. $E = 4\xi^2/U^2 - U_0$, where ξ are the roots of the equation

$$\cos \xi = \pm \gamma \xi, \quad \sin \xi = \pm \gamma \xi \quad \left(\gamma = \frac{1}{a}\sqrt{\frac{4}{U_0}} \right)$$

($\tan \xi > 0$ for the first equation, and $\tan \xi < 0$ for the second).

Problem. Prove that the wave functions Ψ_j corresponding to bound states with distinct energy levels are orthogonal:

$$\int \Psi_1 \overline{\Psi}_2 \, dx = 0.$$

Problem.* What is the connection between stationary energy levels of bound states and the poles of the S-matrix on the imaginary axis?

§ 6. Geometry of a Second-Order Differential Equation and Geometry of a Pair of Direction Fields in Three-Dimensional Space

Here we discuss those local properties of solutions of a second-order differential equation which are geometric, i.e., invariant under diffeomorphisms of the plane of the dependent and independent variables.

With every second-order diferential equation there is associated a pair of direction fields in three-dimensional space. The problem of local classification of second-order equations up to diffeomorphisms of the plane is equivalent to the problem of local classification of generic pairs of direction fields in three-dimensional space up to diffeomorphisms of the space. Below, invariants and "normal forms" are considered for these two equivalent problems.

A. Configuration Properties of Solutions of Linear Equations

The graphs of solutions of the equation $d^2y/dx^2 = 0$ (straight lines) satisfy the con-figuration theorems (Papp, Desargues, etc.) of projective geometry.

Theorem. *The family of graphs of solutions of any second-order homogeneous linear equation*

$$\frac{d^2y}{dx^2} + a(x)\frac{dy}{dx} + b(x)y = 0$$

is locally (in the neighborhood of any point $x = x_0$) diffeomorphic* to the family of graphs of solutions of the simplest equation $d^2y/dx^2 = 0$.

Corollary. *The configuration theorems of projective geometry hold (locally) for the graphs of solutions of any second-order linear equation, for example, for the family of curves $y = A\sin x + B\cos x$ or $y = Ae^x - Be^{-x}$.*

◀ Consider a solution y_1 not vanishing at x_0 and another solution y_2 vanishing at x_0 but not identically zero. The formulas

$$Y = \frac{y}{y_1}, \qquad X = \frac{y_2}{y_1}$$

give the desired diffeomorphism. ▶

Remark 1. The coordinates (X, Y) are determined up to a linear-fractional transfor-mation. (They undergo this transformation under the substitution of the solutions y_1 and y_2 by their linear combinations.) In particular, on the x-axis, the coordinate X induces the structure of a locally homogeneous projective manifold (an atlas in which the transition functions are projective transformations of a straight line, i.e., linear-fractional functions.)

Similarly, in the (x, y)-plane, a second-order homogeneous linear equation defines the structure of a locally projective plane.

Remark 2. Two locally projective manifolds are said to be *equivalent* if there exists a diffeomorphism mapping one locally projective structure into the other.

---

*That is, there exists a diffeomorphism of the neighborhood of the straight line $x = x_0$ onto the (x, y)-plane turning the graphs of solutions into straight lines.

Problem. List all different structures of a locally projective manifold up to equivalence (a) on a straight line, (b) on a circle.

Hint. All locally projective structures on a straight line are induced by a mapping into the projective line (i.e., the circle) with nonvanishing derivative; the number of inverse images of points of the circle under this mapping is an invariant of the structure.

On the circle the two-sheeted covering of the projective line defines a locally projective manifold structure, not equivalent to the structure of the projective line. However, not every locally projective structure on the circle is induced from the structure of the projective line. The classification of locally projective structures on the circle is connected with the classification of Hill's equations (second-order linear equations with periodic coefficients). Even equations with constant coefficients introduce structures not induced from the projective line.

B. Normal Form of the Quadratic Part of a Second-Order Equation in the Neighborhood of a Given Solution

Now we consider an arbitrary second-order nonlinear equation.

$$\frac{d^2y}{dx^2} = \Phi\left(x, y, \frac{dy}{dx}\right),$$

We will study the geometry of the two-parameter family of curves given by this equation in the (x, y)-plane. In particular, we are interested in finding out whether the configuration theorems hold for this family and whether this family can be rectified (turned into a family of straight lines) by an appropriate diffeomorphism of the plane. We shall see that such a rectification is not always possible, and find invariants measuring "infinitesimal nondesarguesness" (violation of the configuration theorems).

Theorem. *In the neighborhood of every linear element (x, y, p) of the plane of the dependent and independent variables, every second-order differential equation can be reduced to*

$$\frac{d^2y}{dx^2} = A(x)y^2 + O(|y|^3 + |p|^3), \qquad p = \frac{dy}{dx}$$

in the neighborhood of the element $(x = 0, y = 0, p = 0)$ by a local diffeomorphism of this plane.

◀ *1. Annihilation of the Linear Terms.* A given linear element defines a unique solution whose graph can be taken as the x-axis. We linearize the equation in the neighborhood of this solution. By the theorem in § 6A, the linear equation thus obtained can be locally rectified (reduced to the form $d^2y/dx^2 = 0$) by an appropriate choice* of the coordinate system. In this coordinate system, $\Phi(x, y, p)$, the right-hand side of our equation vanishes for $y = 0$, $p = 0$, together with its y- and p-derivatives. Con-

---

* It is useful to observe that if the right-hand side Φ is a polynomial of degree $n \geq 1$ in the argument p, then it is also such a polynomial in the new coordinates constructed in § 6A.

sequently, the Taylor series of Φ in y and p begins with terms of not less than second order:

$$\frac{d^2 y}{dx^2} = A(x)y^2 + B(x)yp + C(x)p^2 + O(|y|^3 + |p|^3).$$

2. Transformation of the Independent Variable. We consider the local diffeomorphism of the (x, y)-plane given by the substitution

$$x = F(X, Y), \qquad y = Y$$

(converting the point with coordinates (x, y) into the point with coordinates (X, Y)).

Lemma. *This substitution transforms the equation*

$$\frac{d^2 y}{dx^2} = \Phi(x, y, p), \qquad p = \frac{dy}{dx}$$

into the equation

$$\frac{d^2 Y}{dX^2} = \hat{\Phi}(X, Y, P), \qquad P = \frac{dY}{dX},$$

where

$$\hat{\Phi}(X, Y, P) = \frac{\Delta^3}{F'} \Phi\left(F, Y, \frac{P}{\Delta}\right) + \frac{F'' + 2PF'_Y + P^2 F_{YY}}{F'} P;$$

here $\Delta = F' + PF_Y$; the prime denotes the partial derivative with respect to X, and the arguments of F and of its derivatives X and Y.

◀ Let $y = u(x)$ be a solution of the initial equation and $Y = U(X)$ its image. Then

$$P = \frac{dU}{dX} = \left.\frac{du}{dx}\right|_F (F' + PF_Y), \qquad \left.\frac{du}{dx}\right|_{F(X, U(X))} = \frac{P}{\Delta}.$$

Consequently,

$$\frac{d^2 U}{dX^2} = \Delta \left.\frac{d}{dX}\frac{du}{dx}\right|_{F(X, U(X))} + \frac{P}{\Delta}\left(F'' + 2F'_Y P + F_{YY} P^2 + F_Y \frac{d^2 U}{dX^2}\right).$$

Moreover,

$$\left.\frac{d}{dX}\frac{du}{dx}\right|_{F(X, U(X))} = \left.\frac{d^2 u}{dx^2}\right|_F \Delta = \Phi\left(F(X, U(X)), U(X), \frac{P}{\Delta}\right)\Delta.$$

Hence,

$$\left(1 - \frac{PF_Y}{\Delta}\right)\frac{d^2U}{dX^2} = \Phi\left(F, U, \frac{P}{\Delta}\right)\Delta^2 + \frac{P}{\Delta}(F'' + 2PF_Y' + P^2 F_{YY}),$$

which implies the above formula. ▶

The formula just proved implies the following corollaries.

Corollary 1. *Let* $\Phi = 0$. *Then* $\hat{\Phi}$ *is a polynomial in P of degree not greater than 3.*

Corollary 2. *Let* Φ *be a polynomial in p of degree not greater than 3. Then* $\hat{\Phi}$ *is a polynomial in P of degree not greater than 3.*

Remark. The polynomials in p of degree $n \geqslant 4$ are not converted into polynomials in P by the transformation $\Phi \mapsto \hat{\Phi}$. Indeed, $\Delta^3(P/\Delta)^n$ is not a polynomial in P for $n > 3$.

Corollary 3. *A differential equation of the second order defines the structure of a local projective line on the graph of every solution, and the structure of a local projective plane on the normal bundle of the graph.*

◀ We consider a linearization of our equation along its solution. It is a second-order homogeneous linear equation; therefore, on the plane of the dependent and independent variables, one obtains the structure of a local projective plane, the structure for the initial solution being a local projective line (cf., § 6A).

With the notation of the lemma, the linearized equation has the form $d^2y/dx^2 = \Phi_1$, where $\Phi = \Phi_1 + \Phi_2 + \cdots$ is the expansion in powers of y and p.

Now we consider the normal bundle of the graph of the solution in the plane. The normal space at a point on the submanifold is the quotient of the tangent space to the ambient manifold to the tangent space of the submanifold at that point. The normal bundle of the submanifold is the union of the normal spaces at all points of the submanifold (equipped with the natural projection onto the submanifold).

The solution of the linearized equation assumes values in the normal bundle of the graph of the solution of the initial equation.

Indeed, the linearized equation is defined by means of the coordinate system (x, y), in which the x-axis is the graph of the solution under consideration. The value of the solution of the equation at a point is a vector in the y-direction tangent to the ambient plane at the point on the x-axis under consideration. Its projection in the normal space to the x-axis at this point defines a vector of the normal bundle. The solutions of the linearized equation thus define curves in the space of the normal bundle. It turns out that *these curves do not depend on the choice of the coordinate system in their construction;* in this sense, we say that the linearized equation can be considered as an equation in the normal bundle.

The proof of this statement easily follows, for example, from the lemma above.

In the notation of the lemma, the assertion to be proved means that if $F(X, 0) = X$ and Φ_1 is the linear term in the Taylor series of Φ in y and p, then $\hat{\Phi}_1(X, Y, P) = \Phi_1(X, Y, P)$. This equality easily follows from the formulas of the lemma.

Hence, the structure of a locally projective plane defined by the varied equation is given on the normal bundle. ▶

3. Transformation of the Dependent Variable. Consider the local diffeormorphism of the (x, y)-plane given by the substitution $y = G(X, Z)$, $x = X$, converting the point with coordinates (x, y) to the point with coordinates (X, Z).

Lemma. *The indicated substitution transforms the equation*

$$\frac{d^2 Y}{d^2 X} = \Phi(x, y, p), \qquad p = \frac{dy}{dx}$$

into the equation

$$\frac{d^2 Z}{dX^2} = \tilde{\Phi}(X, Z, \Pi), \qquad \Pi = \frac{dZ}{dX},$$

where

$$\tilde{\Phi}(X, Z, \Pi) = \frac{1}{G_Z} [\Phi(X, G(X, Z), G' + \Pi G_Z) - G'' - 2\Pi G'_Z - \Pi^2 G_{ZZ}];$$

here the prime indicates partial derivation with respect to X, the arguments of G and its derivatives are always X and Z.

◀ Let $y = v(x)$ be a solution of the initial equation, and $Z = V(X)$ its image. Then

$$\frac{dv}{dx} = G' + V' G_Z.$$

(Here and in the following the arguments of G and its derivatives are X and $V(X)$.) Moreover,

$$\frac{d^2 v}{dx^2} = G'' + 2G'_Z V' + G_{ZZ} V'^2 + G_Z V'' = \Phi(X, G, G' + XG_Z).$$

Determining V'' from this equation, we obtain the formula of the lemma. ▶

The lemma just proved implies the following corollaries.

Corollary 1. *Let $\Phi = 0$. Then $\tilde{\Phi}$ is a polynomial in Π of degree not greater than 2.*

Corollary 2. *Let Φ be a polynomial in p of degree not greater than n, $n \geqslant 2$. Then $\tilde{\Phi}$ is a polynomial in Π of degree not greater than n.*

4. Calculation of the Quadratic Terms. Consider the local diffeomorphism of the plane given by the substitution

$$x = F(X, Z) = X + f(X)Z + O(|Z|^2),$$
$$y = G(X, Z) = Z + g(X)Z^2 + O(|Z|^3).$$

Lemma. *This substitution transforms the equation*

$$\frac{d^2 y}{dx^2} = \Phi(x, y, p), \qquad p = \frac{dy}{dx}, \qquad \Phi = O(|y|^2 + |p|^2)$$

into

$$\frac{d^2 Z}{dX^2} = \Psi(X, Z, \Pi), \qquad \Pi = \frac{dZ}{dX},$$

where

$$\Psi(X, Z, \Pi) = \Phi(X, Z, \Pi) + \Omega(X, Z, \Pi) + O(|Z|^3 + |\Pi|^3),$$

$$\Omega = \alpha Z^2 + \beta Z\Pi + \gamma\Pi^2,$$

$$\alpha = -g'', \qquad \beta = -4g' + f'', \qquad \gamma = -2g + 2f'.$$

(X is the argument of f and g and their derivatives.)

◀ This can be proved by applying the lemmas concerning the transformations of the independent and then the dependent variable. Expanding the right-hand sides of the formulas obtained there in a series in (Z, Π) and retaining only the quadratic terms, we obtain the expression Ω indicated above as an addendum to the quadratic terms of Φ. ▶

    5. Reduction of the Quadratic Terms. Denote the quadratic terms of the initial right-hand side by

$$\Phi_2 = Ay^2 + Byp + Cp^2.$$

Then the quadratic terms of the transformed equation will be

$$\Psi_2 = (A - g'')Z^2 + (B - 4g' + f'')Z\Pi + (C - 2g + 2f')\Pi^2.$$

(X is the argument of the functions A, B, C, f, g, and their derivatives. It coincides with x along the solution under consideration). The following lemma is immediate.

Lemma. *The expression*

$$I = 6A - 2B' + C''$$

does not change in the transition from Φ_2 to Ψ_2.

    Choosing arbitrary functions f and g, we can annihilate two of the coefficients A, B, and C (preserving the value of I). In particular, choose f and g from the conditions $4g' - f'' = B$, $2g - 2f' = C$. Then we obtain $\Psi_2 = \bar{A}Z^2$, $\bar{A} = I/6$, which proves the theorem. ▶

C. Infinitesimal Nondesargueness

The coordinate system in which the second-order differential equation has the form

$$\frac{d^2 y}{dx^2} = A(x)y^2 + O(|y|^3 + |p|^3), \qquad p = \frac{dy}{dx}$$

near the graph of a fixed solution is not determined uniquely. We study the extent to which the coefficient A is invariant, i.e., independent of the method of reduction to normal form. A obstructs the rectifiability of the family of solutions and measures the infinitesimal nondesargueness at the point and in the direction in question.

Theorem. *The differential form of order 5/2*

$$\omega = A(x)|dx|^{5/2}$$

is invariantly determined up to a multiplicative constant along the graph of the zero solution.

In other words, if (X, Y) is another coordinate system in which the equation is also in normal form and $y = 0$ corresponds to the solution $Y = 0$ and the coefficient $A(x)$ is changed to $\bar{A}(X)$, then

$$\bar{A}(X) = CA(x)\left|\frac{dx}{dX}\right|^{5/2},$$

where C does not depend on x.

We shall call the form ω *the form of nondesargueness along the solution under consideration.*

◀ The most general diffeomorphism leaving the axis $y = 0$ invariant turns the point (x, y) into the point

$$X = f_0(x) + yf_1(x) + \cdots, \qquad Y = yg_1(x) + y^2g_2(x) + \cdots.$$

The vector of the normal bundle at the point x with y-component ξ is converted to a vector at the point $f_0(x)$ with the y-component $g_1(x)\xi$.

The projective structure of the normal bundle is defined in an invariant manner (cf., § 6B); therefore, the transformation $(x, \xi) \mapsto (f_0(x), g_1(x)\xi)$, constructed by means of the diffeomorphism $(x, y) \mapsto (X, Y)$, converting an equation in normal form into another equation in normal form must be projective. Thus, we find

$$f_0 = \frac{ax + b}{cx + d}, \qquad g_1 = \frac{C}{cx + d}.$$

Every diffeomorphism leaving the axis $y = 0$ invariant and preserving the projective structure of its normal bundle can therefore be represented as the product of a special transformation

$$X = f_0(x), \qquad Y = yg_1(x),$$

and a diffeomorphism preserving normal fibering pointwise,

$$(X, Y) \mapsto (X + Yf_1(X) + \cdots, \qquad Y + Y^2g_2(X) + \cdots).$$

The latter diffeomorphism can be represented as a product of transformations of the dependent and independent variables which we considered in § 6B. Therefore, the invariant I consisting of those terms of the right-hand side of the equation which are quadratic in y and p (cf., § 6B5) does not change under this diffeomorphism.

We study the behavior of I under a special projective transformation.

Every projective transformation of the line splits into the product of translations, expansions, and the transformation $x \mapsto 1/x$.

The expression I is invariant with respect to translations, and expansions of x and y only multiply I by a constant. Therefore, it is sufficient to consider the behavior of I under the substitution $x = 1/X, y = Y/X$.

Calculating the derivatives $P = dY/dX$ and $dP/dX = d^2 Y/dX^2$, we find that $P = y - px$, $dP/dX = X^{-3} dp/dx$ (where $p = dy/dx$). Consequently,

$$\frac{d^2 Y}{dX^2} = X^{-3} A(1/X)(Y/X)^2 + O(|y|^3 + |P|^3).$$

Therefore, the coefficient of Y^2 is equal to $X^{-5} A(x)$. ▶

D. Construction of Scalar Invariants

The differential form ω introduced above yields scalar functions connected with the equation in an invariant manner.

First of all, we note that with any differential form (of arbitrary order) on a one-dimensional manifold, one can associate a vector field in an invariant manner. The value of the form on a vector of this field is equal to 1 at every point.

For example, the vector field $v(x)\partial/\partial x$, where $v = A^{-2/5}$, is invariantly connected with the form $A(x)(dx)^{5/2}$.

Theorem. *Let $v(x)\partial/\partial x$ be a vector field on the line. Then the following scalar functions are connected with this field invariantly with respect to projective transformations of the line:*

$$I_2 = 2v''v - v'^2, \quad I_3 = 2v'''v^2, \ldots, \quad I_n = vI'_{n-1}, \ldots.$$

Here the prime indicates the derivation with respect to x.

◀ The invariance of I_2 can easily be verified by straightforward calculation: it is sufficient to consider the substitution $x = 1/X$, since invariance with respect to translations and expansions is obvious. The derivative of the function along a vector field connection with the function and the field is invariant not only with respect to projective transformations, but also all diffeomorphisms of the line. Therefore, the invariance of all I_n follows from that of I_2. ▶

Remark 1. The invariant I_2 is constructed by the following procedure. The Lie algebra of the projective group of the line is generated by the fields $\partial/\partial x$, $x\partial/\partial x$, and $x^2\partial/\partial x*$.

---

*In affine coordinates, the corresponding one-parameter groups of projective transformations have the form $g^t x = x + t$, $g^t x = e^t x$, $g^t x = x/(1 - tx)$, and consequently, in homogeneous coordinates they are given by second-order unimodular matrices

$$\begin{pmatrix} 1 & t \\ 0 & 1 \end{pmatrix}, \quad \begin{pmatrix} \exp(t/2) & 0 \\ 0 & \exp(-t/2) \end{pmatrix}, \quad \begin{pmatrix} 1 & 0 \\ -t & 1 \end{pmatrix}.$$

Therefore, every vector field can be approximated by a projective field (a field from the Lie algebra of the projective group) up to quadratic terms at every point.

Under projective transformations, the projective field approximating the initial field at the initial point turns into a new projective field approximating the transformed field at the image point. The action of the projective transformations of the line on the three-dimensional space of projective fields is the adjoint action of the projective group on its Lie algebra. This action preserves the quadratic form on the algebra. Indeed, if we express projective transformations by second-order matrices, and the projective fields by matrices of the infinitesimal generators of the one-parameter groups corresponding to them, then the action of the transformation g on the field v is described as the matrix product $gv^{-1}g$. On the other hand, $\det gvg^{-1} = \det v$. Therefore, the determinant of v is a quadratic form on the Lie algebra of the projective fields invariant with respect to the adjoint representation. Consequently, this determinant, calculated for the projective field approximating the vector field under consideration is a scalar connected with the field in an invariant manner with respect to projective transformations.

In the above basis of the Lie algebra, the approximating projective field has the components $(v, v', v''/2)$. Therefore, the matrix corresponding to the field has the form

$$\begin{pmatrix} v'/2 & v \\ -v''/2 & -v'/2 \end{pmatrix};$$

its determinant is I_2 (up to an inessential factor).

Remark 2. It seems clear that every function (polynomial, series, etc.) of the values of v and a finite number of derivatives of v which is invariantly connected with v with respect to projective transformations can be expressed in terms of the invariants I_k.

Remark 3. The projective invariants of a function on the projective line can be constructed in the following way: let its differential (1-form) correspond to the function, let its field correspond to its form, and let the invariants I_k correspond to the field. In particular, the simplest invariant of f with respect to projective transformations of the independent variable is

$$I_2[f] = \frac{2f'f''' - 3f''^2}{f'^4}.$$

(This differs from Schwartz' derivative which is invariant with respect to projective transformations of the axis of the values of the function by the factor f'^2 in the denominator).

Remark 4. The invariants $I_2, I_3, \ldots$ are multiplied by $\lambda^2, \lambda^3, \ldots$ if the vector field is multiplied by λ. It is easy to construct combinations of them which are not sensitive to multiplication of the field by a number; for example, $J = I_2^3/I_3^2$.

Therefore, the matrices of the generating operators are

$$\begin{pmatrix} 0 & 1 \\ 0 & 0 \end{pmatrix}, \quad \begin{pmatrix} \frac{1}{2} & 0 \\ 0 & -\frac{1}{2} \end{pmatrix}, \quad \begin{pmatrix} 0 & 0 \\ -1 & 0 \end{pmatrix}.$$

The quantity J corresponding to v constructed from the (5/2)-form ω is a scalar function on the space of linear elements in the plane entirely independent of the choice of coordinates and only depending on the initial differential equation.

E. Equations Cubic with Respect to the Derivative

The vanishing of the form ω of nondesargueness along any solution is necessary for the rectifiability of the equation (its reduction to the form $d^2y/dx^2 = 0$), but, as we shall immediately see, it is not sufficient.

Theorem 1. *Assume that the differential equation*

$$\frac{d^2y}{dx^2} = \Phi(x, y, p), \qquad p = \frac{dy}{dx},$$

can be reduced to the form $d^2y/dx^2 = 0$ by a diffeomorphism of the plane. Then Φ is a polynomial in p of degree not greater than 3.

In other words, a differential equation of the family of all lines on the plane described in an arbitrary system of curvilinear coordinates has, for its right-hand side, a polynomial in the first derivative of degree not greater than 3.

Theorem 1 follows from the (curious) fact below.

Theorem 2. *Assume that the right-hand side of the differential equation*

$$\frac{d^2y}{dx^2} = \Phi(x, y, p)$$

is a polynomial in p of degree not greater than 3. *Then every diffeomorphism of the plane transforms the equation into an equation of the same kind*, i.e., *the right-hand side remains a polynomial in the derivative of degree not greater than* 3.

◀ Theorem 2 follows from the lemmas in § 6B on the effect of the change of the independent and dependent variables on the right-hand side of the equation (cf., the corollaries in § 6B), since every local diffeormorphism of the plane can be obtained by successive applications of these changes of variables. ▶

◀ Theorem 1 follows from Theorem 2 and the fact that zero is a polynomial in p of degree not greater than 3. ▶

Exercise. *In the equation*

$$\frac{d^2y}{dx^2} = a_0 y'^3 - a_1 y'^2 + b_1 y' - b_0$$

(where a_i and b_i are functions of x and y), make the change of variables $(x, y) \mapsto (y, x)$.

Solution

$$\frac{d^2x}{dy^2} = b_0 x'^3 - b_1 x'^2 + a_1 x' - a_0.$$

Remark. It can be shown that the conditions $\omega = 0$ and $d^4\Phi/dp^4$ are independent. Therefore, the condition $\omega = 0$ is not sufficient for the reducibility of the equation to the form

$$\frac{d^2 y}{dx^2} = 0.$$

The two conditions $\{\omega \equiv 0,\ d^4\Phi/dp^4 \equiv 0\}$ together are sufficient for the reducibility of the equation to the form $d^2y/dx^2 = 0$. This can be seen from the formulas of § 6B (after some calculation).

Exercise. Prove that every second-order equation can be reduced to the form $d^2y/dx^2 = p^2\ B(x, y, p)$ locally (in the neighborhood of the point $x = 0$ of the solution $y = 0$).

F. The Geometry of a Pair of Direction Fields in Three-Dimensional Space

We consider a pair of direction fields in three-dimensional space. It turns out that the local classification (up to a diffeomorphism of the space) of such pairs in general position is equivalent to the local classification of second-order differential equations (up to diffeomorphisms of the plane of the dependent and independent variables; local, meaning near a given point with direction).

First of all, we associate a two-parameter family of curves in the plane with the pair of direction fields in three-dimensional space.

To do this, we rectify the first field by a local diffeomorphism of the space, converting the family of integral curves of the first field into a family of parallel vertical lines. After this, we project the integral curves of the second field onto the horizontal plane in the direction of the vertical lines. On the horizontal plane (on the quotient plane of the space modulo the integral curves of the first field), we obtain a two-parameter family of curves.

From this two-parameter family of curves, we construct a second-order differential equation for which these curves are graphs of solutions.

To this end, we note that in a generic local two-parameter family of curves in the plane, a unique curve of the family passes through every point of the plane in every direction near every linear element (a point and a direction) of every curve of the family. (For this, only a certain Jacobian has to be different from zero).

If the family is generic in the indicated sense and (x, y) are the coordinates in the plane, then d^2y/dx^2 is a smooth function of the element under consideration i.e., $(x, y, dy/dx)$. In this way we obtain a second-order differential equation $d^2y/dx^2 = \Phi(x, y, dy/dx)$. The graphs of the solutions of this equation are curves of the family (by the uniqueness theorem).

Consequently, to a pair of direction fields in three-dimensional space we assigned a second-order differential equation under the nondegeneracy condition that the corresponding Jacobian is different from zero.

Definition. A direction field in three-dimensional space is said to be *nondegenerate with respect to vertical lines* if the direction of the field is nowhere vertical and the horizontal projection of the direction of the field rotates with nonzero velocity as a point moves along a vertical line.

A pair of direction fields in three-dimensional space is said to be nondegenerate if, after applying a diffeomorphism making the first field vertical, the second field becomes nondegenerate with respect to the first one.

Remark. It is easy to see that the definition is unambiguous: if the second field becomes nondegenerate with respect to the first after some diffeomorphism rectifying the integral curves of the first field, then it becomes so after any other diffeomorphism rectifying the first field. It is also not difficult to see that the nondegeneracy of a pair is preserved if the order of the fields is changed (i.e., it is immaterial which one of the two fields is rectified).

Above we associated a second-order differential equation with each nondegenerate pair of local direction fields in three-dimensional space. Indeed, our nondegeneracy condition coincides with the condition that the Jacobian $\partial(y, y')/\partial(u, v)$ does not vanish, where (u, v) are the parameters of the family. Now we prove that this correspondence establishes a complete equivalence of the problems of local classification of pairs of direction fields in $\mathbb{R}^3$ and of differential equations of the second order.

Theorem. *Every second-order equation can be obtained from an appropriate nondegenerate pair of direction fields in three-dimensional space by the construction described above.*
If two pairs of fields can be converted into each other by a diffeomorphism of the space, then the corresponding equations can be converted into each other by a diffeomorphism of the plane. Conversely, if two second-order equations can be converted into each other by a diffeomorphism of the plane, then any two pairs of fields corresponding to them can be converted into each other by a diffeomorphism of the space.

Consequently, the correspondence between pairs of fields and equations is one-to-one (up to diffeomorphisms).
◀ For every equation $d^2y/dx^2 = \Phi(x, y, dy/dx)$ we consider, in the three-dimensional space of 1-jets with coordinates (x, y, p), the family of "vertical" lines in the p-direction (x = const, y = const) and the family of integral curves of the system $dy/dx = p$, $dp/dx = \Phi$ which is equivalent to the equation.

By projection in the vertical direction, we obtain the family of graphs of solutions of our equation. Hence, for every equation, we have produced a pair of direction fields from which it is obtained.

A diffeomorphism of the space mapping vertical lines into vertical lines defines a diffeomorphism of the horizontal plane. Therefore, a diffeomorphism of two pairs of fields in $\mathbb{R}^3$ induces a diffeomorphism of the corresponding equations.

Conversely, diffeomorphisms of the horizontal plane act on linear elements (directions) of curves in the plane. Therefore, a diffeomorphism converting the first equation to the second defines a diffeomorphism in three-dimensional space from the first pair of fields to the second. (To a point of the first space there corresponds an element of a curve in the plane; the diffeomorphism takes it to a new place, the element thus obtained is the projection of the direction of the second field of the second pair at a unique point of the second space; this point is made to correspond to the initial one.) ▶

It follows from the theorem just proved that all our results concerning the geometry of the family of solutions of a second-order equation can be reformulated in terms of the geometry of a nondegenerate pair of direction fields in three-dimensional space (or, if one prefers, in terms of the geometry of the simplest case of a system of two first-order implicit differential equations, where the derivatives are two-valued functions of the coordinates).

We also remark that a nondegenerate pair of direction fields defines a contact structure in three-dimensional space (a completely nonintegrable field of planes spanned by the given directions; for more details, cf., Chapter 2). In the sense of this structure, the integral curves of our fields form Legendre bundles*). Therefore, we may reformulate the preceding results as results on the geometry of a pair of Legendre bundles in $\mathbb{R}^3$.

G. Duality

Above, for every nondegenerate pair of direction fields in three-dimensional space, we constructed a second-order differential equation by rectifying the first field and by projecting the integral curves of the second. However, it would have been possible to proceed conversely: to rectify the second field and project the integral curves of the first one. As a result, we generally obtain another second-order equation.

Hence, a dual equation exists for every second-order differential equation.

In terms of pairs of direction fields in three-dimensional space, the passage to the dual equation amounts to reversing the order of the fields. Another description of the dual equation is as follows. We assume that the family of solutions of the second-order equation for $y(x)$ depending on two parameters is written in the form $F(x, y; u, v) = 0$. Let us now regard x and y as parameters and u and v as variables. Then this relation defines a two-parameter family of functions $v(u)$. This family is the family of solutions of a second-order equation; this is the equation dual to the initial one.

Exercise. Prove that the nondesargueness form ω of a second-order equation (cf., § 6B) is equal to zero if and only if the right-hand side of the dual equation is a polynomial in the derivative of degree not greater than 3.

Hence, the rectifiability condition for a second-order equation can be formulated as follows:

An equation $d^2y/dx^2 = \Phi(x, y, dy/dx)$ can be reduced to the form $d^2y/dx^2 = 0$ if and only if the right-hand side is a polynomial in the derivative of order not greater than 3 both for the equation and for its dual.

H. Survey

The geometry of a second-order equation is the source of several mathematical theories.

1. Tresse, a student of Lie, in his prize-winning paper "Determination des invariants punctuels de l'equation differentielle ordinaire de second ordre", Leipzig, 1896 (cf. also his article, *Sur les invariants differentiels des groupes continues des transformations*, Acta Math. **18** (1894), 1–88.) constructed all semi-invariants of an equation.

A semi-invariant of order k is, by definition, a function of the value of the right-hand

---

*A Legendre submanifold of a contact structure is, by definition, an integral submanifold of greatest possible dimension. A Legendre bundle is a bundle with Legendre fibers.

side of the equation and its derivatives of order not greater than k at point (x, y, p). If it vanishes for the equation at the given point of this three-dimensional space of elements, then it will vanish for the equation transformed by a diffeomorphism of the plane at the transformed element.

It turns out that there are exactly two functionally independent semi-invariants of order 4. One of them is $d^4\Phi/dp^4$. The other is the scalar invariant I_2 constructed from the form of infinitesimal nondesargueness (cf., § 6D). For an equation in the normal form $d^2y/dx^2 = A(x)y^2 + O(|y|^3 + |p|^3)$, we have

$$I_2 = 2v''v - v'^2, v = A^{-2/5}.$$

Consequently, $5AA'' - 12A'^2$ is a semi-invariant.

Tresse also found three semi-invariants of order 5, and for $k > 5$ also $(k^2 - k - 8)/2$ semi-invariants of order k; all other semi-invariants are functions of these. Tresse also mentions that all semi-invariants "can be deduced from the three simplest ones" by differentiations.

2. The problem of the geometry of a second-order differential equation lead Cartan to the theory of manifolds of projective connection (See Cartan, *Sur les variétés à connection projective*, Bull. Soc. Math. France **52** (1924), 205–241; *Oeuvres* III, **1**, N 70, 825–862, Paris, 1955.)

By definition, projective connection on a manifold assigns to every smooth path (on the manifold) a projective mapping of the tangent space at the initial point into the tangent space at the endpoint depending on the path smoothly.* In particular, a projective transformation of the initial tangent space corresponds to a closed path. An "infinitely small projective transformation" corresponding to a circuit along an "infinitely small parallelogram" is called the curvature form of the connection.

A projective connection is called a torsion-free connection if the origin of the tangent space remains at its place if we translate the space along a closed path. From among all torsion-free connections, Cartan chooses the normal connections. In the two-dimensional case, the normalcy of a connection means that every straight line issued from the origin of the tangent space maps onto itself after the traversal of a closed path.

A geodesic line of a connection is, by definition, a line on the manifold whose tangent turns into tangent after displacement along this line.

It turns out that the geodesics of a projective, torsion-free connection in the plane are graphs of solutions of a second-order differential equation whose right-hand side is a polynomial in the derivative of degree not greater than 3. Conversely, to every second-order equation whose right-hand side is a polynomial of degree not greater than 3, there corresponds exactly one normal projective, torsion-free connection for which the geodesics are the graphs of solutions.

To general second-order equations, Cartan also assigns a unique normal projective, torsion-free connection, but the translation of the two-dimensional plane is then defined along paths in the three-dimensional space of elements (see the details in the cited work of Cartan).

3. A translation of Tresse's theory into the language of pairs of direction fields in space was done by Bol [Bol, *Über topologische Invarianten von zwei Kurvenscharen im Raum*, Abh. Math. Sem. Univ. Hamburg **9**, 1 (1932), 15–47.]

---

* And satisfying certain natural conditions which we do not give here.

In this theory, the problem of orbit space has apparently remained unresolved, in particular, the question of determining how many parameters are needed to enumerate the orbits of a k-jet of a second-order equation (under the action of the group of diffeomorphisms of the plane on the space of k-jets of the equation), in the neighborhood of a generic point in the space of k-jets.

It seems that the Poincaré series counting these parameters are rational (in this problem, as in many other analytical classification problems, where functional moduli occur) and that these series provide some rigorous meaning to the expression "depends on p arbitrary functions in q variables".

Chapter 2

First-Order Partial Differential Equations

Partial differential equations have been studied much less than ordinary differential equations. Part of the theory—that of a single partial differential equation of the first order—can be reduced to the study of special ordinary differential equations, the so-called characteristic equations. This is possible because, physically speaking, non-interacting particles can be described by either the partial differential equation of the field or ordinary equations for the particles.

In this chapter we develop this theory (which mathematically reduces to the geometry of the so-called contact structures). We also consider the problem of integrability of a field of hyperplanes (the theorem of Frobenius).

§ 7. Linear and Quasilinear First-Order Partial Differential Equations

The integration of a first-order partial differential equation reduces to the integration of a system of ordinary differential equations, the so-called characteristic equations. The basis of this reduction is a simple geometric analysis of the formation of surfaces from families of curves. We begin with these geometric arguments, and then apply them to partial differential equations.

A. Integral Surfaces of Direction Fields

Let X be a smooth manifold and let V be a direction field on X.

Definition. A smooth submanifold $Y \subset X$ is called an *integral surface* of V if the tangent plane of Y contains the direction of V at every point (Fig. 41).

Theorem. *A submanifold $Y \subset X$ is an integral surface of a field V if and only if it contains, together with every point of it, a segment of the integral curve passing through this point.*

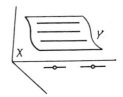

Figure 41.   Figure 42.

The assertion of the theorem is local and invariant with respect to diffeo-morphisms. Therefore, it is sufficient to prove it for a standard field of parallel directions in a linear space. In this case, the assertion of the theorem is obvious. [It reduces to the fact that a function given on an interval is constant if and only if its derivative is equal to zero (Fig. 42)].

Let Γ be a k-dimensional submanifold in an n-dimensional manifold X (Fig. 43). Γ is called a hypersurface if $k = n - 1$.

Definition. The *Cauchy problem* for the direction field V with initial manifold Γ is the problem of finding a $(k + 1)$-dimensional integral submanifold of V containing the initial submanifold Γ.

We note that the integral curves passing through Γ do not always form a submanifold even locally in the neighborhood of Γ, cf., Fig. 44.

Definition. A point of the initial manifold is said to be *characteristic* in the direction field V if the direction of V at this point is tangent to the initial manifold.

Theorem. *Let a point of the k-dimensional initial manifold be given which is not characteristic in V. There exists a $(k + 1)$-dimensional integral surface of the field containing a neighborhood of this point on the initial manifold. This surface is unique in the sense that any two integral surfaces containing a neighborhood of the point under consideration on the initial manifold coincide in some neighborhood of this point.*

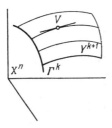

Figure 43.   Figure 44.

◀ The assertion is local and invariant with respect to diffeomorphisms. Therefore, it is sufficient to prove it for a standard field of parallel directions in a linear space. In this case the assertion is obvious. ▶

B. The Homogeneous Linear First-Order Equation

Definition. A *homogeneous linear first-order equation* is an equation

$$L_a u = 0, \tag{1}$$

where a is a given vector field on a manifold M.

Let the field a have the components $(a_1, \cdots, a_n)$ in the coordinates $(x_1, \cdots, x_n)$; every component is a function of the coordinates. Equation (1) becomes

$$a_1 \frac{\partial u}{\partial x_1} + \cdots + a_n \frac{\partial u}{\partial x_n} = 0. \tag{2}$$

Definition. The field a is called the *characteristic vector field* of Eq. (1) and its phase curves are called *characteristics*. The equation $\dot{x} = a(x)$ is called the *characteristic equation* of the partial differential equation (1).

Theorem. *A function u is a solution of a linear first-order equation if and only if it is a first integral of the characteristic equation.*

◀ This is the definition of first integrals. ▶

Definition. The *Cauchy problem* for Eq. (1) is the problem of finding a solution u satisfying the condition $u|_\gamma = \varphi$, where γ is some smooth hypersurface in M and φ is a prescribed smooth function on this hypersurface.

The hypersurface γ is called the *initial hypersurface* and the function φ the *initial condition*.

We note that such a problem does not always have a solution. Indeed, on each characteristic, the solution u is constant. However, a characteristic may intersect the initial surface γ several times. If the values of the prescribed function φ are different at these points, then the corresponding Cauchy problem does not have a solution in any domain containing the characteristic in question (Fig. 45).

Definition. A point x on the initial hypersurface γ is said to be *noncharacteristic* if the characteristic passing through this point is transversal (not tangent) to the initial hypersurface.

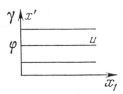

Figure 45.   Figure 46.

Theorem. *Let x be a noncharacteristic point on the initial hypersurface. There exists a neighborhood of x such that the Cauchy problem for Eq. (1) has a unique solution in this neighborhood.*

◀ By the rectifiability theorem, we may choose the coordinates near x so that the components of the field a have the form $(1, 0, \cdots, 0)$ and the equation of γ takes the form $x_1 = 0$ (Fig. 46).

The Cauchy problem in these coordinates becomes

$$\frac{\partial u}{\partial x_1} = 0, \qquad u\big|_{x_1 = 0} = \varphi. \tag{3}$$

The unique solution (in a convex domain) is $u(x_1, x') = \varphi(x')$, where $x' = (x_2, \cdots, x_n)$. ▶

Remark. The solutions of any ordinary differential equation form a finite-dimensional manifold: every solution can be given by a finite collection of initial conditions. We see that in the case of a homogeneous linear first-order partial differential equation in a function of n variables, "there are as many solutions as there are functions of $n - 1$ variables".

Below, we shall observe an analogous phenomenon in the case of general first-order partial differential equations.

C. The Nonhomogeneous Linear Equation of the First Order

Definition. A nonhomogeneous linear equation of the first order is an equation

$$L_a u = b, \tag{4}$$

where a is a given vector field on a manifold M, and b is a given function on M. In coordinates,

$$a_1 \frac{\partial u}{\partial x_1} + \cdots + a_n \frac{\partial u}{\partial x_n} = b, \tag{5}$$

where $a_k = a_k(x_1, \ldots, x_n)$, $b = b(x_1, \ldots, x_n)$.

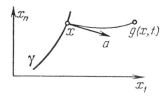

Figure 47.

The Cauchy problem for Eq. (4) can be formulated in the same way as for the homogeneous equation (1).

Theorem. *In a sufficiently small neighborhood of a noncharacteristic point x_0 of the initial surface γ, the Cauchy problem for Eq. (4) has a unique solution; it is given by the formula*

$$u(g(x, t)) = \varphi(x) + \int_0^t b(g(x, \tau))d\tau,$$

where $g(x, t)$ is the value of the solution of the characteristic equation (with the initial condition $g(x, 0) = x$ on the initial surface) at time t.

◀ Upon rectifying the field a, we obtain the problem

$$\frac{\partial u}{\partial x_1} = b, \qquad u\big|_{x_1 = 0} = \varphi(x').$$

Its unique solution is

$$u(x_1, x') = \varphi(x') + \int_0^{x_1} b(\xi, x')d\xi. \ \blacktriangleright$$

In other words, Eq. (4) means that the derivative of the solution along the characteristic is the known function b. The increment of the solution on a segment of the characteristic has to be equal to the integral of b over this segment (Fig. 47).

D. The First-Order Quasilinear Equation

Definition. A *first-order quasilinear equation* is an equation

$$L_\alpha u = \beta, \tag{6}$$

where $\alpha(x) = a(x, u(x))$, $\beta(x) = b(x, u(x))$.

Here x is a point of the manifold M, u is an unknown function on M, a is a given vector field tangent to M depending on the point $u \in \mathbb{R}$ as on a parameter, and b is a given function on $M \times \mathbb{R} = J^0(M, \mathbb{R})$.

In coordinates, the quasilinear equation (6) has the form

$$a_1(x, u) \frac{\partial u}{\partial x_1} + \cdots + a_n(x, u) \frac{\partial u}{\partial x_n} = b(x, u). \tag{7}$$

This differs from a linear equation in that the coefficients a_k and b may depend on the unknown function.

EXAMPLE

We consider a homogeneous medium consisting of particles moving along straight lines in an inertial system, so that the velocity of each particle remains constant. We denote by $u(x, t)$ the velocity of the particle at point x at time t. We write the Newton equation: the acceleration of a particle is equal to zero. If $x = \varphi(t)$ is the motion of the particle, then $\dot{\varphi} = u(\varphi(t), t)$ and

$$\ddot{\varphi} = \frac{\partial u}{\partial x} \dot{\varphi} + \frac{\partial u}{\partial t} = u \frac{\partial u}{\partial x} + \frac{\partial u}{\partial t}.$$

Consequently, the field of the velocities u of noninteracting particles satisfies the quasilinear equation

$$u u_x + u_t = 0. \tag{8}$$

Problem. Construct the graph of the function $u(\,\cdot\,, t)$ if the graph of the function $u(\,\cdot\,, 0)$ has the form shown in Fig. 48.

Answer. Cf., Fig. 49. For $t \geqslant t_1$ there exists no smooth solution. Beginning at $t = t_1$, the particles collide in the medium. [The physical hypothesis of inertial motion, i.e., the lack of interaction between the particles, becomes unrealistic and has to be replaced by other physical hypotheses, the description of the nature of the collision. Thus arise so-called shock waves, functions of the form shown in Fig. 50, satisfying Eq. (8) outside the discontinuity and some additional conditions of physical origin at the discontinuity.]

Figure 48.

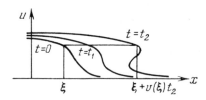

Figure 49.

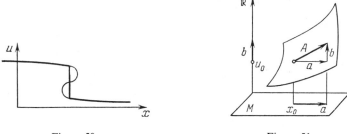

Figure 50.   Figure 51.

E. Characteristics of First-Order Quasilinear Equations

We have just seen how useful it is to pass from the field of velocities to the motion of the particles for the special quasilinear equation (8). Something similar can be done in the case of the general equation (6). In this case, the role of the motion of the particles is played by some curves in the direct product of the domain and range of the unknown function; these curves are called the characteristics of the quasilinear equation.

The quasilinear equation (6) for an unknown function $u: M \to \mathbb{R}$, i.e.,

$$L_{a(x, u(x))}u = b(x, u(x)) \qquad (6)$$

means that if the point x leaves x_0 and begins to move on M with velocity $a(x_0, u_0)$, then the value of the solution $u = u_0$ begins to change with velocity $b(x_0, u_0)$.

In other words, the vector $A(x_0, u_0)$ applied at the point (x_0, u_0) of the space $M \times \mathbb{R}$, which has components $a(x_0, u_0)$ along M and $b(x_0, u_0)$ along $\mathbb{R}$, is tangent to the graph of the solution (Fig. 51).

Definition. The vector $A(x_0, u_0)$ is called the *characteristic vector* of the quasilinear equation (6) at the point (x_0, u_0). The characteristic vectors at all points of the space $M \times \mathbb{R}$ form a *vector field A*. This field is called the *characteristic vector field* of the quasilienear equation (6). The phase curves of the characteristic vector field are called the *characteristics* of the quasilinear equation.

The differential equation given by the field A of phase velocities is called the *characteristic equation*.

EXAMPLE

Let M be equal to $\mathbb{R}^n$ with coordinates $(x_1, \ldots, x_n)$. The characteristic field is given by its components; their values at the point (x, u) are equal to

$$a_1(x, u), \ldots, a_n(x, u); b(x, u).$$

The characteristic equation has the form

$$\dot{x}_1 = a_1(x, u), \ldots, \dot{x}_n = a_n(x, u); \dot{u} = b(x, u).$$

Problem. Determine the characteristics of the equation $uu_x + u_t = 0$ of the medium of noninteracting particles.

Answer. $\dot{x} = u$, $\dot{t} = 1$, $\dot{u} = 0$. The characteristics are the straight lines $x = a + bt$, $u = b$.

Remark. A linear equation is a special case of a quasilinear one. However, the characteristics of a linear equation are not the same as the characteristics of the same equation considered as a quasilinear equation: the characteristics of the linear equation lie in M, and those of the quasilinear one in $M \times \mathbb{R}$. The characteristics of a linear equation are the projections from $M \times \mathbb{R}$ to M of the characteristics of the same equation considered as a quasilinear equation.

F. Integration of a First-Order Quasilinear Equation

Let A be the characteristic vector field of the quasilinear equation (6). We assume that this field vanishes nowhere. Then it defines a direction field.

Definition. The direction field of the characteristic vector field of a quasilinear equation is called the *characteristic direction field* of the equation.

The characteristics of a quasilinear equation are the integral curves of the characteristic direction field.

EXAMPLE

In the case $M = \mathbb{R}^n$ with coordinates $(x_1, \ldots, x_n)$, the characteristic equation can be written in the so-called *symmetric form*

$$\frac{dx_1}{a_1} = \frac{dx_2}{a_2} = \ldots = \frac{dx_n}{a_n} = \frac{du}{b},$$

expressing the collinearity of the tangent to the characteristic with the characteristic vector.

Theorem. *A function u is a solution of a quasilinear equation if and only if its graph is an integral surface of the characteristic direction field.*

◄ This is obvious, since Eq. (6) says that a characteristic vector is tangent to the graph. ►

Corollary. *A function u is a solution of a quasilinear equation if and only if its graph contains, together with every point of it, a segment of the characteristic passing through the point.*

◀ This follows from § 7A. ▶

Consequently, the determination of the solutions of a quasilinear equation reduces to finding its characteristics. If the characteristics are known, it only remains to construct a surface from them which is the graph of a function: this function will be a solution of the quasilinear equation, and all solutions are obtained in this way.

G. The Cauchy Problem for a First-Order Quasilinear Equation

Let $\gamma \subset M$ be a hypersurface (a submanifold of codimension 1) in a manifold M and let $\varphi: \gamma \to \mathbb{R}$ be a smooth function (Fig. 52).

Definition. The *Cauchy problem* for the quasilinear equation (6) *with initial condition φ on γ* consists of determining a solution u which is equal to φ on γ.

The solution of this problem reduces to the solution of the Cauchy problem for the characteristic direction field.

We consider the graph of the function $\varphi: \gamma \to \mathbb{R}$. This graph is a hypersurface in the direct product $\gamma \times \mathbb{R}$. Since γ is imbedded in M, we may regard the graph Γ of φ as a submanifold (of codimension 2) of $M \times \mathbb{R}$ (Fig. 52).

Definition. The submanifold $\Gamma \subset M \times \mathbb{R}$ which is the graph of φ on γ is called the *initial submanifold* for the initial condition φ on γ.

Consequently, the initial manifold Γ determines both the hypersurface γ in M and the initial condition φ on γ.

Definition. An initial condition (φ, γ) is said to be *noncharacteristic for the quasilinear equation* (6) at the point x_0 from γ if the vector $a(x_0, u_0)$ $(u_0 = \varphi(x_0))$ is not tangent to the surface γ at this point (Fig. 52).

Remark. If the equation is linear, then the vector $a(x_0, u_0)$ does not depend on u_0. Therefore, noncharacteristic points of the surface γ can be defined.

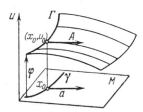

Figure 52.

On the other hand, for a quasilinear equation, only a point $(x_0, u_0) \in \Gamma$ can be characteristic or noncharacteristic, but we cannot speak of a point $x_0 \in \gamma$ being characteristic.

Theorem. *For an initial condition noncharacteristic at a point x_0, the quasilinear equation* (6) *has a locally unique solution in the neighborhood of this point.*

◀ If the initial condition is not characteristic at the point x_0, then we have:

(1) The characteristic field A does not vanish in the neighborhood of the point (x_0, u_0). Therefore, a smooth characteristic direction field is defined in the neighborhood of this point.

(2) The characteristic direction is not tangent to the initial manifold Γ at the point under consideration, and consequently, in its neighborhood. Therefore, the local integral surface, containing the initial manifold Γ, of the characteristic direction field exists and is unique (cf., § 7A).

(3) The tangent plane to the integral surface at the point (x_0, u_0) is nonvertical (does not contain the u-axis). Therefore, the integral surface is the graph of a function. This function is the desired solution (cf., § 7E). ▶

Remark. The proof also contains a procedure for constructing the solution of the Cauchy problem for a quasilinear equation.

§ 8. The Nonlinear First-Order Partial Differential Equation

Nonlinear first-order partial differential equations, like linear ones, can be integrated by means of characteristics. However, while the characteristics of a linear equation with respect to a function on M lie in M, and those of a quasilinear equation lie in $M \times \mathbb{R}$, the characteristics of a general nonlinear first-order partial differential equation lie in the manifold $J^1(M, \mathbb{R})$ of 1-jets of functions.

The manifold of 1-jets of functions has a natural contact structure.

The integration of nonlinear first-order partial differential equations depends on simple geometric facts concerning this contact structure, with which we begin this section.

A. Contact Manifolds

A contact manifold is, by definition, a manifold equipped with a field of hyperplanes in the tangent spaces satisfying the condition of "maximal nonintegrability".

A field of planes (in contrast to a field of one-dimensional directions)

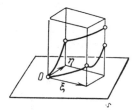

Figure 53.

may not have integral surfaces whose dimension is equal to the dimension of the planes. In order to measure the obstruction to the existence of integral surfaces of a field of hyperplanes, we carry out the following construction.

In the neighborhood of the point O of the manifold under consideration, we introduce coordinates so that the plane of the field at O becomes a coordinate hyperplane; we shall call the corresponding coordinates *horizontal* and the remaining coordinate *vertical*.

For any path in the horizontal plane, we construct a vertical cylinder over the path. The trace of our field of planes on the lateral surface of the cylinder gives a direction field. The integral surfaces (if they exist) intersect the cylinder in integral curves of the direction field (Fig. 53).

Consequently, we may lift any path from the horizontal plane to the desired surface.

Now let (ξ, η) be two vectors in the horizontal coordinate plane at the point under consideration. Take the parallelogram spanned by (ξ, η). There are two paths along the sides of the parallelogram from the point under consideration to the opposite vertex of the parallelogram. By lifting these paths, we obtain two points above the opposite vertex which are different in general. Their difference is the obstruction to the construction of an integral surface, or, as is said, to the "integrability" of the field of hyperplanes.

We consider the difference between the vertical coordinates of the two points obtained above. The principal bilinear part (with respect to ξ and η) of this difference measures the degree of nonintegrability of the field. In order to give a formal definition, we carry out the following construction.

The field of tangent hyperplanes to the manifold can be given locally by a differential 1-form α which is nowhere equal to the zero form. It is defined up to multiplication by a function which vanishes nowhere: the planes of the field are the null spaces of the form (those subspaces of the tangent space on which the form vanishes).

EXAMPLE

In the space $\mathbb{R}^{2n+1}$ with coordinates $(x_1, \ldots, x_n; u; p_1, \ldots, p_n)$, consider the 1-form $\alpha = du - p\,dx$ (where $p\,dx = p_1\,dx_1 + \cdots + p_n\,dx_n$). This 1-form is not equal to the

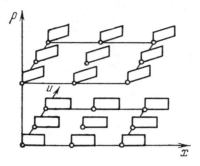

Figure 54.

zero form at any point of $\mathbb{R}^{2n+1}$ and, consequently, defines the field of $2n$-dimensional planes $\alpha = 0$ in $\mathbb{R}^{2n+1}$ (Fig. 54).

Definition. A differential 1-form α which is nowhere equal to the zero form on a manifold M is called a *contact* form if the exterior derivative $d\alpha$ of α defines a nondegenerate exterior 2-form in every plane $\alpha = 0$.

[A bilinear form $\omega: L \times L \rightarrow \mathbb{R}$ is nondegenerate if $\forall \xi \in L \backslash 0 \exists \eta \in L$: $\omega(\xi, \eta) \neq 0$.]

A nondegenerate skew-symmetric 2-form on a linear space is also called a *symplectic structure*.

EXAMPLE

The form constructed in the preceding example is a contact form. Indeed, the exterior derivative of the form α is equal to

$$dx_1 \wedge dp_1 + \cdots + dx_n \wedge dp_n.$$

In the plane $\alpha = 0$, $(x_1, \ldots, x_n; p_1, \ldots, p_n)$ may serve as coordinates. The matrix of the form $\omega = d\alpha|_{\alpha=0}$ has the form $\begin{pmatrix} 0 & -E \\ E & 0 \end{pmatrix}$ in these coordinates, where E is the identity matrix of order n. The determinant of this matrix is equal to 1. Consequently, the 2-form ω is nondegenerate.

Remark. Nondegenerate skew-symmetric bilinear forms can be defined only on *even-dimensional* spaces. Therefore, contact forms exist only on odd-dimensional manifolds.

Theorem. *Let α be a contact form and f a function which vanishes nowhere. The form $f\alpha$ is also a contact form. Moreover, the symplectic structures*

$$d\alpha|_{\alpha=0}, \qquad d(f\alpha)|_{f\alpha=0}$$

differ by a nonvanishing factor on the plane $\alpha = 0$.

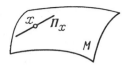

Figure 55.

◀ Using the product rule, we obtain

$$d(f\alpha) = df \wedge \alpha + f\,d\alpha.$$

However, $df \wedge \alpha$ is the zero 2-form on the plane $\alpha = 0$. Consequently, the 2-forms $d\alpha$ and $d(f\alpha)$ differ by the nonzero factor f on the plane $\alpha = 0$. In particular, the 2-form $d(f\alpha)|_{\alpha=0} = f\,d\alpha|_{\alpha=0}$ is nondegenerate. Therefore, $f\alpha$ is a contact form. ▶

Definition. A *contact structure* on a manifold M is a field of tangent hyperplanes which are given locally as the set of zeros of a contact 1-form. The hyperplanes of the field are called *contact hyperplanes*. We shall denote by Π_x the contact hyperplane at the point x (Fig. 55).

Remark 1. From the preceding theorem, it follows that it does not depend on the choice of the form whether the 1-form which defines the field of planes is a contact form. This is determined by the field of contact planes itself. Indeed, if β is another form defining the same field, then (locally) β differs from α by a nowhere-vanishing factor; consequently, the forms β and α are either both contact forms or neither of them is.

Remark 2. Contact structures exist only on odd-dimensional manifolds.

Definition. A submanifold of a contact manifold is said to be an *integral manifold of the field of contact planes* if the tangent plane of the submanifold belongs to the contact plane at every point.

Problem. Prove that *the dimension of an integral manifold of a field of contact planes on a contact manifold of dimension $2n + 1$ does not exceed n.*

Solution. On such a manifold Y, the form $i^*\alpha$ (where $i: Y \to M^{2n+1}$ is the imbedding) is equal to zero. Therefore, $i^*\,d\alpha = di^*\alpha = 0$. Hence, every two vectors of the tangent space are skew-orthogonal: $\omega(\xi, \eta) = 0$. From this it follows that the dimension of the tangent space is not greater than n (cf., Problem 2 in § 8D).

Remark. There exist integral manifolds of dimension n for a contact field in M^{2n+1}. They are called *Legendre submanifolds*. We are going to see now that a Legendre submanifold in the space of 1-jets corresponds to every function.

B. Contact Structure on a Manifold of 1-Jets of Functions

Let V be an n-dimensional manifold. Consider the manifold $J^1(V, \mathbb{R})$ of 1-jets of functions on V.

A 1-jet of a function f on V is defined by a point $x \in V$, the value $u = f(x)$ of f at x, and the first differential of f at x. Consequently, the manifold of 1-jets of functions on V has dimension $2n + 1$. If $(x_1, \ldots, x_n)$ are local coordinates on V, then a 1-jet of a function f on V is defined by a collection $(x_1, \ldots, x_n; u; p_1, \ldots, p_n)$ of $2n + 1$ numbers, where $p_i = (\partial f / \partial x_i)(x)$.

Definition. The *standard contact form* on the manifold $J^1(V, \mathbb{R})$ of 1-jets of functions is the 1-form

$$\alpha = du - p\, dx \qquad (p\, dx = p_1\, dx_1 + \cdots + p_n\, dx_n).$$

We have seen above that this is indeed a contact 1-form on $\mathbb{R}^{2n+1}$.

It is easy to see that the form α defined by means of coordinates does not depend on the choice of $(x_1, \ldots, x_n)$. It is defined globally.

Definition. The 1-*graph* of a function $f: V \to \mathbb{R}$ is the submanifold consisting of the 1-jets of f at all points of V.

Consequently, the 1-graph of a function of n variables is an n-dimensional surface in a $(2n + 1)$-dimensional space.

Theorem. *The standard contact form on the manifold of 1-jets of functions of n variables vanishes on all tangent planes of 1-graphs of functions. The closure of the union of the planes tangent to all 1-graphs of functions coincides with the null space of this form (at every point of the space of jets).*

◀ The first part follows from the definition of the total differential $du = p\, dx$ of a function. The second part follows from the existence of a function with prescribed partial derivatives at a given point. ▶

From the theorem just proved, it follows that the field of planes defined by the standard contact form in the space of 1-jets does not depend on the choice of the coordinate system in the definition of the standard contact form.

Definition. *The standard contact structure* on the manifold of 1-jets of functions on V is the field of hyperplanes which are unions of planes tangent to graphs of functions on V.

Problem. The groups of diffeomorphisms of the spaces V and $\mathbb{R}$ act naturally on the manifold $J^1(V, \mathbb{R})$ of 1-jets of functions. Prove that *the standard contact structure of the space of 1-jets is invariant under this action.*

Moreover, the standard contact 1-form α is invariant under the action of the group of diffeomorphisms of V and is multiplied by a nonvanishing function under the action of diffeomorphisms of the real line $\mathbb{R}$.

C. Geometry on a Hypersurface in a Contact Manifold

We pass to a general contact manifold M^{2n+1}. Let E^{2n} be a smooth hypersurface in M^{2n+1} (Fig. 56).

Definition. A surface E^{2n} in a contact manifold M^{2n+1} is said to be *noncharacteristic* if its tangent plane and contact plane are transversal at every point x (i.e., their sum is the whole tangent space to M^{2n+1} or, equivalently, they intersect in a $(2n - 1)$-dimensional space).*

Definition. The intersection of a plane tangent to a noncharacteristic hypersurface with the contact plane at the point under consideration of a hypersurface in a contact manifold is called the *characteristic plane* at this point:

$$P_x = T_x E \cap \Pi_x.$$

*A generic hypersurface in a contact space may have some isolated points, where the contact plane is tangent to the hypersurface and the transversality does not hold, that is, there may be singular points of the equation for partial derivatives defined by the hypersurface.

For a generic surface at some neighborhood of a point we can reduce both the surface and the contact structure to the normal form

$$dt = \sum p_k \, dq_k - q_k \, dp_k, \qquad t = Q(p, q),$$

where Q is a nondegenerate quadratic form in p and q (which may be chosen in the form $\sum \lambda_k p_k q_k$ in the complex case).

This theorem for smooth (C^∞) functions is proved in V. V. Lychagin, *The local classification of first-order partial differential equations*, Uspekhi Mat. Nauk (Russian Math. Surveys) **30**, 1 (1975), 101–171 (see also: V. I. Arnold, A. B. Givental, *Symplectic Geometry*, Itogi Nauki i Techniki Sovremennye Problemy Mathematiki, Fundamentalnye Napravleniia Dynamicheskie Systemy, Vol. 4, pp. 5–140, Mosc. VINITI, 1985 (Springer translation, 1988). The analytical version of the Lychagin theorem has been proved by M. Ya. Zhitomirskii, Funct. Anal. Appl. **20**, 2 (1986), pp. 65–66.

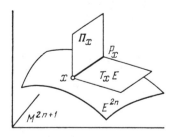

Figure 56.

Consequently, the characteristic planes on a hypersurface in M^{2n+1} form a field of $(2n - 1)$-dimensional planes on a $2n$-dimensional manifold: this is the field of planes defined by the intersections of the contact planes with the tangent spaces of the hypersurface.

It turns out that the contact structure defines a straight line in each of these $(2n - 1)$-dimensional planes, the so-called *characteristic direction*.

D. Skew-Orthogonal Complement

In order to define the characteristic direction, we recall that in a linear space with a nondegenerate bilinear form, orthogonal complements are defined (the orthogonality of a pair of vectors is defined as the vanishing of the form on the pair).

EXAMPLES

1. Let (L, ω) be a Euclidean space, i.e., let L be a linear space and ω a scalar product. To every vector ξ there corresponds a 1-form $\omega(\xi, \cdot)$, the scalar product with the vector ξ. The value of this 1-form at the vector η is equal to $\omega(\xi, \eta)$.
    · For example, grad f is the vector corresponding to the 1-form df.
    To a straight line in L there corresponds the orthogonal complement of this line (the plane of zeros of the 1-form corresponding to the vector of the line). Every plane of codimension 1 is the orthogonal complement of a line in L (Fig. 57). We note that multiplication of ω by a nonzero number preserves orthogonality of vectors. Therefore, the correspondence between straight lines and their orthogonal complements does not change if we multiply the scalar product by a nonzero number.

2. Let (L, ω) be a symplectic space, i.e., let L be a linear space and ω a skew-scalar product (a bilinear skew-symmetric nondegenerate form).
    To every vector ξ there corresponds a 1-form $\omega(\xi, \cdot)$, the scalar product with ξ. The value of this 1-form at the vector η is equal to $\omega(\xi, \eta)$ (Fig. 58).
    For example, a Hamiltonian field with Hamiltonian function H is a field corresponding to the 1-form dH.
    To a straight line in L there corresponds its *skew-orthogonal complement* (the plane of zeros of the 1-form corresponding to the vector of the line). Every plane of codimension 1 is the skew-orthogonal complement of a unique line in L.

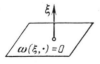

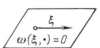

Figure 57.   Figure 58.

We note that orthogonality is preserved under multiplication of ω by a nonzero number. Therefore, the correspondence between straight lines and their skew-orthogonal complements does not change if we multiply the symplectic structure ω by a nonzero number.

Problem. Prove that *every straight line lies in its skew-orthogonal complement.*

Solution. $\omega(\xi, \xi) = -\omega(\xi, \xi) = 0.$

Problem. Prove that *if all vectors in a subspace of a 2n-dimensional sympletic space are skew-orthogonal to each other, then the dimension of this subspace does not exceed n.*

Solution. The skew-orthogonal complement of a k-dimensional subspace has dimension $2n - k$ (indeed, choose a basis $(e_1, \ldots, e_k)$; then the equations $\omega(e_1, \xi) = 0, \ldots, \omega(e_k, \xi) = 0$ form k independent equations in ξ, since a dependence between the equations would imply a dependence between the vectors e_i in view of the condition that ω is nondegenerate).

If $k > n$, then the dimension of the orthogonal complement is smaller than n and the space cannot be skew-orthogonal to itself.

Remark. In a $2n$-dimensional symplectic space there exist n-dimensional subspaces which are skew-orthogonal to themselves. They are called Lagrange subspaces.

E. Characteristics on a Hypersurface in a Contact Space

We return to the geometry on a noncharacteristic hypersurface in a contact manifold M^{2n+1}.

At every point of the hypersurface E^{2n}, we have defined the $(2n - 1)$-dimensional characteristic plane P: the intersection of the contact plane with the tangent plane of E^{2n} (Fig. 59).

Definition. The *characteristic direction* at a noncharacteristic point x on a hypersurface in a contact space is the skew-orthogonal complement of the characteristic plane in the contact plane (the skew-scalar product is defined as $d\alpha|_{\alpha=0}$): l_x is the skew-orthogonal complement of

$$P_x^{2n-1} = T_x E^{2n} \cap \Pi_x^{2n}$$

in Π_x^{2n}.

Figure 59.

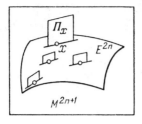

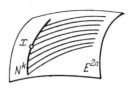

Figure 60.   Figure 61.

We note that the characteristic direction lies in the characteristic plane (cf., § 8D); in the case $n = 1$ the characteristic direction coincides simply with the characteristic plane.

In the general case, the characteristic directions at all points of a non-characteristic smooth hypersurface in a contact space form a smooth field of directions on the hypersurface; at every point, the characteristic direction belongs to the contact hyperplane at this point (Fig. 60).

Definition. The integral curves of the field of characteristic directions on a noncharacteristic hypersurface E in a contact manifold are called the *characteristics of the hypersurface E*.

Let N^k be a k-dimensional integral manifold, lying in E^{2n}, of a field of contact planes.

Definition. A point $x \in N^k$ is said to be *noncharacteristic* if the tangent plane of N at this point does not contain the characteristic direction (Fig. 61).

Problem. Prove that if x is a noncharacteristic point of the manifold N^k, then $k \leqslant n - 1$, and give an example of an n-dimensional integral surface of a contact field of planes in M^{2n+1} lying in E^{2n}.

It turns out that the characteristics passing through a point of an integral noncharacteristic manifold N^k themselves form (locally) an *integral* $(k + 1)$-dimensional submanifold, lying on the hypersurface E^{2n}, of the field of contact planes. In order to prove this, we need a simple general lemma on the invariance of a field of planes with respect to a one-parameter group of diffeomorphisms.

F. Digression: A Condition for the Invariance of a Field of Planes

Let a be a nonvanishing differential 1-form. This form defines a field of hyperplanes. Let v be a nonvanishing vector field. This field defines a direction field. (i.e., a field of straight lines).

We *assume that the direction v of the field at each point belongs to the hyperplane of zeros of the form a at this point:*

$$a(v) \equiv 0.$$

Lemma. *In order that the field of zeros of a be invariant with respect to the phase flow of the field v, it is necessary and sufficient that $da(v, \xi) = 0$ for all ξ in the plane $a(\xi) = 0$ of the field.*

◀ The assertion of the lemma is local and invariant with respect to diffeomorphisms. Therefore, it is sufficient to prove it for the standard field $v = \partial/\partial x_1$ in a Euclidean space with coordinates $(x_1, \ldots, x_n)$ (in view of the theorem on the rectification of a vector field). Let $a = a_1 \, dx_1 + \cdots + a_n \, dx_n$. By assumption, we have $a_1 \equiv 0$ (since $a(v) \equiv 0$).

Problem. Prove that the value of the exterior derivative of the form a at the pair (v, ξ) (where $v = \partial/\partial x_1$, $a(v) \equiv 0$, and ξ is an arbitrary vector) coincides with the value of the partial derivative $\partial a/\partial x_1$ at the vector ξ.

Solution

$$da \equiv \sum \frac{\partial a_i}{\partial x_j} dx_j \wedge dx_i \quad \Rightarrow \quad \left(da(v, \xi) = \sum \frac{\partial a_i}{\partial x_i} \xi_i - \sum \frac{\partial a_1}{\partial x_j} \xi_j \right).$$

However, $a_1 \equiv 0$.

[A perhaps clearer solution can be obtained by applying Stokes' formula to the parallelograms with sides v, ξ.]

The condition for the invariance of the field of zeros of the form a with respect to translations along the x_1-axis is that the partial derivative $\partial a/\partial x_1$ must vanish on the plane $a = 0$ of the field.

Knowing that $\partial u/dx_1(\xi) = da(v, \xi)$, we see that the condition for invariance has the form

$$(a(\xi) = 0) \Rightarrow (da(v, \xi) = 0). \blacktriangleright$$

G. The Cauchy Problem for the Field of Characteristic Directions

We return to the noncharacteristic hypersurface E^{2n} in a contact manifold M^{2n+1}. Let N^k be an integral submanifold, lying in E^{2n}, of the field of contact planes.

Definition. The *Cauchy problem* for a hypersurface E^{2n} in a contact manifold M^{2n+1} with initial manifold N^{n-1} consists of determining an integral manifold Y^n of the field of contact planes, lying in E^{2n} and containing the initial manifold N^{n-1} (Fig. 62).

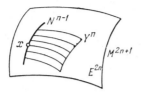

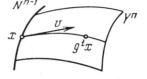

Figure 62.   Figure 63.

Theorem. *Let x be a noncharacteristic* point of the initial manifold N^{n-1}. There exists a neighborhood U of x such that the solution of the Cauchy problem for $E^{2n} \cap U$ with initial condition $N^{n-1} \cap U$ exists and is locally unique (i.e., any two solutions with the same initial condition coincide in some neighborhood of x).*

The manifold Y^n consists of characteristics passing through the points of the initial manifold N^{n-1}.

◀ The family of characteristics passing through the points of the initial manifold forms a smooth manifold Y^n of dimension n in E^{2n} in the neighborhood of the point x. We prove that this manifold is the integral manifold for the field of contact hyperplanes.

We consider a contact 1-form α defining the contact field of hyperplanes in the neighborhood of x. We denote by a the restriction of α to the neighborhood of x in the hypersurface E^{2n}.

The form a is nonvanishing, since E^{2n} is noncharacteristic (cf., § 8C). This form defines the field of tangent hyperplanes on E^{2n} (they are the traces of the contact hyperplanes on E^{2n}). The field of characteristic directions on E^{2n} lies in the field of null planes of a (cf. § 8E).

In the neighborhood of x on E^{2n}, we consider a vector field v whose direction is characteristic everywhere. We denote by $\{g^t\}$ the local one-parameter group of diffeormorphisms on E^{2n} defined by v (g^t is defined in the neighborhood of x for t near 0), Fig. 63.

Every point of Y^n can be obtained from some point of the initial manifold by means of an appropriate diffeormorphism g^t.

Lemma A. *The form a vanishes on the planes tangent to Y at the points of the initial manifold.*

Lemma B. *The diffeomorphisms g^t convert null planes of a into null planes.*

◀ A. The form a is equal to zero at vectors tangent to N^{n-1}, since N is an integral manifold. The form a is equal to zero at v, since the characteristic direction lies in the contact hyperplane. Consequently, the form a is equal to 0 on $T_x N + \mathbb{R}v$. ▶

---

*The definition of noncharacteristic points is given in § 8E.

◀ *B*. We consider a vector ξ at which *a* vanishes. (ξ is not necessarily based at a point of the initial manifold). We calculate the value of *da* at the pair (v, ξ) at this point. By the definition of the form $a = \alpha|_E$, this value is equal to the value of the derivative of the contact form $d\alpha(v, \xi)$. By the definition of characteristic direction, the latter is equal to 0 for any vector ξ in the contact plane. By the lemma in § 8F, the condition $a = 0$ is invariant with respect to $\{g^t\}$. ▶

Lemmas A and B imply that the form *a* vanishes at all vectors tangent to Y^n. Consequently, Y^n is an integral manifold.

Hence, we have constructed an integral submanifold $Y^n \subset E^{2n}$ of the field of contact hyperplanes, passing through the initial manifold N^{n-1}.

H. Proof of Uniqueness

Lemma. *The tangent plane to any n-dimensional integral manifold in E^{2n} of the field of contact planes contains the characteristic direction.*

◀ The restriction of the contact 1-form α to its integral manifold is equal to zero. The restriction of the 2-form $d\alpha$ to this manifold is equal to the derivative of the restriction of the form α, i.e., equal to zero. Consequently, any two vectors tangent to the integral manifold are skew orthogonal (in the sense of the skew-scalar product $d\alpha|_{\alpha=0}$).

The vector of a characteristic direction on E^{2n} is skew orthogonal to all vectors in E^{2n} lying in the plane $\alpha = 0$. Assume that it does not belong to the tangent plane of an *n*-dimensional integral manifold, lying in E^{2n}, of the field of contact planes. Then the subspace spanned by this vector and this tangent plane has dimension $n + 1$. However, all vectors of the subspace just constructed are mutually orthogonal. According to Problem 2 in § 8D, the dimension of this subspace does not exceed *n*. ▶

It follows from the lemma that every *n*-dimensional integral manifold in E^{2n} of the field of contact planes contains, together with every point of it, a segment of the characteristic passing through this point. This implies the uniqueness of the integral manifold containing a given initial manifold. ▶

I. Application to a Nonlinear First-Order Partial Differential Equation

We now interpret nonlinear first-order partial differential equation for a function $u: V^n \to \mathbb{R}$ as a hypersurface E^{2n} in the manifold $M^{2n+1} = J^1(V^n, \mathbb{R})$ of 1-jets equipped with the standard contact structure.

Let $(x_1, \ldots, x_n)$ be local coordinates on V^n and let *u* be the coordinate in $\mathbb{R}$. Denote by $(x_1, \ldots, x_n; u; p_1, \ldots, p_n)$ the corresponding local coordinates in the space of 1-jets. Then the differential equation can be written

locally in the form

$$\Phi(x, u, p) = 0. \tag{1}$$

Solving this equation reduces to determining the integral surfaces of the field of contact planes in E^{2n} which are 1-graphs of functions (cf., § 8B).

Our general theorems reduce the problem of solving this equation to the construction of characteristics on E^{2n}, for which we need to find the integral curves of a field of directions on E^{2n} (i.e., we have to solve a system of $2n - 1$ ordinary differential equations).

Theorem. *The solutions of Eq. (1) are the functions whose 1-graphs consist of characteristics on E^{2n}.*

◀ Cf., § 8B, § 8G, and § 8I. ▶

J. The Cauchy Problem for a Nonlinear First-Order Partial Differential Equation

Let $\gamma^{n-1} \subset V^n$ be an $(n - 1)$-dimensional submanifold of a manifold V^n, $\varphi : \gamma^{n-1} \to \mathbb{R}$ a smooth function, and $E^{2n} \subset J^1(V^n, \mathbb{R})$ the smooth non-characteristic hypersurface given by Eq. (1).

Definition. The *Cauchy problem* for Eq. (1) is the problem of determining a solution $u : V \to \mathbb{R}$ which is equal to φ on γ.

Definition. The *initial manifold N* constructed from the initial condition (γ, φ) is the set consisting of all 1-jets of functions on V^n satisfying the following conditions (Fig. 64):

(1) The base point of the jet lies on γ^{n-1}.
(2) The value of the function at this point is equal to φ.

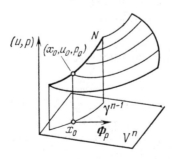

Figure 64.

(3) The value of the total differential of the function at this point is such that its restriction to the plane tangent to γ^{n-1} is equal to the total differential of the initial condition φ.

(4) The jet belongs to E^{2n}.

Definition. A point of the initial manifold is said to be *noncharacteristic* for Eq. (1) if the projection of the characteristic direction at this point onto V is transversal to γ. (This definition is different from the one given in § 8E.)

Remark. We say that "the derivatives of the unknown function are determined by the initial condition in $n - 1$ directions on γ, and the derivative in the last direction (transversal to γ) is determined by Eq. (1)."

EXAMPLE

Let γ be given by the equation $x_1 = 0$ in the space with coordinates (x_1, x'). Then N is given by the conditions

$$x_1 = 0, \qquad u = \varphi(x'), \qquad p' = \frac{\partial \varphi}{\partial x'};$$

p_1 is determined from the equation $\Phi(x, u, p) = 0$.

Theorem. *Assume that the point (x_0, u_0, p_0) of the space of 1-jets is a noncharacteristic point of the initial manifold N. Then the solution of Eq. (1) with initial condition N exists in some neighborhood U of the initial point x_0 and is locally unique (in the sense that every two solutions of the equation satisfying the initial condition $u|_{U \cap \gamma} = \varphi|_{U \cap \gamma}, u(x_0) = u_0, du(x_0) = p_0$ coincide in some neighborhood of x_0).*

◀ This follows from the theorem of § 8G, which also provides a means of constructing the solution. ▶

K. Explicit Formulas

Problem. Calculate explicitly the differential equation of the characteristics for the equation $\Phi(x, u, p) = 0$.

Answer.

$$\dot{x} = \Phi_p,$$
$$\dot{p} = -\Phi_x - p\Phi_u,$$
$$\dot{u} = p\Phi_p.$$

Solution. Consider a vector tangent to the manifold $\Phi = 0$. For such a vector with components (X, U, P) we have

$$\Phi_x X + \Phi_u U + \Phi_p P = 0.$$

This vector lies in the contact plane if the form $du - p\,dx$ vanishes for it, i.e., $U = pX$.

Hence, a condition necessary and sufficient for the vector (X, pX, P) to lie in the intersection of the contact plane with the tangent to the manifold $\Phi = 0$ is that

$$(\Phi_x + \Phi_u p)X + \Phi_p P = 0. \tag{2}$$

The characteristic vector $(\dot{x}, \dot{u} = p\dot{x}, \dot{p})$ is defined by the condition that the skew-scalar product with all vectors (X, pX, P) satisfying the preceding equality vanishes.

This skew-scalar product is equal to the value of $d\alpha = dx \wedge dp$ at the pair of the vectors $(\dot{x}, \dot{u} = p\dot{x}, \dot{p})$, $(X, U = pX, P)$, i.e., equal to $\dot{x}P - \dot{p}X$.

Hence, Eq. (2) for X and P must be equivalent to the equation

$$\dot{x}P - \dot{p}X = 0. \tag{3}$$

Consequently, the coefficients of X and P in Eqs. (2) and (3) are proportional. This provides the above answer in view of the equality $\dot{u} = p\dot{x}$.

L. Conditions for a Noncharacteristic Point

Problem. Give explicit conditions for γ, φ, and Φ under which the point (x_0, u_0, p_0) is noncharacteristic for the equation $\Phi(x, u, p) = 0$ with the initial condition φ on γ.

Answer. $\Phi_p(x_0, u_0, p_0)$ is not tangent to γ at x_0 (cf., Fig. 62).

Solution. We project the plane tangent to the surface formed by the characteristics passing through the initial manifold onto the x-space. If the surface is the 1-graph of the solution and the point (x_0, u_0, p_0) is noncharacteristic, then the tangent plane is generated by the tangent to N and the characteristic direction. It is projected isomorphically. Consequently, the x-component of the characteristic vector must be transversal to γ at x_0. But this component is Φ_p (cf. § 8K).

Conversely, let Φ_p be not tangent to γ at x_0. Then:

(1) *In the neighborhood of* (x_0, u_0, p_0), *the hypersurface* $\Phi = 0$ *is smooth.* Indeed, $\Phi_p \neq 0$, and consequently, $d\Phi|_{(x_0, u_0, p_0)} \neq 0$.

(2) *At* (x_0, u_0, p_0), *the equation* $\Phi = 0$ *is noncharacteristic.* Indeed, the vector $(0, 0, \Phi_p)$ lies in the contact plane and is not tangent to the surface $\Phi = 0$ at (x_0, u_0, p_0), since $\Phi_p(x_0, u_0, p_0) \neq 0$.

(3) *Near* (x_0, u_0, p_0) *the initial manifold N is smooth.*

Indeed, we choose coordinates $(x_1, \ldots, x_n) = (x_1, x')$ so that the local equation of γ takes the form $x_1 = 0$. Then the condition of solvability of the equation

$$\Phi\left(0, x', \varphi(x'), p_1, \frac{\partial \varphi}{\partial x'}\right) = 0$$

for $p_1(x')$ takes the form $\partial\Phi/\partial p_1|_{(x_0, u_0, p_0)} \neq 0$; the condition that the vector Φ_p is not tangent to γ has the same form.

(4) *The point (x_0, u_0, p_0) of the initial manifold is noncharacteristic.*

Indeed, if the characteristic vector were tangent to the initial manifold N, then its projection Φ_p would be tangent to the projection γ of N onto the x-space.

(5) *The characteristics intersecting the initial manifold in the neighborhood of $(x_0,$ $u_0, p_0)$ form a smooth manifold in this neighborhood which is projected diffeomorphically onto the x-space (and thus is a 1-graph of a function).*

Indeed, the image of the plane tangent to this manifold at (x_0, u_0, p_0) under projection onto the x-space contains the tangent plane of γ and a vector transversal to it. Consequently, at (x_0, u_0, p_0), the derivative of the mapping under consideration is an isomorphism, and the projection mapping itself is a local diffeomorphism (by the inverse function theorem).

Hence, all five conditions for being noncharacteristic are satisfied at (x_0, u_0, p_0) if Φ_p is not tangent to γ at this point.

M. The Hamilton–Jacobi Equation

Definition. *The Hamilton–Jacobi equation is the equation*

$$H(x, u_x) = 0. \tag{1}$$

The difference between this and a general first-order partial differential equation is that here the unknown function does not appear explicitly in the equation.

EXAMPLE

Let γ be a smooth hypersurface in the Euclidean space $\mathbb{R}^n$ and let $u(x)$ be the distance of the point x from γ (Fig. 65). The function u satisfies the Hamilton–Jacobi equation

$$\left(\frac{\partial u}{\partial x_1}\right)^2 + \cdots + \left(\frac{\partial u}{\partial x_n}\right)^2 = 1 \tag{2}$$

(at the points of smoothness of the function u).

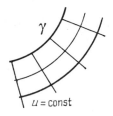

Figure 65.

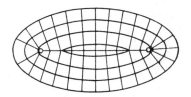

Figure 66.

Indeed, the modulus of the gradient of this function is equal to the modulus of the derivative of the distance from γ in the direction normal to γ, i.e., it is equal to 1.

Globally, the function u may not be smooth. For example, let γ be an ellipse in the plane. The singularities of u form an interval inside the ellipse (Fig. 66).

Problem. Prove that every solution of a Hamilton–Jacobi equation (2) is locally the sum of the distance from a hypersurface and a constant.

When studying the Hamilton–Jacobi equation, it is useful to consider the cotangent bundle T^*V^n instead of the manifold $J^1(V^n, \mathbb{R})$ of 1-jets of functions on V^n. In mechanics, the space T^*V^n is called the *phase space of the configuration space V^n*. A *vector cotangent* to V^n at x is, by definition, a linear homogeneous function on the space tangent to V^n at x. All vectors cotangent to V^n at x form a linear space called the *space cotangent to V^n* at x and denoted by $T_x^*V^n$. The vectors cotangent to V^n at all points form a smooth manifold of dimension $2n$. This is called the *cotangent bundle of V^n* (or simply the cotangent bundle) and is denoted by T^*V^n.

Let $(x_1, \ldots, x_n)$ be local coordinates on V^n. A cotangent vector to V^n at x can be given by a collection $(p_1, \ldots, p_n)$ of n numbers in the following way. To the collection p_k of numbers, there corresponds the 1-form $p_1 \, dx_1 + \cdots + p_n \, dx_n$ on the tangent space to V^n at x. The collection $(p_1, \ldots, p_n; x_1, \ldots, x_n)$ of $2n$ numbers forms a system of local coordinates in the cotangent bundle of V^n.

There exists a natural projection π of the space $J^1(V^n, \mathbb{R})$ of 1-jets of functions onto the cotangent bundle of V^n:

$$\pi: J^1(V^n, \mathbb{R}) \to T^*V^n.$$

The mapping π "eliminates the value of a function"; in coordinates, it can be given as

$$(x_1, \ldots, x_n; u; p_1, \ldots, p_n) \to (p_1, \ldots, p_n; x_1, \ldots, x_n).$$

Definition. *The characteristics of the Hamilton–Jacobi equation* (1) *are the projections of the characteristics of the first-order partial differential equation* (1) *onto the cotangent bundle.*

Problem. Determine the differential equation of the characteristics of the Hamilton–Jacobi equation (1).

Answer. $\dot{x} = H_p, \dot{p} = -H_x$.

Remark. This system of differential equations is called the *system of Hamilton's canonical equations*. The corresponding vector field is defined on not only the surface $H = 0$ but on the whole phase space.

Problem. Determine the characteristics of the Hamilton–Jacobi equation (2).

Answer. $x = 2at + b, p = a$ (a and b are constant vectors and $a^2 = 1$).
Consequently, the projections of the characteristics onto V^n are straight lines.

In geometrical optics, the Hamilton–Jacobi equation (2) is called the *eikonal equation* and the projections of the characteristics onto V^n are called *rays*. The function u is called the *optical distance* and its level surfaces are called *fronts*. Besides these objects, in geometrical optics, *caustics* play an essential role. We consider, for example, a wall lighted by rays reflected from a concave surface (for example, from the interior of a cup). On the wall, more intense lines are visible with singular points, which lines are the caustics.

The definition of caustics is as follows. Consider the Cauchy problem for a first-order partial differential equation. Even if the corresponding characteristics can be extended unboundedly and do not intersect, forming a global integral manifold, the projection of this manifold onto V^n is generally not a diffeomorphism.

The set of critical values of the projection of the integral manifold onto V^n is then called a caustic.

In the special case of the Hamilton–Jacobi equation (2) with initial condition $u = 0$ on γ, the caustic is the locus of focal points or centers of curvature of the hypersurface γ.

Problem 1. Draw the locus of the centers of curvature of an ellipse in the plane.

Problem 2. On every interior normal of an ellipse, an interval of length t is drawn. Determine the curve thus obtained and study its change as t increases.

Answer. Cf., Fig. 67.

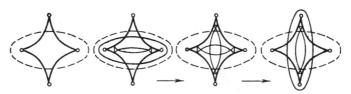

Figure 67.

§ 9. A Theorem of Frobenius

A direction field in the plane always defines a family of integral curves and is locally rectifiable (it reduces to a field of parallel lines by a diffeomorphism). In a three-dimensional space this is not true: a field of planes in $\mathbb{R}^3$ may in general not have integral surfaces.

In this section we discuss conditions for the local rectifiability of a field of hyperplanes, i.e., conditions under which the field is a field of tangent hyperplanes to a family of smooth hypersurfaces.

A. Completely Integrable Fields of Hyperplanes

Let M^n be a smooth manifold on which a field of hyperplanes is given.

In the neighborhood of any point, such a field is given by a differential 1-form α which is not degenerated to the zero form and is determined up to multiplication by a function which vanishes nowhere.

Definition. A field of hyperplanes is said to be *completely integrable* if the form $d\alpha$ is identically zero on any plane of the field.

Remark. The property of local integrability does not depend on the choice of the form α which defines it locally, since under the multiplication of α by a nonvanishing function, the form $d\alpha|_{\alpha=0}$ is multiplied by this function. (cf., § 8A).

Proposition. *In order that a field of hyperplanes $\alpha = 0$ be completely integrable it is necessary and sufficient that*

$$\alpha \wedge d\alpha \equiv 0.$$

◀ In the tangent space of M^n at the point under consideration, we choose a basis (Fig. 68) of $n - 1$ "horizontal" vectors $(e_1, \ldots, e_{n-1})$ in the plane $\alpha = 0$ and one "vertical" vector f. The value of the 3-form $\alpha \wedge d\alpha$ is equal to zero at the three horizontal vectors, since $\alpha = 0$. Moreover, $(d\alpha \wedge \alpha)$ $(e_i, e_j, f) = 0$ is a sum in which every term contains a factor of the form either $\alpha(e_i)$ or $d\alpha(e_i, e_j)$; these quantities are equal to zero.

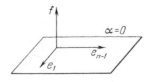

Figure 68.

Conversely, if $\alpha \wedge d\alpha = 0$, then $d\alpha(e_i, e_j) = 0$. Indeed, the only term in $(\alpha \wedge d\alpha)(e_i, e_j, f)$ not containing a factor $\alpha(e_i)$ or $\alpha(e_j)$ is $d\alpha(e_i, e_j)\alpha(f)$; on the other hand, $\alpha(f) \neq 0$. ▶

Remark. The condition $\alpha \wedge d\alpha \equiv 0$ is called the *integrability condition of Frobenius.* From the proposition just proved, it follows that it is a constraint on the field of planes: it is satisfied or not simultaneously for all forms α defining the field.

B. Existence of Integral Manifolds

Theorem. *In order that a field of hyperplanes $\alpha = 0$ be a field of tangent hyperplanes to a family of hypersurfaces it is necessary and sufficient that the field satisfy the integrability condition $\alpha \wedge d\alpha \equiv 0$ of Frobenius.*

◀ On the surfaces of the family, we have $\alpha = 0$, and therefore, $d\alpha = 0$. Conversely, let $d\alpha = 0$ on the planes $\alpha = 0$. In the neighborhood of a point x, we construct a family of integral surfaces in the following way. Let v be a vector field for which $\alpha(v) \equiv 0$ (i.e., the vector of the field lies in the plane of the field of planes at every point). Let Γ^k be an integral submanifold of the field of planes (Fig. 69) and let $v(x)$ not lie in the tangent plane to Γ^k at x.

Lemma. *The phase curves of the field v passing through points of the integral manifold Γ^k form a smooth integral manifold Y^{k+1} of a completely integrable field of planes $\alpha = 0$.*

◀ Denote by $\{g^t\}$ the local phase flow of the field v. Then (a) *the diffeomorphisms g^t convert the planes of our field $\alpha = 0$ into planes of the field.*

Indeed, $d\alpha(\xi) = 0$ for every vector ξ of a plane of the field. Therefore, the field of planes is invariant under the diffeomorphisms g^t according to the lemma of § 8F.

Moreover, (b) *the tangent space to Y^{k+1} lies in the plane of the field at the points of the initial manifold Γ^k.*

Indeed, both the tangent plane of the integral manifold Γ^k and the vector v belong to a plane of the field, and the tangent space to Y^{k+1} at x is spanned by $v(x)$ and $T_x\Gamma^k$.

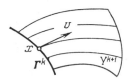

Figure 69.

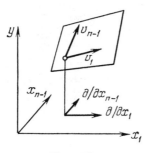

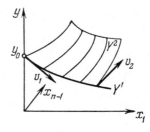

Figure 71.

Figure 70.

From (a) and (b) it follows that the manifold Y^{k+1} is an integral manifold for the field of planes $\alpha = 0$. ▶

Now the integral manifolds of dimension $n - 1$ are constructed by increasing the dimension successively.

We consider a coordinate system $(x_1, \ldots, x_{n-1}; y)$ in which the coordinate plane $y = 0$ belongs to the field of planes $\alpha = 0$ at zero.

In the neighborhood of zero, the projection in the direction of the y-axis from a plane of the field onto the coordinate plane $(x_1, \ldots, x_{n-1})$ is an isomorphism (Fig. 70). Consider basis vector fields in the coordinate plane: $(\partial/\partial x_1, \ldots, \partial/\partial x_{n-1})$. Their inverse images in the planes of the field form smooth vector fields in the neighborhood of the point O. Denote these fields by $(v_1, \ldots, v_{n-1})$.

As initial (zero-dimensional) integral manifold Γ^0, we take a point y_0 of the y-axis (Fig. 71).

Applying the lemma to Γ^0 and v_1, we obtain a one-dimensional integral manifold Y^1. On Y^1 we have $x_2 = \ldots = x_{n-1} = 0$, and therefore, $v_2 \notin T_0 Y^1$.

Applying the lemma to Y^1 and v_2, we obtain a two-dimensional integral manifold Y^2. Continuing in the same way, we begin with an integral manifold Y^k on which $x_{k+1} = \ldots = x_{n-1} = 0$. Applying the flow of the field v_{k+1}, we obtain an integral manifold Y^{k+1} on which $x_{k+2} = \ldots = x_{n-1} = 0$.

The process concludes with the construction of the desired manifold Y^{n-1}. ▶

Chapter 3

Structural Stability

In every mathematical investigation, the question will arise whether we can apply our mathematical results to the real world. Indeed, let us assume that the result is very sensitive to the smallest change in the model. In such a case, an arbitrarily small change in the model (say, a small change of the vector field defining a differential equation) leads to another model with essentially different properties. A result like this cannot be transferred to the real process under consideration, because, when constructing the model, the real situation was idealized and simplified, the parameters were determined only approximately, etc. Consequently, the question arises of choosing those properties of the model of a process which are not very sensitive to small changes in the model, and thus may be viewed as properties of the real process.

One of the attempts to choose such properties led to the notion of robustness or structural stability (Andronov and Pontrjagin, 1937). The significant success of the theory of structural stability for phase spaces with small dimension (1 or 2) gave birth to hopes broken in the 1960's by the work of Smale: He showed that for phase spaces of large dimension, systems exist in the neighborhood of which there is no structurally stable system. For the qualitative theory of differential equations this result has approximately the same significance as Liouville's theorem on the impossibility of solving differential equations by quadrature for the integration theory of differential equations. It shows that the problem of the complete topological classification of differential equations with high-dimensional phase space is hopeless even if restricted to generic equations and nondegenerate cases.

In this chapter, we give a short survey of the basic notions, methods, and results of the theory of structural stability.

§ 10. The Notion of Structural Stability

In this section structural stability is defined and structurally stable vector fields are studied on a one-dimensional phase space.

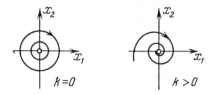

Figure 72.

A. The Naive Definition of Structural Stability

We shall consider the differential equation

$$\dot{x} = v(x), \qquad x \in M,$$

given by the vector field v on the manifold M. We will say that the field v
defines a *dynamical system* (or briefly, a system). We shall assume (as a rule)
that the solutions of the equation can be extended infinitely; this is always
so if M is compact.

EXAMPLE

The equation of a pendulum with friction is (Fig. 72):

$$\dot{x}_1 = \dot{x}_2, \qquad \dot{x}_2 = -x_1 - kx_2.$$

   If $k = 0$, then all phase curves are closed. If $k > 0$, they spiral towards
the singular point O, which is of focal type. Consequently, a small change
in the friction coefficient changes the behavior of the phase curves qualitatively if the coefficient had been zero before the change. It does not change
the qualitative picture if the coefficient had been positive.
   The definition of structural stability given below formalizes this difference: a pendulum without friction turns out to be a structurally unstable
system, and a pendulum with friction is a structurally stable one.

Definition. A system is said to be *structurally stable* if for any small change
in the vector field the system thus obtained is equivalent to the initial system.

   In order to give this definition a meaning, we have to define small change
of a field and what systems are considered to be equivalent.

B. Topological Equivalence

The finest classification of differential equations is based on the notion of
diffeomorphism. Two systems (M_1, v_1) and (M_2, v_2) are said to be *diffeo-*

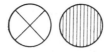

Figure 73.

morphic if there exists a diffeomorphism $h\colon M_1 \to M_2$ converting the vector field v_1 into the vector field v_2.

Diffeomorphic systems are indistinguishable from the point of view of the geometry of smooth manifolds. The following example shows that classification up to diffeomorphisms is too fine (too many systems turn out to be inequivalent).

EXAMPLE

Consider the equations $\dot{x} = x$ and $\dot{x} = 2x$ with a one-dimensional phase space.

In both cases, 0 is the only (repelling) equilibrium position. However, the two systems are not diffeomorphic.

◀ If a diffeomorphism converts a singular point of a vector field into a singular point of another vector field, then the derivative of the diffeomorphism converts the operator of the linear part of the first field at the singular point into the operator of the linear part of the second field at its singular point. Consequently, these two linear operators are similar and, in particular, have the same eigenvalues. Hence, the eigenvalues of the linearization of a vector field at a singular point are invariants with respect to diffeomorphisms and change continuously with the field. Such invariants are called *moduli*. Due to the existence of moduli, the decomposition of the set of vector fields into classes of diffeomorphic vector fields is continuous rather than discrete (Fig. 73).

In particular, the two fields indicated above are not diffeomorphic, since $1 \neq 2$. ▶

In order not to distinguish these two fields, a coarser equivalence is introduced, the so-called topological equivalence. We note that homeomorphisms (one-to-one mappings continuous in both directions) do not act on vector fields. Therefore, the topological equivalence of vector fields is defined in the following way.

We consider the phase flows defined by given vector fields. The phase flow of a field v on M consists of the transformations $g^t\colon M \to M$ converting every initial condition x_0 of the equation $\dot{x} = v(x)$ at time 0 into the value $g^t x_0$ of the solution at time t; it is obvious that $g^{t+s} = g^t g^s$, $g^0 = 1$. If M is compact, then $g^t x$ are defined for all $t \in \mathbb{R}$ and $x \in M$.

Definition. Two systems are *topologically equivalent* if there exists a homeomorphism of the phase space of the first system onto the phase space of the second, which converts the phase flow of the first system into the phase flow of the second:

$$hg_1^t x \equiv g_2^t hx.$$

In other words, the diagram

$$\begin{array}{ccc} M_1 & \xrightarrow{g_1^t} & M_1 \\ h \downarrow & & \downarrow h \\ M_2 & \xrightarrow{g_2^t} & M_2 \end{array}$$

is commutative.

For example, the systems $\dot{x} = x$ and $\dot{x} = 2x$ are topologically equivalent.

Remark. Various application of homeomorphisms for the expulsion of moduli, as introduced above, provided the fundamental motivation for introducing homeomorphism and continuous (not differential) topology.

C. Orbital Equivalence

Unfortunately, the topological equivalence does not save us from moduli.

EXAMPLE

Consider a vector field having a closed phase curve, say, a limit cycle. Then every topologically equivalent system also has a limit cycle with the same period. A small change in the field may cause a small change in the period. Consequently, the period of motion on the cycle is a continuously changing invariant (modulus) with respect to topological equivalence as well. In order to get rid of such moduli, a classification of systems is introduced which is still coarser than the classification up to a homeomorphism.

Definition. Two systems are said to be *topologically orbitally equivalent* if there exists a homeomorphism of the phase space of the first system onto the phase space of the second, converting oriented phase curves of the first system into oriented phase curves of the second. No coordination of the motions on corresponding phase curves is required.

The *structural stability* conjecture consists of the assumption that the decomposition of systems into classes of orbital equivalence does not have moduli, at least if we are restricted to generic cases and neglect degeneracies.

D. The Final Definition of Structural Stability

Let M be a compact manifold (of class $C^{r-1}, r \geq 1$). Let v be a vector field of class C^r (if M has a boundary, then it is assumed that v is not tangent to it).

The system (M, v) is said to be structurally stable if there exists a neighborhood of v in the space C^1 such that every vector field in this neighborhood defines a system topologically orbitally equivalent to the initial one, and the homeomorphism realizing the equivalence is close to the identity homeomorphism.

E. The One-Dimensional Case

Let M be a circle. A vector field on the circle is defined by a periodic function. The singular points of the field correspond to the zeros of this function. A singular point is said to be *nondegenerate* if the derivative of the function is different from zero at that point.

Theorem. *A vector field on the circle defines a structurally stable system if and only if it has only nondegenerate singular points.*

Two vector fields on the circle with nondegenerate singular points are topologically orbitally equivalent if and only if they have the same number of singular points.

The structurally stable vector fields form an everywhere dense open set in the space of vector fields on the circle.

◀ Let all singular points of the field be nondegenerate. There are a finite number of them and they are alternately stable and unstable. Every non-constant solution of the equation $\dot{x} = v(x)$ tends to a stable equilibrium position as $t \to +\infty$ and to an unstable one as $t \to -\infty$. This easily implies all assertions of the theorem except one: it remains to be proven that all singular points can be made nondegenerate by means of an arbitrarily small perturbation of the system.

It is convenient to prove the last assertion by means of Sard's theorem.

Lemma. *The measure of the set of critical values of a smooth function defined on the interval* $[0, 1]$ *is equal to zero.*

◀ We divide the interval into N equal parts and single out those which contain critical points. If N is sufficiently large, then the derivative of the function does not exceed C/N on each of the parts singled out (C is a constant independent of N). Therefore, the length of the image of each of the parts singled out does not exceed C/N^2. Cover this image by an interval of length $2C/N^2$. We have obtained a cover of the set of critical values by intervals of total length not greater than $2C/N$. ▶

We consider a family of vector fields with parameter ε on the circle given by the formula $v(x, \varepsilon) = v(x) - \varepsilon$. A point x is a degenerate singular point of the field corresponding to the value ε of the parameter if and only if ε is a critical value of the function v at the point x.

All critical points form a set of measure zero. Therefore, there exist arbitrarily small noncritical values. Fix a noncritical value ε. All singular points of the field corresponding to this value of the parameter are nondegenerate. ▶

F. Digression: Sard's Theorem

Let $f: M \to N$ be a smooth mapping of manifolds of any dimension. A point of the source space is said to be *critical* if the dimension of the image of the differential of the mapping at the point is smaller than the dimension of the image space. The value of the mapping at a critical point is said to be a *critical value.*

Theorem. *The measure of the set of critical values of every sufficiently smooth mapping is equal to zero.*

◀ *1.* If the dimension of the source space is equal to zero, then the theorem is obvious; if the dimension is equal to 1, then it has already been proved above. We assume that the theorem is proved in all cases where the dimension of the source space is equal to $m - 1$, and prove it in the case where it is equal to m.

2. We divide the set K of critical points of the mapping into several parts. A point of the source space is called a *point of flattening of order r* if all partial derivatives of orders $1, \ldots, r$ are equal to zero at this point. We denote by K_r the set of all points of flattening of order r.

3. First we consider the set of critical points of $K \backslash K_1$. We prove that *the measure of the corresponding set of critical values* (i.e., the measure of the set $f(K \backslash K_1)$) *is equal to zero.*

At every point of $K \backslash K_1$, one of the partial derivatives is different from zero, say, $\partial f_1 / \partial x_1$. In the neighborhood of such a point, f_1 can be chosen as a local coordinate instead of x_1, retaining the other coordinates $x_2, \ldots, x_m$ in the source space. In these coordinates, f can be written as a one-parameter family of smooth mappings of $(m - 1)$-dimensional spaces into $(n - 1)$-dimensional ones:

$$(f_1 ; x_1, x_2, \ldots, x_m) \mapsto (f_1 ; f_2, \ldots, f_n).$$

We fix a value c of the parameter of f_1. The mapping f induces a mapping f_c of the $(m - 1)$-dimensional plane $f_1 = c$ in the source space into the $(n - 1)$-dimensional plane $f_1 = c$ in the image.

The set of critical values of the mapping f_c has $(n - 1)$-dimensional measure zero in the plane $f_1 = c$ by the induction hypothesis (the theorem is proved for $(m - 1)$-dimensional source spaces). By Fubini's theorem, the n-dimensional measure of the union of the sets of critical values of the mappings f_c is equal to zero.

On the other hand, the image of the set of critical points in $K \backslash K_1$ lying in the neighborhood of the point under consideration is contained in this union. This implies that the measure of $f(K \backslash K_1)$ is equal to zero.

4. We consider the set $K_r \backslash K_{r+1}$ of points of r-flattening. We prove that *the measure of the corresponding set $f(K_r \backslash K_{r+1})$ of critical values is equal to zero.*

At every point of $K_r \backslash K_{r+1}$, one of the partial derivatives of order $r + 1$, say $\partial g / \partial x_1$, is different from zero, where g is one of the partial derivatives of f_1 of order r (in an appropriate local coordinate system).

In the neighborhood of such a point, the set $K_r \backslash K_{r+1}$ is contained in the smooth $(m - 1)$-dimensional hypersurface $g = 0$. The points of $K_r \backslash K_{r+1}$ are critical for the restriction of f to this hypersurface, since $df = 0$ on K_r. By assumption, the measure of the image of the set of critical values of the restriction of f to this hypersurface is equal to zero. Consequently, the measure of $f(K_r \backslash K_{r+1})$ is equal to zero.

5. Finally, we consider the set K_r of r-flat critical points for sufficiently large r. We prove that *the measure of the corresponding set $f(K_r)$ of critical values is equal to zero if r is sufficiently large.*

To this end, we divide each side of the m-dimensional cube in the source space (choosing local coordinates) into N equal parts, divide the cube into N^m equal small cubicles, and single out those which contain points from K_r. The diameter of the image of any cubicle singled out does not exceed $c(1/N)^{r+1}$ (where the constant c does not depend on N). Therefore, the images of all cubes singled out are covered by open cubes with total measure not greater than

$$c_1 N^m \left(\frac{1}{N} \right)^{n(r+1)},$$

even if all N^m cubicles are singled out.

This number converges to zero as $N \to \infty$, and, therefore, the measure of K_r is equal to 0 for $r > (m/n) - 1$.

Let us represent the whole set K of critical points as the union of the sets $K \backslash K_1$, $K_i \backslash K_{i+1}$, and K_r. We have proved that the measure of the image of each of these sets is equal to zero. Therefore, the measure of the whole set of critical values is equal to zero. ▶

G. Structurally Stable Systems on the Two-Dimensional Sphere

Passing to systems with a phase space of dimension greater than one, we encounter singular points and closed phase curves.

Definition. A singular point of a vector field is said to be *degenerate* if zero is an eigenvalue of the linearization of the field at that point.

Remark. A nondegenerate singular point of the field does not disappear for a small perturbation of the field, and moves only slightly (by the implicit function theorem). In contrast to this, a degenerate singular point, in general, bifurcates under a small perturbation of the field (splits into several nondegenerate ones) or disappears. Therefore, all singular points of a structurally stable system are nondegenerate.

Definition. A closed phase curve (cycle) of a vector field is said to be *degenerate* if 1 is an eigenvalue of the linearization of the Poincaré mapping. [The

Figure 74.   Figure 75.

Poincaré mapping is the mapping of a transversal of the cycle into itself assigning to every point of the transversal close to the cycle the next point of intersection of the phase curve emanating from this point of the transveral with the transversal, cf., Fig. 74.]

Remark. A nondegenerate cycle does not disappear under a small perturbation of the field but moves a little (in view of the implicit function theorem). In contrast to this, a degenerate cycle, in general, bifurcates (splits into several nondegenerate ones) or disappears under a small perturbation of the field. Therefore, all cycles of a structurally stable system are nondegenerate.

We consider a vector field on a two-dimensional surface. In the two-dimensional case, the generic singular points are, topologically, either saddle or nodal points. A phase curve converging to a saddle point as $t \to +\infty$ is called an *incoming separatrix* of the saddle point and a phase curve converging to the saddle as $t \to -\infty$ is called an *outgoing separatrix* (Fig. 75).

Theorem. *A vector field on the two-dimensional sphere defines a structurally stable system if and only if the following conditions are satisfied:*

(1) *All singular points of the field are nondegenerate.*
(2) *The real parts of the eigenvalues of the linear parts of the field at all singular points are nonzero.*
(3) *No outgoing separatrix of a saddle point is incoming.*
(4) *All closed phase curves are nondegenerate cycles.*

Remark. The proof of structu:al instability if at least one of conditions (1)–(4) is violated is not complicated (cf., Fig. 76). The proof of the fact that conditions (1)–(4) imply

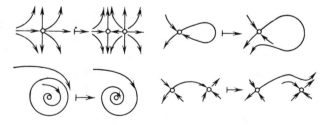

Figure 76.

structural stability is more complicated; it is given in detail by De Baggis (cf., H. F. De Baggis, *Dynamical systems with stable structures*, Contrib. Theory Nonlinear Oscillation, **2** (1952), 37–59; M. M. Peixoto, *Structural stability on two-dimensional manifolds*, Topology, **1** (1962), 101–120; **2** (1963), 179–180); M. S. Peixoto, M. M. Peixoto, *Structural stability in the plane with enlarged boundary conditions*, Anals Acad. Brasileira Ciências, **31**, 2 (1959), 135–160.

Theorem. *The structurally stable vector fields form an everywhere dense open set in the space of vector fields on the two-dimensional sphere.*

◀ This theorem follows from the preceding one. ▶

Remark. Analogous results hold for a vector field on the disk not tangent to the bounding circle.

§ 11. Differential Equations on the Torus

In this section we discuss the theory of vector fields without singular points on the two-dimensional torus, due to Poincaré and Denjoy. In particular, all structurally stable fields are described.

A. The Two-Dimensional Torus

The n-dimensional torus is the direct product of n copies of the circle. The two-dimensional torus $T^2 = S^1 \times S^1$ can be represented as the square

$$\{x, y: 0 \leqslant x \leqslant 2\pi, 0 \leqslant y \leqslant 2\pi\}$$

with opposite sides pasted together [the points $(0, y)$ and $(2\pi, y)$ as well as the points $(x, 0)$ and $(x, 2\pi)$ are identified (Fig. 77)].

The torus can also be viewed as the set of cosets of the group $\mathbb{R}^2$ modulo, the subgroup $2\pi\mathbb{Z}^2$ of vectors with integral components multiplied by 2π:

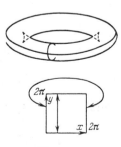

Figure 77.

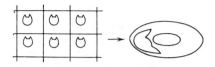

Figure 78.

$$T^2 = \mathbb{R}^2/2\pi\mathbb{Z}^2 = \{(x, y) \in \mathbb{R}^2 \bmod 2\pi\}.$$

Consequently, the plane $\mathbb{R}^2$ covers the torus in a locally diffeomorphic fashion. The covering $\mathbb{R}^2 \to T^2$ (Fig. 78) enables us to carry over every picture from the torus to the plane (where it is reproduced infinitely many times). Smooth functions on the torus correspond to 2π-periodic smooth functions on the plane.

To every closed curve in the plane there corresponds a closed curve on the torus. The converse is not true: closed curves on the torus correspond to not only closed curves in the plane but also to mappings $\varphi: [0, 1] \to \mathbb{R}^2$ for which $\varphi(0) = \varphi(1) \bmod 2\pi$.

Moreover, if the coordinates of $\varphi(1) - \varphi(0)$ are equal to $(2\pi p, 2\pi q)$, then we say that the curve on the torus *closes after p revolutions along the parallel and q revolutions along the meridian.*

B. Vector Fields on the Torus

Every vector field on the torus defines a field on the plane periodic of period 2π in both coordinates. Conversely, to every field 2π periodic in both coordinates in the plane, there corresponds a vector field on the torus.

EXAMPLE

The equations

$$\dot{x} = \alpha, \qquad \dot{y} = \beta,$$

where α and β are constant, define a vector field on the torus without singular points.

Theorem. *If the ratio $\lambda = \beta/\alpha$ is rational, then all phase curves are closed on the torus. If it is irrational, then they are everywhere dense.*

◀ (1) Let $\lambda = p/q$. The phase curve passing through the point (x_0, y_0) has the equation $y - y_0 = (x - x_0)p/q$. If $x - x_0 = 2\pi q$, then $y - y_0 = 2\pi p$ and, consequently, $(x, y) = (x_0, y_0) \bmod 2\pi$, i.e., the phase curve is closed.

(2) Below we prove that (in the case where λ is irrational) any phase curve is uniformly distributed on the torus, i.e., in any part of the torus*, it spends a time proportional to the area of that part. This implies that, in particular, a sufficiently long segment of the phase curve is as close to any point of the torus as we wish, i.e., the phase curve is everywhere dense. ▶

C. Uniform Distribution

The general definition of uniform distribution is as follows.

Let v be a vector field on a smooth compact manifold M with a fixed volume element (for example, a field on the torus with the area element $dx\,dy$). We shall denote the volume (area) of a domain D by $\mu(D)$.

Consider the solution φ of the equation $\dot{x} = v(x)$ with initial condition z. We denote by $\tau(D, T, z)$ the measure of the set of those values of the time $t \in [0, T]$ for which $\varphi(t)$ belongs to D.

Definition. A solution to the equation $\dot{x} = v(x)$ is uniformly distributed if for every domain D with piecewise smooth boundary, we have

$$\lim_{T \to \infty} \frac{\tau(D, T, z)}{T} = \frac{\mu(D)}{\mu(M)}.$$

Theorem. *For irrational β/α, the solutions of the equation $\dot{x} = \alpha$, $\dot{y} = \beta$ are uniformly distributed on the torus.*

Uniform distribution can also be defined in terms of *time averages* of functions.

Let f be a function (complex valued, in general) on M.

Definition. The limit

$$\lim_{T \to \infty} \frac{1}{T} \int_0^T f(g^t z)\, dt = \hat{f}(z)$$

is called the *time average of the function f* (here g^t is the phase flow).

Remark. Of course, this limit does not always exist, and even if it does, it may, in general, depend on the initial point.

The theorem on the uniform distribution on the torus follows from the theorem below.

---

*By "a part of the torus" we mean a Jordan measurable domain, for example, a domain with piecewise smooth boundary.

Theorem (On the Coincidence of Averages). *For irrational* $\lambda = \beta/\alpha$, *the time average of any continuous (or at least Riemann integrable) function* $f: T^2 \rightarrow \mathbb{C}$ *exists on a solution of* $\dot{x} = \alpha$, $\dot{y} = \beta$ *on the torus. It is independent of the initial point, and coincides with the space average*

$$f_0 = \frac{1}{4\pi^2} \oiint f\,dx\,dy.$$

In order to obtain the uniform distribution theorem from this theorem, it is sufficient to take the characteristic function of a set D (equal to 1 on D and zero outside D) as f.

D. Proof of the Theorem on the Coincidence of Averages

We denote the vector with components (α, β) by ω. Then the solution with initial condition z has the form $z + \omega t$. The assertion of the theorem reads as follows:

$$\lim_{T \to \infty} \frac{1}{T} \int_0^T f(z + \omega t)\,dt = \frac{1}{4\pi^2} \oiint f(z)\,dx\,dy.$$

◀ We note that among the functions on the torus, there are the harmonics of the form $e^{i(k, z)}$, where k is a vector with integral components. For harmonics, the theorem can be verified by direct calculation of the integral. Let $k \neq 0$. We have

$$\int_0^T e^{i(k, z+\omega t)}\,dt = e^{ikz} \int_0^T e^{i(k, \omega)t}\,dt = \frac{e^{i(k, z)}}{i(k, \omega)}\left[e^{i(k, \omega)T} - 1\right].$$

The function in square brackets is bounded. Therefore, the time average of a harmonic with nonzero index k is equal to zero. The space average is also equal to zero. For $k = 0$ the harmonic equals 1. Both averages of the function identical to 1 are equal to 1. The coincidence of the averages is proved for harmonics.

The coincidence of the averages for harmonics implies their coincidence for trigonometric polynomials: the average of a linear combination is equal to the linear combination of the averages with the same coefficients. In particular, the theorem is proved for $f = \cos(k, z)$ and $\sin(k, z)$.

Now we prove the theorem for real functions; then (in view of the linearity of averages) it will be proved for complex-valued functions, as well. We approximate a given function f from above and from below by continuous functions P and Q so that

$$P < f < Q, \quad \frac{1}{4\pi^2} \oiint (Q - P)\,dx\,dy < \varepsilon$$

(the possibility of such an approximation for any $\varepsilon > 0$ characterizes the Riemann integrable functions). Then we approximate the functions P and Q by trigonometric polynomials p and q so that

$$|p - P| < \varepsilon, |q - Q| < \varepsilon.$$

We denote by p_0 and q_0 the constant terms of these polynomials. The numbers p_0 and q_0 are both space and time averages of the polynomials p and q (since for trigonometric polynomials, time averages coincide with space averages). Consequently, the space average f_0 of f is between $p_0 - \varepsilon$ and $q_0 + \varepsilon$:

$$p_0 - \varepsilon < f_0 < q_0 + \varepsilon, \qquad q_0 - p_0 < 3\varepsilon.$$

We denote by p_T, f_T, and q_T the averages of p, f, and q in time T:

$$p_T(z) = \frac{1}{T} \int_0^T p(z + \omega t)\, dt, \quad \text{etc.}$$

Then $p_T(z) - \varepsilon < f_T(z) < q_T(z) + \varepsilon$ for any T. For T sufficiently large, we have

$$|p_T(z) - p_0| < \varepsilon, \qquad |q_T(z) - q_0| < \varepsilon.$$

Consequently, for suficiently large T,

$$|f_T(z) - f_0| < 6\varepsilon. \ \blacktriangleright$$

E. Some Consequences

1. The fact that the torus was two-dimensional did not play any role in the preceding. We consider the equation $\dot{z} = \omega, z \in T^n$, on the n-dimensional torus. The frequency vector ω is said to be a resonant vector if there exists a nonzero vector k with integral components such that $(\omega, k) = 0$.

If the vector ω is nonresonant, then the time and space averages of continuous (or even Riemann integrable) functions coincide and the solutions are uniformly distributed.

2. From the uniform distribution theorem, it follows that the first digit of the number 2^n is equal to 7 more often than to 8. More precisely, denote by $N_k(n)$ the number of positive integers $m \leqslant n$ for which 2^m begins with the digit k. Then

$$\lim_{n \to \infty} \frac{N_7(n)}{N_8(n)} = \frac{\log 8 - \log 7}{\log 9 - \log 8}.$$

The same distribution governs the first digits of the population figures of the world's countries, but not those of the lengths of rivers (N. N. Konstantinov). One possible explanation is the exponential growth of populations with time. Such growth implies the desired distribution for the population figures of every fixed country in different years. To obtain the required explanation we have only to substitute temporal averaging for "spatial" averaging (averaging over the set of countries).

3. The discovery of the uniform distribution theorem was motivated by the following problem of Lagrange: determine $\omega = \lim_{t \to \infty} (1/t) \arg f(t)$, where

$$f(t) = \sum_{k=1}^{n} a_k e^{i\omega_k t}.$$

We give the answer for the nonresonant vectors $\omega = (\omega_1, \cdots, \omega_n)$. Let $n = 3$. If a triangle can be constructed from the three line segments (a_1, a_2, a_3), then $\omega = \sum \alpha_k \omega_k / \pi$, where α_k is the angle opposite a_k.

For an arbitrary n, the solution has the form of a weighted average of the frequencies ω_k: $\omega = \sum W_k \omega_k$. The weights W_k can be calculated in the following way. Denote by $W(a_1, \cdots, a_s; b)$ the probability of the event that the distance from the origin to the endpoint of the planar polygon with s sides of lengths $a_1, \cdots, a_s$ with random directions is smaller than b. We then have $W_k = W(\hat{a}_k, a_k)$ ($\hat{a}_k$ is the collection of all numbers a_i except a_k).

For a proof, cf., for example, H. Weyl, *Mean motion*, Am. J. Math. **60**, (1938), 889–896; **61** (1939), 143–148.

Lagrange encountered the above problem (the so-called problem of mean motion) in the following way. Consider the vector connecting the Sun with the center of the ellipse on which a planet moves (it is called the Laplace vector). In the first approximation of the perturbation theory, the evolution of the Laplace vector under the influence of mutual attraction of planets has the form of a motion of a sum of uniformly rotating vectors (their number equals the number of planets).

If for the planets of the solar system we calculate the frequencies ω_k and amplitudes a_k, it turns out that for all planets except the Earth and Venus one of the amplitudes a_k is greater than the sum of the others. Therefore, Lagrange managed to determine the mean motion of the perihelia of all planets except the Earth and Venus. In the case of the Earth and Venus, however, several terms have approximately the same amplitude. The problem was solved only in the Twentieth Century by Bohl, Sierpinski, and H. Weyl.

F. Poincaré Mapping and Angular Function

Consider a general differential equation

$$\dot{z} = \omega(z), \qquad z \in T^2$$

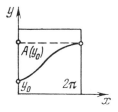

Figure 79.

on the torus. Assume that the field ω has no singular points and $\omega_1 \neq 0$. [If there are no singular points or cycles, then in an appropriate coordinate system, the first component of the field is everywhere different from zero (cf., Siegel, *Note on differential equations on the torus*, Ann. Math. **46**, 3 (1945), 423–428); it is easy to construct a field without singular points but with cycles which do not admit such a coordinate system].

Now we pass to the study of integral curves of a nonautonomous equation with doubly periodic right-hand side:

$$\frac{dy}{dx} = \lambda(x, y), \qquad \lambda = \frac{\omega_2}{\omega_1}.$$

All solutions of this equation can be infinitely extended, since the right-hand side is bounded.

Definition. The *Poincaré mapping* of the equation under consideration on the torus is the mapping A of the y-axis into itself assigning to every initial point $(0, y_0)$ the value of the solution at $x = 2\pi$ with this initial condition (Fig. 79).

The Poincaré mapping is differentiable (by the theorem on the differentiability of the solution with respect to the initial conditions) and has the periodicity property: $A(y + 2\pi) = A(y) + 2\pi$; the inverse mapping A^{-1} is also differentiable. Consequently, A defines a diffeomorphism of the circle onto itself. The Poincaré mapping may be thought of as that diffeomorphism of the meridian of the torus which takes every point of the meridian into the following point of intersection of the integral curve passing through this point with the same meridian.

The study of properties of integral curves on the torus thus reduces to that of properties of diffeomorphisms of the circle. For example, assume that a diffeomorphism of the circle has a fixed point. Then, on the torus there exists a closed integral curve. The converse is not true (for example, take a rotation of the circle by angle π). For an integral curve passing through a given point of the meridian of the torus to be closed, it is necessary and sufficient that this point be a periodic point of the diffeomorphism, i.e., that it be mapped onto itself after several applications of the diffeomorphism.

The Poincaré mapping defines an orientation-preserving diffeomorphism of the circle. Therefore, it can be written in the form

$$Ay = y + a(y), \text{ where } a(y + 2\pi) = a(y), a'(y) > -1.$$

The function a will be called the *angular function*.

G. Rotation Number

The rotation number characterizes the average slope of integral curves of an equation on the torus; for the simplest equation $dy/dx = \lambda$ with constant right-hand side, the rotation number is equal to λ.

Definition. The *rotation number* of the equation $dy/dx = \lambda(x, y)$ on the torus is the number

$$\mu = \lim_{x \to \infty} \frac{\varphi(x)}{x},$$

where φ is the solution of the corresponding equation in the plane.

The rotation number can be expressed by the angular function:

$$\mu = \frac{1}{2\pi} \lim_{k \to \infty} \frac{a(y) + a(Ay) + \cdots + a(A^{k-1}y)}{k}.$$

In this form, the definition can be carried over to any orientation-preserving diffeomorphism of the circle.

Theorem. *The limit in the definition of the rotation number exists and does not depend on the initial point; it is rational if and only if some power of the diffeomorphism has a fixed point (i.e., the differential equation has a closed phase curve).*

◀ *1.* Consider the angle of rotation of a point y under k-fold application of the diffeomorphism. Denote it by

$$a_k(y) = a(y) + a(Ay) + a(A^2y) + \cdots + a(A^{k-1}y).$$

For any two points y_1 and y_2, we have

$$|a_k(y_1) - a_k(y_2)| < 2\pi.$$

Indeed, the inequality holds for $|y_1 - y_2| < 2\pi$, since the mappings A and A^k of the line convert intervals of length 2π into intervals of length 2π.

On the other hand, the function a_k is 2π-periodic, and hence y_2 can be changed by an integral multiple of 2π, so that $a_k(y_2)$ will not change and the distance of y_1 from y_2 will become smaller than 2π.

2. Denote by m_k the integer for which

$$2\pi m_k \leqslant a_k(0) < 2\pi(m_k + 1).$$

We prove that for any y and for any integer l,

$$\left| \frac{a_{kl}(y)}{2\pi kl} - \frac{m_k}{k} \right| < \frac{2}{k}.$$

Indeed, $|a_k(y) - 2\pi m_k| < 4\pi$ for any y according to § 11G1, and therefore,

$$\left| \frac{a_k(y)}{2\pi k} - \frac{m_k}{k} \right| < \frac{2}{k}.$$

However, $a_{kl}(y)/2\pi kl$ is the arithmetic mean of the l quantities $a_k(y_i)/2\pi k$, where $y_i = A^i y$, $i = 0, \ldots, l - 1$.

3. Denote by σ_k the interval $[(m_k - 2)/k, (m_k + 2)/k]$. We have proved that $a_{kl}(y)/2\pi kl$ belongs to σ_k for all l. We prove that the intervals σ_k with distinct k intersect each other.

Indeed, $a_{kl}(y)/2\pi kl$ belongs to both σ_k and σ_l.

4. Thus, the intervals σ_k have lengths converging to 0 and they intersect each other. Consequently, they have a unique common point: it is the rotation number. We have proved that the limit defining the rotation number exists and does not depend on the initial point.

5. Assume that A^q has a fixed point y on the circle; then the corresponding point on the line translates by an integral multiple of 2π under q-fold application of the corresponding mapping, i.e., $a_q(y) = 2\pi p$. In this case for any l we have $a_{ql}(y) = 2\pi pl$, and, therefore, the rotation number $\mu = p/q$ is rational.

6. Let $\mu = p/q$. If for all y we have $a_q(y) > 2\pi p$, then for some $\varepsilon > 0$ we have $a_q(y) > 2\pi p + \varepsilon$ for all y.

Then $\mu > p/q$. If we had $a_q(y) < 2\pi p$ for all y, then we would have $\mu < p/q$. Hence, $a_q - 2\pi p$ changes sign. Consequently, y exists such that $a_q(y) = 2\pi p$. ▶

Remark. If the rotation number μ is irrational, then the order of the points $(y, Ay, A^2 y, \ldots, A^N y)$ on the circle is the same for any y as in the case of a rotation by angle $2\pi\mu$. Indeed, $a_q(y) > 2\pi p$ if and only if $\mu > p/q$.

We also note that the rotation number of an equation on the torus *depends* on the choice of the circle transversal to phase curves (the y-axis in our notation).

H. Structurally Stable Equations on the Torus

The simplest equation $\dot{z} = \omega$ on the torus is structurally unstable, for both resonant and nonresonant values of ω.

Theorem 1. *The differential equation $dy/dx = \lambda(x, y)$ on the torus is structurally stable if and only if the rotation number is rational and all periodic solutions are nondegenerate*.*

◀ This theorem follows from the analogous assertion proved below for orientation-preserving diffeomorphisms of the circle. ▶

Definition. A *cycle of order q* of a diffeomorphism $A: M \to M$ is a set of q distinct points $(y, Ay, \ldots, A^{q-1}y)$ with $A^q y = y$. A cycle is said to be *nondegenerate* if the point y is a nondegenerate fixed point of A^q (i.e., 1 is not an eigenvalue of the derivative of A^q at y).

Remark. The derivatives of A^q are similar at the points of the same cycle and, therefore, all points of a cycle are degenerate or nondegenerate at the same time.

Theorem 2. *An orientation-preserving diffeormorphism of the circle is structurally stable if and only if the rotation number is rational and all cycles are nondegenerate. The structurally stable diffeomorphisms form an everywhere dense open set in the space C^2 of all twice differentiable orientation-preserving diffeomorphisms of the circle.*

Consequently, generic diffeomorphisms with rational rotation number have a rather simple structure: the topological type of the mapping is determined by the number of cycles, which must be even (since the points of stable and unstable cycles alternate). The order of all cycles is equal to q provided that the rotation number $\mu = p/q$. The ordering of points of each cycle on the circle is the same as for a rotation by the angle $2\pi\mu$.

Theorem 2 is proved in § 11J. The proof is simple modulo the following nontrivial theorem of Denjoy (1932).

---

*A. G. Maier, *Robust transformations of the circle into the circle*, Učenye Zapiski GGU, **12** (1939), 215–229; V. A. Pliss, *On the robustness of differential equations given on the torus*, Vestnik LGU, Ser. Mat. **13**, 3 (1960), 15–23.

Theorem 3. *If an orientation-preserving diffeomorphism of the circle of class* C^2 *has the irrational rotation number* μ, *then it is topologically equivalent to a rotation of the circle by the angle* $2\pi\mu$.

The preceding theory is due to Poincaré (1885); Denjoy's theorem was stated by Poincaré in the form of a conjecture (for equations whose right-hand sides are trigonometric polynomials). Denjoy also gave examples showing that C^2 cannot be replaced by C^1.

I. Proof of Denjoy's Theorem

◀ *1.* The ordering of the points $\ldots, A^{-1}y, y, Ay, \ldots$ of the orbit of the mapping A on the circle is the same as the ordering of the points of the orbit of rotation by the angle $2\pi\mu$ (cf., § 11G). Therefore, *to prove the theorem it is sufficient to establish that the orbit of A is everywhere dense on the circle.* Indeed, we then would obtain a homeomorphism of the circle converting A to a rotation by continuously extending the mapping which takes the points of the orbit $\ldots, A^{-1}y, y, Ay, \ldots$ into the corresponding points of the orbit of the rotation.

2. If on the circle there is an arc free from points of the orbit of A, then the images of this arc under the powers of A are mutually disjoint. Indeed, consider the maximal arc containing the given arc and free from points of the orbit. Its images are also maximal arcs. The endpoints of a maximal arc belong to the closure of the orbit. Therefore, the endpoints of a maximal arc cannot lie inside maximal arcs. Hence, any two intersecting maximal arcs must coincide. On the other hand, if a maximal arc coincides with its image, then its boundary point belongs to a cycle, in contridiction with the irrationality of μ.

3. The sum of the lengths of the images of a maximal arc is bounded. Therefore, the lengths of the consequent images of such an arc under the action of both A^N and A^{-N} converge to zero as $N \to \infty$. Consequently, *the integrals of the Jacobians of both the positive and negative iterates of A over a maximal arc converge to zero: in the notation*

$$u_N = \prod_{i=0}^{N-1} \frac{dA}{dy}(A^i y), \qquad v_N = \prod_{i=0}^{N-1} \frac{dA^{-1}}{dy}(A^{-i}y),$$

we have

$$\int u_N \, dy \to 0, \qquad \int v_N \, dy \to 0$$

as $N \to \infty$ (the above integrals are taken over a maximal arc).

4. Consider the sequence $(\alpha_0, \alpha_1, \alpha_2 \ldots)$ of the points of the orbit of rotation by angle $2\pi\mu$. Assume that α_q is the closest to α_0 among the points $(\alpha_1, \ldots, \alpha_q)$. Then the points $\alpha_q, \ldots, \alpha_{2q-1}$ alternate with the points $\alpha_0, \ldots, \alpha_{q-1}$.

Indeed, consider the arc (α_s, α_{q+s}), $s < q$, of length δ, the distance of α_0 from α_q. Assume that α_r lies on this arc. If $r < s$, then α_0 lies on the arc $(\alpha_{s-r}, \alpha_{s-r+q})$. Therefore, the distance from α_{s-r} to α_0 is less than δ, contrary to the choice of α_q. If $r > s$, then α_{r-s} lies on the arc (α_0, α_q), and therefore, $r - s > q$. Then the distance from α_0 to α_{r-s-q} is less than δ. Hence, the arc (α_s, α_{q+s}) contains no points α_r, $r < 2q$, which was to be proved.

5. Consider the points $(y, Ay, \ldots, A^{q-1}y)$ and $(A^{-1}y, \ldots, A^{-q}y)$. These two sets of points alternate (§ 1I4). Therefore, for any function f of bounded variation defined on the circle, for any point y, and for any q defined in § 1I4, *the quantity*

$$\sum f(A^i y) - \sum f(A^{-j}y), \qquad 0 \leqslant i < q, \qquad 0 < j \leqslant q$$

is bounded from below and above by constants not depending on y or q.

6. Let f be $\ln(dA/dy)$. This function is of bounded variation, since A is of class C^2. Consequently, *the quantity*

$$\prod_{i=0}^{q-1} \frac{dA}{dy}(A^i y) / \prod_{j=1}^{q} \frac{dA}{dy}(A^{-j}y) = u_q v_q$$

is bounded from below and above by positive constants not depending on y or q (if q is chosen as in § 1I4).

7. A contradiction to § 1I3 completes the proof: applying the Schwarz inequality to $\sqrt{u_q}$ and $\sqrt{v_q}$, we obtain

$$\left(\int \sqrt{u_q v_q}\, dy \right)^2 \leqslant \int u_q\, dy \int v_q\, dy. \; \blacktriangleright$$

J. Proof of the Theorem on Structurally Stable Diffeomorphisms of the Circle

◀ *1. If for any two orientation-preserving diffeomorphisms of the circle with the same rational rotation number and the same number of cycles, all cycles are nondegenerate, then there exists a homeomorphism converting the first diffeomorphism into the second one.*

For the proof, we first assign to the points of a stable cycle of the first diffeomorphism the points of some stable cycle of the second diffeomorphism, and then, to the point of a neighboring unstable cycle of the first diffeomorphism we assign neighboring unstable points of the second one, and so on for all cycles (the order of points of a cycle on the circle is the same as for the rotation). This correspondence can then be extended to adjacent intervals using the following lemma, which can be proved easily:

Any two homeomorphisms of an interval onto itself without fixed points are topologically conjugate.

2. If the rotation number is rational and all cycles are nondegenerate, the rotation number, the number of cycles, and the nondegeneracy of cycles are preserved under a small perturbation (according to the implicit function theorem). Consequently, *a diffeomorphism with rational rotation number and nondegenerate cycles is structurally stable* (cf., § 11J1).

3. If a diffeomorphism has a degenerate cycle, then a small perturbation of the diffeomorphism in the neighborhood of the points of this cycle, can change the number of cycles. Therefore, *a diffeomorphism with a degenerate cycle is structurally unstable.*

4. *If the rotation number is irrational, then it can be changed by an arbitrarily small perturbation of the diffeomorphism.* Indeed, consider the perturbed diffeomorphism $y \mapsto Ay + \varepsilon$, $\varepsilon > 0$. By Denjoy's theorem, in some (not smooth) coordinate system we have $z \mapsto z + 2\pi\mu + \varphi(z)$, $\varphi > 0$. Therefore, the rotation number of the perturbed diffeomorphism is greater than μ. Hence, *every diffeomorphism with irrational rotation number is structurally unstable.*

5. *The rotation number is a continuous function of the diffeomorphism.* Indeed, $\mu < p/q$ if and only if the q-fold application of the diffeomorphism moves points by less than $2\pi p$. This property is preserved under small perturbation of the diffeomorphism.

6. *The diffeomorphisms with rational rotation numbers constitute a dense set.* This follows from § 11J4 and § 11J5 and the density of the set of rational numbers.

7. *All cycles of a diffeomorphism with rational rotation number can be made nondegenerate by an arbitrarily small perturbation of the diffeomorphism.*

Indeed, any cycle can be made nondegenerate by an arbitrarily small perturbation in the neighborhood of the cycle. Let γ be one of the arcs into which a degenerate cycle divides the circle. Define a smooth function φ equal to 1 on γ except in small neighborhoods of the endpoints of γ and to 0 outside γ. Set $A_\varepsilon(y) = Ay + \varepsilon\varphi(y)$. The rotation number does not change, since the cycle has been preserved. Let q be the order of the cycle. Then $A_\varepsilon^q(y)$ coincides with $A^q(y) + \varepsilon$ on the arc $A\gamma$ outside a neighborhood of the endpoints of $A\gamma$.

We apply Sard's theorem to the function $A^q(y) - y$ on $A\gamma$. We see that for almost all ε, all fixed points of A_ε^q on $A\gamma$ are nondegenerate. On the other hand, every cycle of A_ε has representatives on $A\gamma$. Consequently, the cycles of A_ε are nondegenerate. ▶

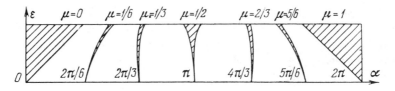

Figure 80.

K. Discussion

1. The preceding theorems give the impression that a generic diffeomorphism of the circle has rational rotation number, and diffeomorphisms with irrational rotation number are exceptional. Nevertheless, numerical experiments usually lead to (at least apparently) everywhere dense orbits. To explain this phenomenon, we consider, for example, the family of the diffeomorphisms

$$A_{a,\varepsilon}: y \mapsto y + \alpha + \varepsilon \sin y, \qquad \alpha \in [0, 2\pi], \qquad \varepsilon \in [0, 1).$$

We shall represent every diffeomorphism by a point in the (α, ε)-plane. As is easily seen, the set of diffeomorphisms with rotation number $\mu = p/q$ is bounded by a pair of smooth curves and approaches the axis $\varepsilon = 0$ with increasingly narrow tongues as q increases. The union of these sets is dense. Nevertheless, it turns out that the measure of the set of points of the parameter plane for which the rotation number is rational is small in the domain $0 \leqslant \varepsilon \leqslant \varepsilon_0$, $0 \leqslant \alpha \leqslant 2\pi$, compared to the measure of the whole domain (Fig. 80).

Consequently, a diffeomorphism chosen randomly from our family with small ε has irrational rotation number with great probability.

Moreover, an analogous result holds for any analytic or sufficiently smooth family of diffeomorphisms close to rotations; for example, for the family $y \mapsto y + \alpha + \varepsilon a(y)$ with an arbitrary analytic function a: for small ε, the orbits are everywhere dense on the circle and the rotation number is irrational with preponderant probability.

Consequently, the idea of structural stability is not the only approach to the notion of a generic system. The metric approach indicated above is more appropriate for the description of the actually observable behavior of the system in some cases.

2. According to Denjoy's theorem, a smooth mapping with irrational rotation number is topologically equivalent to a rotation. The question arises of whether this mapping is *smoothly* equivalent to a rotation.

The answer to this question turns out to be negative in the case where the rotation number can be approximated abnormally rapidly by rational numbers (Finzi). The question of smooth equivalence to a rotation reduces to the question of smoothness of the invariant measure of the transformations. If the rotation number is rational, then the invariant measure is concentrated on separate points. If the rotation number can be approximated very rapidly by rational numbers with not too large denominators, then the invariant measure can be approximated so rapidly by measures concentrated on separate points that it cannot be even absolutely continuous with respect to Lebesgue measure. Therefore, the homeomorphisms in Denjoy's theorem cannot be replaced by diffeomorphisms.

3. From the metric point of view, a randomly chosen number μ is irrational with probability 1; moreover, it does not admit too rapid approximation by rational numbers with small enough denominators. For example, for any $\varepsilon > 0$ with probability 1, there exists $C > 0$ such that

$$\left| \mu - \frac{p}{q} \right| \geqslant \frac{C}{q^{2+\varepsilon}}$$

for any integers $p, q > 0$. This gives rise to the conjecture that the phenomenon in § 11K2 occurs with probability 0. We formulate two results in this direction.

Theorem. *For almost every rotation number μ, every sufficiently smooth (of class C^3 or smoother) diffeomorphism of the circle with rotation number μ is smoothly equivalent to the rotation by the angle $2\pi\mu$ (Herman, 1976).*

Here "almost every" means that the Lebesgue measure of the exceptional set of rotation numbers is equal to zero.

Herman's theorem was preceded by an analogous theorem for mappings close to a rotation and by the following result (proved in the analytic case in 1959* and in the smooth case by J. Moser in 1962).

Theorem. *In a sufficiently smooth family $y \mapsto y + \alpha + \varepsilon a(y)$, the proportion of pairs (α, ε) in the domain $0 \leqslant \alpha \leqslant 2\pi, 0 \leqslant \varepsilon \leqslant \varepsilon_0$ for which the diffeomorphism cannot be reduced to a rotation by a smooth diffeomorphism converges to zero together with ε_0.*

This theorem also holds for mappings of the n-dimensional torus.

The proof of these results goes beyond the scope of the present course; nevertheless, in § 11L we shall consider a technique, due to A. N. Kolmogorov, for proving theorems of this sort in the simplest case of an anlytic diffeomorphism.

---

*See: V. I. Arnold, L. D. Mechalkin, *The Kolmogorov seminar on selected topics in analysis* (*1958–1959*), Uspeki Math. Nauk **15**, 1 (1960) 247–250.

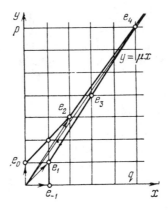

Figure 81.

L. Approximation of Irrational Numbers by Rational Ones

Theorem. *For any irrational number μ, there exist arbitrarily accurate rational approximations whose error is less than the reciprocal value of the square of the denominator:*

$$\left| \mu - \frac{p}{q} \right| < \frac{1}{q^2}.$$

For example, the number π can be approximated with an error of the order of one millionth by a rational fraction with a three-digit numerator and denominator: $\pi \approx 335/113$.

Before proving the theorem, we indicate a geometric method of finding an infinite sequence of such approximations (the method is called the algorithm of *continuous fractions*, or the algorithm of stretching the noses, or simply the *Euclidean algorithm*).

Consider the plane with the coordinate system (x, y) (Fig. 81).

We draw the line $y = \mu x$. For the sake of definiteness, we assume $\mu > 0$. In the first quadrant we mark all points with integral coordinates. Except for the point O, they do not lie on our line, since μ is irrational. We consider the convex hulls of lattice points of the quadrant lying on one side of our line ("below" it) and on the other side ("above" it). [In order to construct these convex hulls, we may visualize a thread fastened at infinity and lying on our line. Let us imagine that a nail is hammered in at every lattice point of our quadrant other than O. Pull the free end O of the thread downward (upward). Then the thread will touch some nails and stretch, forming the boundary of the lower (upper) convex hull.] The vertices of the convex polygonal lines thus constructed give the required approximations of the irrational number μ. If the integers (q, p) are coordinates of a vertex, then the fraction p/q corresponding to the vertex is called a *convergent fraction for* μ. It turns out that for any convergent fraction we have

$$\mu - \left(\frac{p}{q}\right) < \frac{1}{q^2}.$$

To prove this inequality, we describe the construction of our convex polygonal lines by another method. Denote by e_{-1} the basis vector $(1, 0)$ and by e_0 the vector $(0, 1)$. These vectors lie on different sides of the line $y = \mu x$. We construct a sequence of vectors $e_1, e_2, \cdots$ in the following way. Let e_{k-1} and e_k be already constructed and lie on different sides of our line. We add e_k to e_{k-1} as many times as we can in such a way that the sum lies on the same side of the line $y = \mu x$ as e_{k-1}.

In this way, we obtain a sequence of natural numbers a_k and a sequence of lattice vectors

$$e_1 = e_{-1} + a_0 e_0, \ldots, \qquad e_{k+1} = e_{k-1} + a_k e_k, \ldots.$$

The vectors e_k are the vertices of our two convex hulls (the upper one for even k and the lower one for odd k).

Lemma. *The oriented area of the parallelogram spanned by the vectors (e_{k+1}, e_k) is equal to $(-1)^k$ (taking into account the orientation).*

◀ For the initial parallelogram (e_0, e_{-1}), this is evident. Every following parallelogram has a common side with the preceding one and equal altitude, and gives an opposite orientation of the plane. ▶

Corollary. *Denote by q_k and p_k the coordinates of the point e_k. The difference of two subsequent convergent fractions is equal to*

$$\frac{p_k}{q_k} - \frac{p_{k+1}}{q_{k+1}} = \frac{(-1)^k}{q_k q_{k+1}}.$$

◀ In bringing the fractions to a common denominator, it turns out that the numerator is the determinant of the components of e_{k+1} and e_k, which is equal to the oriented area of the parallelogram. ▶

◀ *Proof of Theorem.* The vectors e_k lie alternately on one or the other side of the line $y = \mu x$.

Therefore, the convergent fractions are alternately larger or smaller than μ. Consequently, the difference between μ and a convergent fraction is smaller than the modulus of the difference between the convergent fraction and the subsequent convergent fraction. By the corollary, the absolute value of this difference is equal to $1/q_k q_{k+1}$, which is not larger than $1/q_k^2$, since $q_{k+1} \geq q_k$ for $k \geq 0$. ▶

Remark. The numbers a_k are called *partial quotients*. The convergent fractions can be expressed in terms of the partial quotients in the following way:

$$\frac{p_k}{q_k} = a_0 + \cfrac{1}{a_1 + \cfrac{\ddots}{\quad + \cfrac{1}{a_{k-1}}}}$$

The expression $a_0 + \cfrac{1}{a_1 + \cdots}$ is called an infinite continuous fraction. The number μ is expanded in an infinite continuous fraction in the sense that $\lim p_k/q_k = \mu$.

§ 12. Analytic Reduction of Analytic Circle Diffeomorphisms to a Rotation

In this section, a theorem on analytic diffeomorphisms of the circle close to a rotation and having an almost arbitrary rotation number is proved by means of Kolmogorov's modification of the Newton method.

A. Formulation of the Theorem

Denote by Π_ρ the strip $|\operatorname{Im} y| < \rho$. For a holomorphic function a bounded in this strip, we shall write

$$\|a\|_\rho = \sup |a(y)|, \qquad y \in \Pi_\rho.$$

Let μ be an irrational number, and let $K > 0$, $\sigma > 0$. We say that μ is a *number of type* (K, σ) if for any integers p and $q \neq 0$,

$$\left| \mu - \frac{p}{q} \right| \geq \frac{K}{|q|^{2+\sigma}}$$

Theorem. *There exists $\varepsilon > 0$ depending only on K, ρ, and σ such that if a is a 2π-periodic analytic function, real on the real axis with $\|a\|_\rho < \varepsilon$ and such that the transformation*

$$y \mapsto y + 2\pi\mu + a(y)$$

defines a diffeomorphism of the circle with rotation number μ of type (K, σ), then this diffeomorphism is analytically equivalent to the rotation by the angle $2\pi\mu$.

B. Homological Equation

Denote by $\mathfrak{A}$ the rotation by the angle $2\pi\mu$ and by H the desired diffeomorphism converting the rotation into A. The following diagram is commutative:

$$S^1 \xrightarrow{A} S^1$$
$$H \uparrow \qquad \uparrow H, \qquad \text{i.e., } H \cdot \mathfrak{A} = A \cdot H.$$
$$S^1 \xrightarrow{\mathfrak{A}} S^1$$

We write H in the form $Hz = z + h(z)$, $h(z + 2\pi) = h(z)$. For h, we obtain the functional equation

$$h(z + 2\pi\mu) - h(z) = a(z + h(z)).$$

If A differs little from a rotation, then a is small. It is natural to expect that h is of the same order of smallness. Then $a(z + h(z))$ differs from $a(z)$ by a quantity of order of smallness higher than that of a. Therefore, "in first approximation", we obtain the equation

$$h(z + 2\pi\mu) - h(z) = a(z)$$

for h. This linear equation is called the *homological equation*.

Remark. We may consider the collection of diffeomorphisms A as an "infinite-dimensional manifold" on which the "infinite dimensional group" of diffeomorphisms H acts. Moreover, the function a can be interpreted as a tangent vector to the manifold of the diffeomorphisms at the point $\mathfrak{A}$ and the function h as a tangent vector to the group at identity.

Using these terms, the homological equation can be interpreted in the following way: a belongs to the tangent space to the orbit of the point $\mathfrak{A}$ under the action of the group if and only if the homological equation is solvable for h.

C. Formal Solution of the Homological Equation

We expand the given function a and the unknown function h in the Fourier series:

$$a(z) = \sum a_k e^{ikz}, \qquad h = \sum h_k e^{ikz}.$$

Comparing the coefficients of e^{ikz}, we find

$$h_k = \frac{a_k}{e^{2\pi i k \mu} - 1}.$$

For the solvability of the equation, it is necessary that the denominators vanish only together with the numerators. In particular, the homological equation is not solvable if $a_0 \neq 0$. If $a_0 = 0$ and the rotation number μ is irrational, then the preceding formulas give a solution of the homological equation in the class of formal Fourier series. In order to obtain the actual solution, it is necessary to study the convergence of the series.

D. Behavior of the Fourier Coefficients of Analytic Functions

Lemma 1. *If f is a 2π-periodic function which is analytic in the strip Π_ρ, continuous in the closure of this strip, and $\|f\|_\rho \leqslant M$, then its Fourier coefficients are decreasing in* *geometric progression*:

$$|f_k| \leqslant Me^{-|k|\rho}.$$

◀ As is known, $f_k = \dfrac{1}{2\pi} \oint f(z)e^{-ikz}\,dz$. Let $k > 0$. Let us shift the path of integration down (by $-i\rho$). The integral does not change, since the integrals on the vertical sides of the rectangle thus obtained are equal. Hence,

$$f_k = \frac{1}{2\pi} \int_0^{2\pi} f(x - i\rho)e^{-ikx - k\rho}\,dx, \qquad |f_k| \leqslant Me^{-k\rho}.$$

For $k < 0$ the path of integration has to be shifted upward (by $i\rho$). ▶

Lemma 2. *If* $|f_k| \leqslant Me^{-|k|\rho}$, *then the function* $f = \Sigma f_k e^{ikz}$ *is analytic in the strip* Π_ρ *and*

$$\|f\|_{\rho - \delta} \leqslant \frac{4M}{\delta} \qquad \text{for} \qquad \delta < \rho,\, \delta < 1.$$

◀ $\|f\|_{\rho - \delta} \leqslant \sum |f_k| |e^{ikz}| \leqslant M \sum e^{-|k|\rho e|k|(\rho - \delta)} = M \sum e^{-|k|\delta} \leqslant \dfrac{2M}{1 - e^{-\delta}} \leqslant \dfrac{4M}{\delta}.$ ▶

Remark. In the case of functions of n variables, Lemma 1 still holds. In Lemma 2, the estimate $4M/\delta$ has to be replaced by CM/δ^n, where $C = C(n)$ is a constant independent of δ and f.

E. Small Denominators

In solving the homological equation, the Fourier coefficients of the right-hand side have to be divided by the numbers $e^{2\pi ik\mu} - 1$. If μ is irrational, then for $k \neq 0$, these numbers are different from 0. Nevertheless, some of them are very close to 0. Indeed, every number μ admits rational approximations p/q with error $|\mu - (p/q)| < 1/q^2$ with arbitrarily large q. For $k = q$, the denominator $e^{2\pi ik\mu} - 1$ will be very small.

It turns out that, with probability 1, all these small denominators admit lower estimates in terms of powers of k.

Lemma 3. *Let* $\sigma > 0$. *For almost every real* μ *there exists* $K = K(\mu, \sigma) > 0$ *such that*

$$\left| \mu - \frac{p}{q} \right| \geqslant \frac{K}{|q|^{2+\sigma}}$$

for all integers p *and* $q \neq 0$.

◀ Consider those numbers μ in the interval $[0, 1]$, for which the above inequality (with fixed p, q, K, and σ) is violated. These numbers form an interval of length not greater than $2K/q^{2+\sigma}$. The union of these intervals for all p (for fixed $q > 0$, K, and σ) has total length not greater than $2K/q^{1+\sigma}$. By summing over q, we obtain a set of measure not greater than CK, where $C = 2\Sigma q^{-(1+\sigma)} < \infty$. Consequently, the set of numbers $\mu \in [0, 1]$ for which the desired K does not exist is covered by sets of arbitrarily small measure. Consequently, this set is of measure zero (on the interval $[0, 1]$ and, therefore, on the entire line). ▶

Remark. The numbers μ satisfying the above inequality are called numbers of type (K, σ) in § 12A.

    For a number μ of type (K, σ), a small denominator admits the following lower estimate:

$$|e^{2\pi i k \mu} - 1| \geqslant \frac{K}{2|k|^{1+\sigma}} \qquad (|k| > 0).$$

◀ Indeed, the distance of $k\mu$ from the closest integer can be estimated from below by the number $K/|k|^{1+\sigma}$; a chord of the unit circle is not shorter than the length of the small arc subtended by it divided by π. ▶

F. Study of the Homological Equation

Let a be a 2π-periodic analytic function with mean value 0.

Lemma 4. *For almost all μ, the homological equation has a 2π-periodic analytic solution (which is real if a is a real function). There exists a constant $\nu = \nu(K, \sigma) > 0$ such that if μ is of type (K, σ), then for any $\delta > 0$ smaller than ρ and any $\rho < \frac{1}{2}$, we have*

$$\|h\|_{\rho-\delta} \leqslant \|a\|_\rho \delta^{-\nu}.$$

Remark. We can see that the passage from a to h worsens the properties of the function not more than differentiation ν times. (It is useful to observe that $\|d^\nu f/dz^\nu\|_{\rho-\delta} \leqslant C\|f\|_\rho \delta^{-\nu}$ according to the Cauchy estimate of Taylor coefficients.) Disregarding the deterioration of the function due to ν differentiations, we may say that the solution h of the homological equation is of the same order of smallness as the right-hand side a.

◀ *1.* By Lemma 1, we have $|a_k| \leqslant Me^{-|k|\rho}$ provided that $\|a\|_\rho \leqslant M$.

    2. Since μ is of type (K, σ), we have $|h_k| \leqslant 2Me^{-|k|\rho}k^{1+\sigma}/K$.

    3. The function $x^m e^{-\alpha x}$, $x \geqslant 0$, has a maximum at the point $x = m/\alpha$. Therefore, $x^m e^{-\alpha x} \leqslant C\alpha^{-m}$, $C = (m/e)^m$ for any $\alpha > 0$ and $x > 0$. Consequently, for any $\alpha > 0$, we have $|k|^{1+\sigma}e^{-\alpha|k|} \leqslant C\alpha^{-m}$, $m = 1 + \sigma$.

    4. Thus, $|h_k| \leqslant Me^{-|k|(\rho-\alpha)}2CK^{-1}\alpha^{-m}$. By Lemma 2, we have $\|h\|_{\rho-\delta} \leqslant DM$, where $D = 8C/K\alpha^m(\delta - \alpha)$. Take $\alpha = \delta/2$. The number D does not exceed $\delta^{-\nu}$ if ν is sufficiently large (because $\delta < \frac{1}{2}$). ▶

G. Construction of Successive Approximations

We solve the homological equation with right-hand side $\tilde{a} = a - a_0$ (a_0 is the mean value of the function a). Denote by h^0 the solution. We define a mapping H_0 by the formula $H_0(z) = z + h^0(z)$. We construct the mapping $A_1 = H_0^{-1} \circ A \circ H_0$ and define a function a^1 by the relation $A_1 z = z + 2\pi\mu + a^1(z)$.

In other words, on the circle we have introduced a new coordinate z_1 (where $z = H_0(z_1)$) and described the mapping A in terms of the new coordinate. We have obtained the mapping $z_1 \mapsto A_1 z_1$ which differs from the rotation by angle $2\pi\mu$ by the "mismatch" a^1.

The next approximation is constructed in the same way, beginning with A_1 in place of A. We construct h^1 and a substitution H_1 converting A_1 into

$$A_2 = H_1^{-1} \circ A_1 \circ H_1.$$

This gives rise to a sequence of substitutions H_n. We consider the substitution $\mathscr{H}_n = H_0 \circ H_1 \circ \ldots \circ H_{n-1}$. We have $A_n = \mathscr{H}_n^{-1} \circ A \circ \mathscr{H}_n$.

It turns out that *the sequence $\mathscr{H}_n$ is convergent if μ is a number of type (K, σ) and if* $\|a\|_\rho$ is sufficiently small. The limit substitution $\mathscr{H}$ converts the initial mapping into $\mathscr{H}^{-1} \circ A \circ \mathscr{H} = \lim A_n$, i.e., rotation by angle $2\pi\mu$.

H. Estimation of the Mismatch after the First Approximation

Lemma 5. *There exist constants $\varkappa, \lambda > 0$ depending only on K and σ such that for every δ in the interval $(0, \rho)$, where $\rho < \frac{1}{2}$, we have*

$$\|a\|_\rho \leqslant \delta^\varkappa \Rightarrow \|a^1\|_{\rho-\delta} \leqslant \|a\|_\rho^2 \delta^{-\lambda}.$$

Remark. This means that the mismatch a^1 remaining after the first substitution of the variable is of second-order smallness compared to the initial difference a from the rotation (up to a deterioration from λ differentiations). Consequently, in the above scheme of successive approximation, the error of every approximation is of the order of the square of the preceding error. After n approximations, we obtain an error of the order ε^{2^n}, where ε is the error of the initial (zeroth) approximation.

This kind of convergence, which is characteristic of Newton's method of tangents (Fig. 82), enables us to overcome the effect of small denominators occurring in every step (i.e., the occurrence of the deteriorating multiplier $\delta^{-\lambda}$); this technique was devised by Kolmogorov in 1954.

◀ *1.* Let Ω be a convex domain in $\mathbb{C}^n$ (or $\mathbb{R}^n$) and let $h: \Omega \to \mathbb{C}^n$ (or $\mathbb{R}^n$) be a smooth mapping with $\|h_*\| = \sup_{x \in \Omega} \|h_*(x)\| < 1$. Then *the mapping H taking x into $x + h(x)$ is a diffeomorphism of Ω onto $H\Omega$.*

◀ The eigenvalues of $H_*(x)$ are different from 0; therefore, H is a local diffeomor-

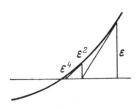

Figure 82.

phism. In view of the condition $|h_*| < q < 1$ and the convexity of Ω, h is a contraction. Consequently, the difference between the displacements of any two distinct points under the mapping H is smaller than the distance between these points and, therefore, their images are distinct, i.e., H is one-to-one. ▶

2. We show that *if $\varkappa$ is sufficiently large, then A_1 is analytic in the strip $\Pi_{\rho-\delta}$.* ◀ Let $\|a\|_\rho < M = \delta^\varkappa$. We have $a_0 < M$ and $\|\tilde{a}\|_\rho \leqslant 2M$. By the theorem in § 12F, we have $\|h^0\|_{\rho-\alpha} \leqslant 2M\alpha^{-\nu}$. Consequently, $\|dh^0/dz\|_{\rho-2\alpha} \leqslant 2M\alpha^{-(\nu+1)}$.

Choose $\alpha = \delta/8$. If $\varkappa$ is sufficiently large, then from the preceding inequalities we obtain

$$\|a\|_\rho < \alpha, \qquad \|h^0\|_{\rho-\alpha} < \alpha, \qquad \|dh^0/dz\|_{\rho-2\alpha} < \alpha.$$

Consequently, according to § 12H1, H_0 is a diffeomorphism of the strip $\Pi_{\rho-2\alpha}$ and the image contains the strip $\Pi_{\rho-3\alpha}$. Since

$$H_0\Pi_{\rho-\delta} \subset \Pi_{\rho-\delta+\alpha}, \qquad A \circ H_0\Pi_{\rho-\delta} \subset \Pi_{\rho-\delta+2\alpha} \subset \Pi_{\rho-3\alpha},$$

the diffeomorphism H_0^{-1} is defined on $A \circ H_0\Pi_{\rho-\delta}$. Hence, the mapping $A_1 = H_0^{-1} \circ A \circ H_0$ is analytic in $\Pi_{\rho-\delta}$ and is a diffeomorphism there. ▶

3. *We estimate the mismatch a^1.* ◀ The commutative diagram defining a^1 gives

$$z + 2\pi\mu + a^1(z) + h^0(z + 2\pi\mu + a^1(z)) \equiv z + h^0(z) + 2\pi\mu + a(z + h^0(z)).$$

By the homological equation, we obtain

$$a^1(z) = [a(z + h^0(z)) - a(z)] - [h^0(z + 2\pi\mu + a^1(z)) - h_0(z + 2\pi\mu)] + a_0.$$

The expression within the first pair of square brackets can be estimated using the mean value theorem and the Cauchy inequality. By § 12H2, we obtain

$$\|a(z + h^0(z)) - a(z)\|_{\rho-\delta} \leqslant \frac{M}{\delta}\|h^0\|_{\rho-\delta} \leqslant M^2\delta^{-u},$$

where the constant u depends only on ν, i.e., only on K and σ.

The expression within the second pair of square brackets can be estimated analogously:

$$\|[\quad]\|_{\rho-\delta} \leqslant 2M\alpha^{-(\nu+1)}\|a^1\|_{\rho-\delta} \leqslant M\delta^{-u_1}\|a^1\|_{\rho-\delta}.$$

Hence,

$$\|a^1\|_{\rho-\delta}(1 - M\delta^{-u_1}) \leqslant |a_0| + M^2\delta^{-u}. ▶$$

4. Now *estimate the quantity $|a_0|$* using the fact that the rotation number of A and, therefore, of A_1 is equal to $2\pi\mu$. ◀ From this it follows that a^1 vanishes at some real point z_0. We substitute the value z_0 into the formula for $a^1(z)$. We obtain $a_0 = a(z_0) - a(z_0 + h^0(z_0))$, and, consequently, $|a_0| \leqslant M^2\delta^{-u}$ (cf., § 12H3). ▶

5. From estimates in § 12H3 and § 12H4, it follows that

$$\|a^1\|_{\rho-\delta} \leqslant 4M^2 \delta^{-u}. \blacktriangleright$$

I. Convergence of the System of Approximations

1. We shall consider the mapping A_n constructed in the nth step in a strip of radius ρ_n decreasing with every step: $\rho_0 = \rho, \rho_n = \rho_{n-1} - \delta_{n-1}$.

The sequence of the numbers δ_n is chosen decreasing in the following way:

$$\delta_n = \delta_{n-1}^{3/2}, \qquad \delta_0 < \tfrac{1}{2}.$$

Then, for sufficiently small δ_0, $\sum \delta_n < \rho/2$.

2. We form a numerical sequence M_n by setting

$$M_n = \delta_n^N.$$

A sufficiently large number N (depending only on K and σ) will finally be chosen below. Note that $M_n = M_{n-1}^{3/2}$.

3. Assume $\|a\|_\rho \leqslant M_0$. *We prove* $\|a^n\|_{\rho_n} \leqslant M_n$.
◀ According to Lemma 5, if $N > \varkappa$, then

$$\|a^1\|_{\rho_1} \leqslant M_0^2 \delta_0^{-\lambda} = \delta_0^{2N-\lambda}.$$

On the other hand,

$$\delta_0^{2N-\lambda} < \delta_1^N = \delta_0^{3N/2},$$

provided that $N > 2\lambda$. We choose N larger than 2λ and $\varkappa$. Then we obtain

$$\|a^1\|_{\rho_1} \leqslant \delta_1^N = M_1.$$

The passage from a^{n-1} to a^n is analogous. ▶

4. We prove the convergence of the products $\mathscr{H}_n = H_0 \circ \ldots \circ H_{n-1}$ *in* $\Pi_{\rho/2}$.

The diffeomorphism H_0 is analytic in Π_ρ and satisfies the inequalities $\|h^0\|_{\rho_1} \leqslant \delta_0$, $\|dh^0/dz\|_{\rho_1} \leqslant \delta_0$ (cf., § 12H2).

In the same way, for H_{n-1} we obtain $\|h^{n-1}\|_{\rho_n} \leqslant \delta_{n-1}$, $\|dh^{n-1}/dz\|_{\rho_n} \leqslant \delta_{n-1}$. Consequently, $\mathscr{H}_n$ is analytic in Π_{ρ_n} and its derivative is bounded from above and below by the numbers

$$C = \prod (1 + \delta_k), c = \prod (1 - \delta_k).$$

This implies that $\mathscr{H}_n$ is a diffeomorphism of Π_{ρ_n} and that the sequence $\mathscr{H}_n$ converges in $\Pi_{\rho/2}$. Indeed,

$$\|\mathscr{H}_n - \mathscr{H}_{n+1}\|_{\rho/2} \leqslant C\|h^n\|_{\rho/2} \leqslant C\delta_n.$$

Denote by H the limit of the sequence $\mathscr{H}_n$. Passing to the limit in $A \circ \mathscr{H}_n = \mathscr{H}_n \circ A_n$, we obtain $A \circ \mathscr{H} = \mathscr{H} \circ \mathfrak{A}$, where $\mathfrak{A}$ is the rotation by $2\pi\mu$. The theorem is proved. ▶

J. Remarks

1. Moser observed that an analogous theorem can be proved in the case of finite smoothness by combining the above approximation with a smoothing process of Nash (cf., J. Moser, *A rapidly converging iteration method and nonlinear partial differential equations*, I, Ann. Scuola Norm. Sup. Pisa (3), **20** (1966), 265–316, II, **20** (1966), 499).

In the first publications of Moser, hundreds of derivatives were required. Subsequent efforts of Moser and Rüssmann reduced the number of derivatives (H. Rüssmann, *Kleine Nenner II: Bemerkungen zur Newtonischen Methode*, Nachr. Acad. Wiss. Göttingen, Math. Phys. Klasse **1** (1972), 1–10). See also the recent works by Sinai, Khanin and others (1987).

2. In the multidimensional case, the rotation number is not defined. Nevertheless, in the family of mappings $y \mapsto y + \alpha + a(y)$ with small a and $y \in T^n$, for most α the mapping is smoothly equivalent to a translation $y \mapsto y + 2\pi\mu$. In particular, for the analytic family $y \mapsto y + \alpha + \varepsilon a_1(y) + \varepsilon^2 a_2(y) + \cdots$, there exists, for almost every μ, an analytic function $\alpha(\varepsilon) = 2\pi\mu + \varepsilon\mu_1 + \cdots$ such that the mapping $y \mapsto y + \alpha(\varepsilon) + \varepsilon a_1(y) + \cdots$ turns into $y \mapsto y + 2\pi\mu$ upon an analytic substitution $y = z + \varepsilon h_1(z) + \cdots$.

The coefficients $h_1, \ldots$ can be found by comparing the terms containing the same power of ε. However, the convergence of the series in ε thus obtained is proved only indirectly, using the Newton approximation.

3. It seems plausible that an analytic diffeomorphism of the circle is analytically equivalent to an irrational rotation if and only if the fixed points of the powers of the diffeomorphism do not accumulate at the real axis. One may also think that for some irrational numbers μ which can be approximated abnormally well by rational numbers, the function $\alpha(\varepsilon)$ described in § 12J2 is not even smooth (even in the one-dimensional case).

§ 13. Introduction to the Hyperbolic Theory

In the present section, we prove Anosov's theorem on the structural stability of an automorphism of the torus and the Grobman–Hartman theorem on the structural stability of a saddle.

A. The Simplest Example: A Linear Automorphism of the Torus

Differential equations with multidimensional phase space define a large class of structurally stable systems in which every phase curve lies between neighboring ones in the same way as an equilibrium position of saddle type between the neighboring hyperbolas. We begin with the simplest example (Fig. 83).

Consider the automorphism A of the torus T^2 which is given by the

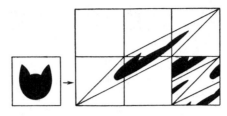

Figure 83.

unimodular (having determinant equal to 1) linear transformation $\hat{A}$ of the plane with the matrix

$$\begin{pmatrix} 2 & 1 \\ 1 & 1 \end{pmatrix}.$$

The lattice $2\pi\mathbb{Z}^2$ is transformed into itself by $\hat{A}$. Therefore, equivalent (equal modulo 2π) points of the plane are mapped by $\hat{A}$ into equivalent ones. Consequently, $\hat{A}$ defines a mapping A of the torus onto itself. The matrix of $\hat{A}^{-1}$ is also integral, since $\det \hat{A} = 1$. Therefore, A is a diffeomorphism of the torus onto itself. In addition, A is an automorphism of the group $T^2 = \mathbb{R}^2/2\pi\mathbb{Z}^2$.

B. Properties of the Torus Automorphism

A finite set of points is called a cycle of the mapping A if A permutes them cyclically.

Theorem 1. *The automorphism A of the torus has a countable number of cycles. All points both of whose coordinates are rational multiples of 2π, and only they, are cyclic points of the automorphism A.*

◀ *1*. Fix an integer N. Then the points of the torus whose coordinates are rational multiples of 2π with denominator N form a finite set. The mapping A maps this set onto itself. Consequently, all points of this set belong to cycles.

2. Let $2\pi\xi$ be a point of a cycle of order $n > 1$. We have $\hat{A}^n\xi = \xi + m$, where m is an integral vector. The linear equation obtained for ξ has nonzero determinant. Therefore, the components of ξ are rational. ▶

Theorem 2. *The iterates of the automorphism A smear every domain F uniformly over the torus: for every domain G we have*

$$\lim_{n\to\infty} \frac{\mathrm{mes}(A^n F) \cap G}{\mathrm{mes}\, F} = \frac{\mathrm{mes}\, G}{\mathrm{mes}\, T^2}.$$

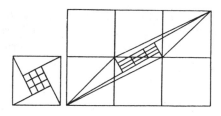

Figure 84.

This property of A is called *mixing*; it holds for any measurable sets F and G.

◀ In terms of functions on the torus, this relation can be written in the following form:

$$\lim_{n \to \infty} (A^{n*}f, g) = \frac{(f, 1)(1, g)}{(1, 1)},$$

where $(u, v) = \int u(x)\bar{v}(x)\,dx$, $(A^{n*}f)(x) = f(A^n x)$.

Now let f be an exponential function: $f = e^{i(p, x)}$. Then $A^{n*}f$ is also an exponential function with wave vector $p' = \hat{A}^n p$. If $p \neq 0$, then the orbit of p under $\hat{A}^n$ is infinite. Therefore, for any exponential function $g = e^{i(q, x)}$, we have $\lim_{n \to \infty} (A^{n*}f, g) = 0$. We obtain the required result by approximating f and g by sums of exponential functions in mean square. ▶

A more instructive (although more complicated in proper execution) proof of the mixing property can be obtained in the following way.

Theorem 3. *On the torus there exist two direction fields invariant with respect to the automorphism A. The integral curves of each of these direction fields are everywhere dense on the torus. The automorphism A converts the integral curves of each field into integral curves of the same field, expanding by $\lambda > 1$ for the first field and contracting by λ for the second (Fig. 84).*

◀ We consider the eigenvalues $\lambda_{1,2} = (3 \pm \sqrt{5})/2$ of the transformation $\hat{A}$. It is obvious that $\lambda_1 > 1 > \lambda_2$ and the numbers λ_1 and λ_2 are irrational. In the plane we consider the family of all straight lines parallel to the first eigenvector of $\hat{A}$. Since λ_1 is irrational, the components of the eigenvector are incommensurate. Therefore, the lines of the family determine an everywhere dense winding line on the torus. The transformation $\hat{A}$ of the plane converts this family of lines into itself, dilating them $\lambda_1 > 1$ times. Therefore, the transformation A of the torus converts the family of winding lines into itself, dilating them the *same number of times*. This family is called the *expanding foliation* of A.

The second eigendirection determines a *contracting foliation* in a similar way. ▶

Now consider the image of a planar domain $\hat{F}$ under the transformation A^n of the plane. This transformation represents a hyperbolic rotation: expansion λ_1^n times in the first eigendirection and contraction λ_1^n times in the other. Therefore, for large n, the image of the domain $\hat{F}$ is a narrow long strip stretched out in the first eigendirection. Consequently, on the torus, the image of F under A^n is a long narrow strip close to a long segment of a phase curve of the equation $\dot{x} = \omega$ with nonresonant vector ω. This curve is uniformly distributed on the torus. This implies that for increasing n, the images $A^n F$ intersect any domain G on the torus; the mixing property can be deduced from these arguments with little effort.

C. Structural Stability of an Automorphism of the Torus

The astonishing fact discovered in the early 1960's and appearing to be one of the most significant results of the theory of differential equations in recent decades is that the automorphism considered above is structurally stable in the class of all diffeomorphisms of the torus. In particular, every diffeomorphism sufficiently close to A has a countable number of cycles and an everywhere dense set of periodic points.

Anosov's Theorem. *The torus automorphism $A: T^2 \to T^2$ defined by the matrix $\begin{pmatrix} 2 & 1 \\ 1 & 1 \end{pmatrix}$ is structurally stable in the C^1-topology. In other words, every diffeomorphism B sufficiently close to A together with its derivative is conjugate with A by means of some homeomorphism H, i.e., $B = H^{-1}AH$.*

Remark. The homeomorphism H can be chosen arbitrarily close to the identity transformation if B is sufficiently close to A, but it cannot be made smooth in general.

Anosov's theorem shows that in the case of systems with multidimensional phase space, behavior of the phase curves other than attraction to stable equilibrium positions and cycles is possible and is preserved under small perturbations, in contrast to the case of vector fields on the two-dimensional sphere or torus. Later we shall discuss the physical meaning of this behavior of dynamical systems, which is more complicated than self-sustained oscillation. The proof of Anosov's theorem is given in § 13D–§ 13G.

D. The Homological Equations

We are looking for a homeomorphism H, $H(x) = x + h(x)$ making the diagram

$$
\begin{array}{ccc}
\mathbb{R}^2 & \xrightarrow{B} & \mathbb{R}^2 \\
H\uparrow & & \uparrow H \\
\mathbb{R}^2 & \xrightarrow{A} & \mathbb{R}^2
\end{array}
$$

commutative, where $B(x) = A(x) + f(x)$ and the functions f and h are 2π-periodic in x.

From the diagram, we obtain the following nonlinear functional equation for h:

$$h(Ax) - Ah(x) = f(x + h(x)).$$

We assume that f is small and that h turns out to be small of the same order. Therefore, we replace the right-hand side by $f(x)$, omitting a "quadratically small quantity". We obtain the following linearized equation:

$$h(Ax) - Ah(x) = f(x).$$

This equation is called the *homological equation*.

E. Solution of the Homological Equation

The left-hand side of the homological equation is linear in h. We denote by L the linear operator assigning the left-hand side of the homological equation to h. The solution of the homological equation has the form $h = L^{-1}f$. We only have to prove that the operator L is invertible.

Lemma 1. *The space of vector fields on the torus splits into the direct sum of two subspaces invariant with respect to L.*

◀ The spaces of vector fields parallel to the first and second eigendirections of the operator A are invariant under the transformation, and every vector field can be represented uniquely as the sum of two fields in the eigendirections.

Let $f = f_1 e_1 + f_2 e_2$, $h = h_1 e_1 + h_2 e_2$, be the decompositions of the fields f and h. Then the homological equation takes the form

$$h_1(Ax) - \lambda_1 h_1(x) = f_1(x),$$
$$h_2(Ax) - \lambda_2 h_2(x) = f_2(x).$$

Here $\lambda_1 = \lambda_2^{-1} > 1 > \lambda_2 = \lambda$ are the eigenvalues.

Consider the operator of applying A to the argument in the space of continuous functions on the torus. We denote it by S. We have

$$(Sg)(x) = g(Ax), \qquad \|S\| = 1, \qquad \|S^{-1}\| = 1.$$

The homological equation can be written in the form (E is the identity operator)

$$(S - \lambda_i E)h_i = f_i, \qquad i = 1, 2.$$

Let $i = 1$. Then

$$(S - \lambda_1 E)^{-1} = -\lambda(E + \lambda S + \lambda^2 S^2 + \cdots).$$

Since $\lambda < 1$ and $\|S\| = 1$, the inverse operator exists and

$$\|(S - \lambda_1 E)^{-1}\| \leq \frac{\lambda}{1 - \lambda}.$$

Analogously,

$$(S - \lambda_2 E)^{-1} = S^{-1}(E - \lambda S^{-1})^{-1} = S^{-1}(E + \lambda S^{-1} + \lambda^2 S^{-2} + \cdots),$$

$$\|(S - \lambda_2 E)^{-1}\| \leq \frac{1}{1 - \lambda}.$$

Consequently, L^{-1} exists and $\|L^{-1}\| \leq 1/(1 - \lambda)$. The homological equation is solved. ▶

F. Construction of the Mapping H

The nonlinear functional equation from § 13D can now be solved by the simple contraction mapping principle. We set

$$\Phi[h](x) = f(x + h(x)) - f(x).$$

Our functional equation has the form

$$Lh = \Phi h + f, \qquad h = L^{-1}\Phi h + L^{-1}f.$$

Lemma 2. *If the C^1-norm of f is sufficiently small, then the operator $L^{-1}\Phi$ is a contraction in the space C^0.*

◀ It is sufficient to verify that the nonlinear operator Φ satisfies a Lipschitz condition with a small constant. Indeed, according to E we have

$$\|L^{-1}\Phi h^1 - L^{-1}\Phi h^2\| \leq \|\Phi h^1 - \Phi h^2\|/(1 - \lambda).$$

On the other hand,

$$\|\Phi h^1 - \Phi h^2\| = \max |f(x + h^1(x)) - f(x + h^2(x))| \leq \|f\|_{C^1}\|h^1 - h^2\|.$$

Hence, $L^{-1}\Phi$ is contractive provided that $\|f\|_{C^1} < 1 - \lambda$. ▶
Under this condition our equation is solved and H is constructed.

G. Properties of the Mapping H

We prove that H is a homeomorphism of the torus.

◀ If h is small in the C^1 metric, then the mapping $H = E + h$ is a homeomorphism. We only know that h is small in the C^0-metric. Nevertheless, from $H(x) = H(y)$, in view of the hyperbolic properties of $\hat{A}$, it follows that $x = y$.

Indeed, $\hat{B}\hat{H} = \hat{H}\bar{A}$ in the plane. Therefore, $\hat{H}\hat{A}\hat{x} = \hat{H}\hat{A}\hat{y}$ and $\hat{H}\hat{A}^n\hat{x} = \hat{H}\hat{A}^n\hat{y}$ in general. In view of the hyperbolicity of $\hat{A}$, the distance between the points $\hat{A}^n\hat{x}$ and $\hat{A}^n\hat{y}$ approaches ∞ as either $n \to +\infty$ or $n \to -\infty$. This contradicts the boundedness of h. We necessarily have $\hat{x} = \hat{y}$ and, consequently, $x = y$ on the torus.

We prove that the range of H is the whole torus. Indeed, the image under $\hat{H}$ of a sufficiently large disk in the plane contains a disk of radius 2π (since h is bounded). Therefore, $HT^2 = T^2$. Hence, H is a homeomorphism of the torus. Moreover, $BH = HA$. ▶

This completes the proof of the theorem of § 13C.

H. Theorem on the Structural Stability of a Saddle Point

The preceding arguments also prove the following proposition.

Theorem of Grobman–Hartman. *Let $A: \mathbb{R}^n \to \mathbb{R}^n$ be a linear transformation without eigenvalues equal to 1 in absolute value. Every local diffeomorphism $B: (\mathbb{R}^n, O) \to (\mathbb{R}^n, O)$ with linear part A at the fixed point O is topologically equivalent to A in a sufficiently small neighborhood of O.*

◀ In the neighborhood of O the local diffeomorphism B coincides with a global diffeomorphism $C: \mathbb{R}^n \to \mathbb{R}^n$ defined in the following way: Let φ be a smooth function equal to 0 outside a 1-neighborhood of O and equal to 1 in a small neighborhood of O. Then C coincides with A outside the ε-neighborhood in which φ_ε is different from 0. Inside this neighborhood, we set $C = A + \varphi_\varepsilon(B - A)$; here $\varphi_\varepsilon(x) = \varphi(x/\varepsilon)$.

Our proof of Anosov's theorem shows that every $\mathbb{R}^n \to \mathbb{R}^n$ diffeomorphism of $\mathbb{R}^1$ C^1-close to A is topologically equivalent to A. On the other hand, the C^1-smallness of the difference $C - A$ can be achieved by an appropriate choice of $\varepsilon > 0$, because

$$|B - A| \leqslant C\varepsilon^2, \qquad |(B - A)'| \leqslant C\varepsilon$$

in the ε-neighborhood of zero. Hence C is topologically equivalent to A.

On the other hand, C, just as A, has O as its only fixed point. Consequently, any homeomorphism converting A into C leaves O fixed. ▶

§ 14. Anosov Systems

In this section, Anosov diffeomorphisms and Anosov flows are defined. We also discuss their applications in the theory of geodesic flows on manifolds of negative curvature and in other problems.

A. Definition of an Anosov Diffeomorphism

An analysis of the above diffeomorphism of the torus shows that in the preceding arguments only the contracting and expanding foliations are essential; therefore, it is possible to introduce a general notion of hyperbolic diffeomorphisms without assuming that M is the torus.

Let $A: M \rightarrow M$ be a diffeomorphism of a compact manifold. We assume that:

(1) The tangent space of M is decomposed into the direct sum of two subspaces at every point of M:

$$T_x M = X_x \oplus Y_x.$$

(2) The fields of planes $X = \{X_x\}$ and $Y = \{Y_x\}$ are continuous and invariant with respect to A.
(3) For some Riemannian metric, A contracts the planes of the first field and expands the planes of the second field: there exists a number $\lambda < 1$ such that for any point x of M,

$$\|A_* \xi\| \leqslant \lambda \|\xi\|, \qquad \forall \xi \in X_x, \qquad \|A_* \eta\| \geqslant \lambda^{-1}\|\eta\|, \qquad \forall \eta \in Y_x.$$

Then we say that A is an *Anosov system*.

EXAMPLE

Let $M = T^2$ be the torus and

$$A = \begin{pmatrix} 2 & 1 \\ 1 & 1 \end{pmatrix}$$

an automorphism of it. A is an Anosov system.

Indeed, the eigendirections of the corresponding automorphism of the plane define invariant direction fields on the torus: a contracting and an expanding one.

Remark 1. Instead of the above inequalities, we may require the apparently weaker conditions

$$\|A^n_*|_X\| \leqslant c\lambda^n, n > 0; \qquad \|A^n_*|_Y\| \leqslant c\lambda^{-n}, n < 0.$$

If this condition is satisfied for one metric, then it is satisfied for any other (possibly with a different c). This condition also implies the above inequalities (possibly in a modified metric).

Remark 2. The definition does not require *smoothness* of the fields X and Y of planes. A torus diffeomorphism close to the automorphism

$$\begin{pmatrix} 2 & 1 \\ 1 & 1 \end{pmatrix}$$

is always an Anosov system, although its contracting and expanding direction fields may not be of class C^2 even in the case where the diffeomorphism is analytic (in the multidimensional case the fields of planes may not be even of class C^1).

Remark 3. The definition of an Anosov system was suggested by Anosov. He called such a system a U-system. This term comes from the first letter of the Russian word for "*condition*". Anosov called conditions (1)–(3) U conditions and proposed that they be called C conditions in English; he also offered to call U-diffeomorphisms U-cascades. Smale has called them "Anosov diffeomorphisms".

B. Properties of Anosov Diffeomorphisms

Theorem (Anosov). *Every Anosov diffeomorphism is structurally stable.*

The proof can be carried out following the method of § 13 for automorphisms of the torus; the details can be found, for example, in J. Mather, *Anosov diffeomorphisms*, Bull. Am. Math Soc. **73** (1967), 747–817 (appendix).

The first proof was related to the following property of Anosov diffeomorphisms.

Theorem. *The contracting and expanding plane fields of an Anosov diffeomorphism are completely integrable.*

In other words, there exist contracting and expanding foliations* whose tangent planes form contractive and expanding fields of planes. We note

---

* A foliation on an n-dimensional manifold is a partition into submanifolds (leaves) of the same dimension k satisfying the following condition: every point of the manifold has a neighborhood whose partition into connected components of leaves is diffeomorphic to the partition of an n-dimensional cube into parallel k-dimensional planes.

that the Frobenius theorem cannot be used, since our fields of planes are not smooth.

The proof is based on the observation that under the diffeomorphism the angle between planes not too different from the planes of the expanding field decreases: the expanding field is an attracting fixed point in the space of fields of planes under the action of the Anosov diffeomorphism on this space.

In order to construct the expanding foliation, we can partition the manifold into sufficiently small domains and take, in every domain, an arbitrary foliation whose leaves have dimension equal to that of the planes of the expanding field and form a not too large an angle with these planes. We apply the Anosov diffeomorphism and its iterates to these foliations. It turns out that the sequence of partial foliations thus obtained converges to the true expanding foliation.

Remark. A special case of the above construction is the construction of stable and unstable invariant manifolds of a fixed point of a diffeomorphism in the case where the absolute values of the eigenvalues of the linear part of the diffeomorphism are all different from 1. For the construction of the unstable manifold, we may apply the iterates of the diffeomorphism to any manifold tangent to the unstable invariant subspace of the linear part of the diffeomorphism.

The procedure described above enables us to construct contracting and expanding foliations for not only any given Anosov diffeomorphism, but for all diffeomorphisms close to it. Consequently, the property of being an Anosov diffeomorphism is preserved under a small perturbation (in the C^1 sense) of the diffeomorphism. Besides, it is clear from the construction that the contracting and expanding foliations (or more precisely, plane fields) depend continuously on the diffeomorphism.

After the contracting and expanding foliations are constructed for the given and perturbed diffeomorphisms, the proof of Anosov's theorem is simple.

Indeed, consider a phase point and the sequence of its images under the initial diffeomorphism. We consider the system of ε-neighborhoods of the image points. Let ε be small. If the distance between the perturbed diffeomorphism and the unperturbed one is sufficiently small, then each of the ε-neighborhoods is foliated into connected components of the leaves of the contracting foliation for both the initial and perturbed diffeomorphisms.

We shall call these components vertical *disks*. Consider the vertical disk of the initial foliation going through the initial point and its images under the action of the positive powers of the initial Anosov diffeomorphism.

There exists a unique vertical disk of the perturbed foliation such that its images under the positive powers of the perturbed diffeomorphism remain inside the above ε-neighborhoods.

Indeed, the initial Anosov diffeomorphism is expanding in the horizontal direction. Therefore, the perturbed diffeomorphism is also expanding in the horizontal direction.

We denote by U_n the neighborhoods described above, by $U_n \to B_n$ their foliations into perturbed vertical disks, and by A the perturbed diffeomorphism. Since A is expanding in the horizontal direction, A^{-1} induces contractive mappings $a_n: B_n \to B_{n-1}$. Now the desired point $b_0 \in B_0$ is defined as

$$b_0 = \bigcap_{n \to +\infty} a_1 a_2 \cdots a_n B_n.$$

In the same way, there exists a unique horizontal perturbed disk whose images under the negative powers of our diffeomorphism do not leave the neighborhoods with negative indices.

The intersection of the perturbed horizontal and vertical disks thus constructed defines the point which is assigned by the conjugating homeomorphism to the initial phase point.

It is not difficult to verify the fact that this construction actually defines a homeomorphism conjugating the unperturbed Anosov diffeomorphism with the perturbed one.

Anosov diffeomorphisms with invariant measure given by a positive density have an everywhere dense set of periodic points (cycles). Quite complete study of the ergodic (mixing, etc.) properties of Anosov diffeomorphisms with invariant measure was carried out by Anosov and J. G. Sinai (cf., D. V., Anosov, *Geodesic flows on closed Riemannian manifolds of negative curvature*, Trudy Steklov, **90** (1967), 3–209: Proceedings of the Steklov Institute of Mathematics, Petrovskii and Mikolskii, ed., American Mathematical Society, 1969, pp. 1–235.

C. Anosov Flows

In passing to one-parameter groups of diffeomorphisms, the definition of hyperbolicity has to be altered, since there are no contractions or expansions along phase curves.

Consider the integral curves in the case of the saddle point $\dot{x} = -x$, $\dot{y} = y$ (Fig. 85). The t-axis is the intersection of two planes consisting of

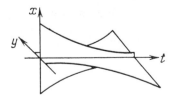

Figure 85.

integral curves approaching it as $t \to +\infty$ [the (x, t) plane] and as $t \to -\infty$ [the (y, t)-plane]; the remaining integral curves move away both for $t \to +\infty$ and $t \to -\infty$.

A one-parameter group of diffeomorphisms is called an *Anosov flow* if the phase curves near a given phase curve behave like the integral curves in the above example. A formal definition follows:

Definition. Let M be a compact smooth manifold, v a vector field on M without singular points, and $\{g^t\}$ the corresponding phase flow. Assume that:

(1) At every point of M, the tangent space of M can be represented as the direct sum of three subspaces:

$$T_x M = X_x \oplus Y_x \oplus Z_x.$$

(2) The fields X, Y, and Z of planes are continuous and invariant with respect to the phase flow.
(3) The field Z is generated by the field of phase velocity.
(4) For some positive constants c, λ, and for some Riemannian metric on M we have

$$\|g^t_* |_X\| \leqslant ce^{-\lambda t} \quad \text{for} \quad t > 0, \qquad \|g^t_* |_Y\| \leqslant ce^{\lambda t} \quad \text{for} \quad t < 0.$$

Then the phase flow is called an *Anosov flow* and the equation $\dot{x} = v(x)$ an *Anosov system*.

EXAMPLE

Consider the three-dimensional manifold M obtained from the direct product of the torus with the interval $[0, 1]$ by gluing the boundary tori by an Anosov automorphism:* $(x, 1)$ is glued with $(Ax, 0)$, where $x \in T^2$ and

$$A = \begin{pmatrix} 2 & 1 \\ 1 & 1 \end{pmatrix}.$$

Consider a vector field directed along the factor $[0, 1]$ in the direct product $T^2 \times [0, 1]$. Upon gluing boundaries in the direct product, this field turns into a smooth (why?) field v on M.

The field v thus obtained defines an Anosov flow on M.

Theorem. *Every Anosov flow is structurally stable.*

◀ This can be proved by the same method as for Anosov diffeomorphisms (cf., the references cited above). ▶

---

*This manifold appears in the "Analysis situs" of Poincaré.

Some Anosov flows have infinite sets of closed phase curves. Consequently, even confined to structurally stable vector fields, in the multi-dimensional case we cannot expect to obtain as simple a picture with a finite number of equilibrium positions and cycles as in the case of systems on the two-dimensional sphere.

In 1961, Smale constructed the first examples of structurally stable systems with an infinite number of cycles. In these examples, an exponential divergence takes place not on the entire phase space, but on a closed subset. Such sets are now called hyperbolic. The general theory of hyperbolic sets was created later, under the influence of the theory of Anosov systems.

The appearance of similar examples led to a sharp change in the interpretation of the behavior of phase curves of multidimensional systems. Some specialists hastened to announce these results as not having any real meaning, since such systems, however structurally stable they are, "cannot describe any real physical processes" in view of the instability of the individual trajectories.

Nevertheless, there are very important real cases where systems with exponential divergence of trajectories apparently describe reality best. We speak of the mathematical description of phenomena of the type of turbulence and of the motion of colliding particles (say, in models of a gas consisting of rigid spheres). Simpler, but quite real, is the problem of motion along geodesics of manifolds of negative curvature. We are now going to analyze the simplest version of this problem, the problem of geodesics on surfaces of constant negative curvature. To do this, we need some information about Lobachevsky geometry.

D. The Lobachevsky Plane

The *Lobachevsky plane* is the upper half-plane $\operatorname{Im} z > 0$ with the metric*

$$ds^2 = \frac{dx^2 + dy^2}{y^2},$$

where $z = x + iy$. The straight line $y = 0$ is called the *absolute*. We note that angles in this metric coincide with Euclidean angles, and the distance from the absolute is infinite.

Theorem. *The geodesics of the Lobachevsky plane are all circles and straight lines orthogonal to the absolute (Fig. 86).*

◀ The metric is invariant under (1) translations in the direction of the absolute; (2) expansions from the origin of the coordinate system; (3)

---

*As well as any Riemannian manifold isometric to it.

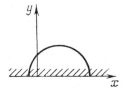

Figure 86.

symmetries $z \mapsto -\bar{z}$ (this is obvious). It is easy to verify that (4) the metric is also invariant under the inversion $z \to 1/\bar{z}$.

It follows from (1)–(4) that the metric is invariant under all fractional linear transformations of the upper half-plane into itself. Besides, it follows from (3) that the y-axis is a geodesic. On the other hand, the y-axis can be converted into any circle or straight line orthogonal to the absolute line by a real fractional linear transformation. Consequently, they all are geodesics.

Conversely, a circle or a straight line orthogonal to the absolute passes through every point in every direction. Consequently, there are no other geodesics. ▶

Remark. At the same time, we have proved that the motions (preserving the metric and orientation) of the Lobachevsky plane are the fractional linear transformations of the upper half-plane into itself.

Theorem. *The circles of the Lobachevsky plane are all Euclidean circles not intersecting the absolute.*

◀ Consider the unit disk. The upper half-plane can be transformed into the unit disk by a fractional linear transformation (cf., Chap. 1, § 5E). Therefore, the interior of the unit disk can be regarded as a model of the Lobachevsky plane (Fig. 34).

Then a fractional linear transformation preserving the upper half-plane turns into a fractional linear transformation leaving the unit disk fixed. Therefore, the metric of the Lobachevsky plane in the model on the disk is invariant with respect to all fractional linear transformations of the disk onto itself.

On the other hand, among these transformations are the rotations around the center. Consequently, all points of a Euclidean circle with the same center as the unit disk are at a constant distance from the center in the sense of the Lobachevsky metric. Hence, a Euclidean circle is a Lobachevsky circle provided that its center is at the center of the disk.

On the other hand, any Euclidean circle not intersecting the absolute can be converted into a Euclidean circle with center at the origin by a motion of the Lobachevsky plane. Consequently, every Euclidean circle not intersecting the absolute is a circle in the sense of the Lobachevsky metric (in

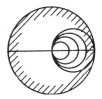

Figure 87.

both the model on the disk and the model in the half-plane). From this, it follows that, conversely, every Lobachevsky circle is a Euclidean circle. ▶

Definition. The limit of a sequence of circles tangent to each other at a given point and of increasing radius in the Lobachevsky plane is called a *horocycle*.

Theorem. *The horocycles of the Lobachevsky plane are exactly the Euclidean circles and straight lines tangent to the absolute.*

◀ Consider a half-geodesic issued from a point of the Lobachevsky plane (Fig. 87). On this half-geodesic, we choose the point at a distance t from the initial point. Then the circle of radius t with center at this point goes through the initial point perpendicularly to the geodesic. Now let t approach infinity. Then, in the Euclidean sense, the circle thus constructed converges to the circle perpendicular to the geodesic being considered and passing through its point on the absolute. This Euclidean circle is tangent to the absolute. ▶

Remark 1. By the same passage to the limit $t \to \infty$, we can construct horocycles on surfaces of negative curvature and horospheres on manifolds of negative curvature.

Remark 2. There are two horocycles with common tangent passing through every point of the Lobachevsky plane; they are obtained from the above construction as $t \to +\infty$ and as $t \to -\infty$.

E. Geodesic Flows on Surfaces of Negative Curvature

Let M be a Riemannian manifold. We shall assume that M is complete as a metric space. For example, any compact manifold is complete; the Lobachevsky plane is complete, since the distance from the absolute is infinite.

We consider the set of vectors tangent to M and of length 1. This set is a manifold of dimension $2n - 1$ if M is of dimension n. It is denoted by $T_1 M$.

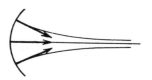

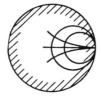

Figure 88.   Figure 89.

Definition. The *geodesic flow* on M is the one-parameter group of diffeomorphisms of the manifold of unit tangent vectors defined in the following way: in time t, every vector moves forward along the geodesic tangent to it, remaining tangent to the geodesic at distance t.

Theorem. *The geodesic flow on the Lobachevsky plane satisfies conditions (1)–(4) of the definition of an Anosov flow.*

◀ *1.* We construct the contracting and expanding foliations. To do this, for every vector we construct a horocycle orthogonal to it which is the limit of circles whose centers are located forward of the base point of the vector. At every point we equip the horocycle with a normal unit vector so that we obtain a continuous field of normal vectors (Fig. 88).

We note that if we had started with any of these vectors, we would have obtained the same horocycle with the same field. This horocycle with the field may be considered as a curve in the three-dimensional space $T_1 M$ of unit tangent vectors of the Lobachevsky plane. Consequently, we have constructed a one-dimensional foliation in $T_1 M$, a decomposition of the space of unit tangent vectors into curves. This decomposition is the contracting foliation.

The expanding foliation can be constructed in the same way, beginning with circles whose centers are located behind the origin of the vector.

2. Conditions (2) and (3) express the invariance of geodesics and horocycles with respect to the geodesic flow and can be verified directly. Indeed, the family of geodesics perpendicular to a given horocycle intersects the absolute at the point of tangency of the absolute with the horocycle, and every horocycle tangent to the absolute at this point is orthogonal to all geodesics of the family (Fig. 89).

Therefore, the geodesic flow converts every horocycle (equipped with the field of normal vectors) into the horocycle tangent to the absolute at the same point (and also equipped with the field of normal vectors).

3. Condition (1) means that the tangent vectors of a geodesic equipped with the tangent field and of both horocycles equipped with normal fields are linearly independent. This can be verified easily: the only important thing is that the tangency of both horocycles is of the first and not higher order.

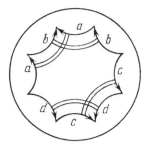

Figure 90.

4. We prove that segments of the contracting horocycle contract exponentially under the action of the phase flow. We assume that the initial horocycle is the straight line $y = 1$ in the upper half-plane. The geodesics are the lines $x = $ const and the geodesic flow converts the line $y = 1$ into the line $y = e^t$ over time t (the distance of the point 1 from the point y on the y-axis is equal to $\ln y$). Consequently, every segment of a horocycle turns into a segment whose length is e^t times smaller. This implies that the phase flow contracts the leaves of the contracting foliation (in the sense of the natural metric on $T_1 M$).

Condition (4) of the definition of an Anosov system is verified analogously for expanding horocycles. ▶

Corollary. *The geodesic flow on a compact surface of constant negative curvature is an Anosov flow.*

◀ A change in the time reduces everything to the case of curvature -1. For a surface of constant negative curvature -1, the Lobachevsky plane is a universal covering; the surface can be obtained from the Lobachevsky plane by indentifying points mapped onto each other by some discrete group of motions of the Lobachevsky plane (Fig. 90).

Under this identification, the geodesics, circles, and horocycles of the Lobachevsky plane are projected onto geodesics, circles, and horocycles of the surface; the geodesic flow on the Lobachevsky plane and its contracting and expanding foliations are projected onto similar foliations of the surface. ▶

In particular, this implies that *the geodesic flow on q compact surface of constant negative curvature is structurally stable and has an everywhere dense set of closed geodesics.*

Remark. For a multidimensional manifold of negative (not necessarily constant) curvature, the geodesic flow is also an Anosov flow. The proof is similar to the one above given for the simplest case: only the proof of the existence of horocycles (horospheres) is somewhat more complicated (cf., the work of Anosov cited earlier).

Figure 91.

F. Billiard Systems

We consider the geodesic flow on the surface of an ellipsoid. We assume that the small axis of the ellipsoid decreases to zero, so that the ellipsoid flattens out and turns into an ellipse. The geodesic flow then turns into the so-called billiard system in a domain bounded by an ellipse: the point moves along a straight line inside the domain and it is reflected from the boundary according to the law that the "angle of incidence is equal to angle of reflection" (Fig. 91).

The billiard trajectory inside the ellipse is never everywhere dense. Nevertheless, for domains bounded by other curves (for example, by nonsmooth curves concave inside), the billiard motion has almost the same properties of exponential instability of trajectories and mixing as Anosov systems.

In particular, we consider the billiard system on the torus with a hole. This system can be considered as a limit of geodesic flows on a pretzel (the pretzel degenerates to the two-sided torus with a hole just as the ellipsoid degenerates to the two-sided ellipse). Moreover, a two-sided torus with a hole and with planar metric can be considered a limiting case of a pretzel of negative curvature (upon degeneration the entire curvature is concentrated on the rim of the hole). Therefore, it is not surprising that this billiard system has the properties of Anosov flows.

There is hope that arguments close to the hyperbolic theory would enable one to prove the ergodicity of a system of rigid balls in a box, which has been postulated in statistical mechanics since the time of Boltzmann. (Ergodicity means that every invariant subset of the phase space has measure zero or full measure; it implies the almost everywhere coincidence of time and space averages. In our case, by phase space we mean a level set of energy.) In the planar case, the proof has been published by Sinai (Ja. G. Sinai, *Dynamical systems with elastic reflection. Ergodic properties of diffracting billiards*, UMN **25**, 2 (1970), 141–192). Concerning billiard systems, cf., also L. A. Bunimovič, *On billiards close to dispersing ones*, Matem. Sbornik **94**, 1 (1974), 49–73. *On the ergodic properties of nowhere dispersing billiards*, Commun. Math. Phys. **65** (1979), 295–312. Sinai and Chernov, UMN **42**, 3 (1987).

G. Anosov Systems and Double Sweeping

The hyperbolic situation arises in problems of numerical mathematics solved by the method of sweeping. As an example, let us try to solve a

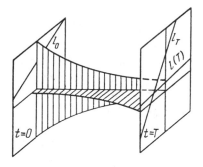

Figure 92.

boundary-value problem for the second-order equation $\ddot{x} = x$ (i.e., for the system $\dot{x} = p$, $\dot{p} = x$, on the interval $[0, T]$. Assume that the following nonhomogeneous boundary conditions are given: the initial point $\varphi(0)$ with coordinates $(x(0), p(0))$ lies on a given straight line l_0 of the phase plane (x, p) and the endpoint $\varphi(T)$ lies on another given line l_T.

If the initial point $\varphi(0)$ were known, then in our attempt to solve the Cauchy problem with initial condition $\varphi(0)$, we would encounter a loss of accuracy increasing exponentially with the length T of the interval of integration. Indeed, the solutions with initial conditions proportional to the expanding vector $(1, 1)$ increase exponentially. Consequently, in the passage from the plane $t = 0$ to the plane $t = T$ there arises an expansion in the direction of the vector $(1, 1)$ (in what follows, this direction is called horizontal) and contraction in the direction of the vector $(1, -1)$ (called vertical, cf., Fig. 92).

Now consider the image of the line l_0 under our transformation. Although the image of each point of the line is determined with an exponentially large loss of accuracy, the image of the line itself is determined quite accurately in general.

Indeed, the direction of this image is, in general, close to the horizontal direction.

Therefore, the error in the computation of a point on this almost horizontal line only slightly affects the position of the line: only the horizontal component of the error can be large; the vertical one is small.

We find the point $\varphi(T)$ as the intersection of the line l_T and the image of the line l_0. For the final determination of the solution, we have to solve the Cauchy problem backwards. Then the horizontal error does not increase, and the vertical component of $\varphi(t)$ is fixed by the fact that $\varphi(t)$ lies on the line $l(t) = g^t l_0$. Consequently, first, going from 0 to T, we determine the lines $l(t)$, and then, going backwards from T to 0, we choose a point on each of them. All this is accomplished without an exponential loss of accuracy.

H. On Applications of Anosov Systems

Presently, Anosov systems and related objects are in the same situation as limit cycles were in the time of Poincaré. The whole mathematical apparatus for the study of limit cycles had been created, but serious application of limit cycles in engineering began only several decades later, when the progress of radio technology turned the theory of nonlinear oscillations into a field of applied mathematics.

Since the beginning of the 1960's, it has been conjectured that a natural area of application of Anosov flows is in the theory of turbulent motions of a fluid. Let us imagine a closed vessel filled with an incompressible viscous fluid brought into motion by some exterior force (stirring). Stirring is necessary; otherwise viscosity would damp motion over time.

The Navier–Stokes equations of hydrodynamics define a dynamical system* in a function space (the points of this infinite-dimensional phase space are divergence free vector fields, the fields of the velocities of the fluid).

The equilibrium positions of this dynamical system are stationary fields of velocities, i.e., those motions of the fluid for which the velocity does not change over time at each point of the space. To the cycles of the system there correspond periodic motions of the fluid, in which the velocity changes periodically at every point of the space. Such a motion may be observed by turning on a water faucet.

The conjecture concerning the mathematical description of turbulence is based on the fact that the phenomenon essentially reduces to a finite-dimensional dynamical system, since viscosity extinguishes high harmonics rapidly. In other words, it is assumed that in the infinite-dimensional phase space, a finite-dimensional manifold or set exists to which all phase curves are attracted; on this very set, the phase flow is an Anosov system or has the similar properties of exponential instability of trajectories and mixing.

In this case, the observable properties of the motion of the fluid have to be as follows: under any initial condition, the motion rapidly assumes a limiting behavior; however, this behavior is neither stationary nor periodic; although the limiting motion is in fact determined by a finite number of parameters ("phase" of the limiting behavior), the parameters themselves are highly unstable (limiting flows with close initial phases diverge exponentially); actually, the statistical characteristics of the flow do not depend on these unstable phases.

The following results have so far been obtained in this direction. If the viscosity is large enough, then the Navier–Stokes system has a unique

---

* Actually, the theory of partial differential equations has not been able to solve the problem of existence and uniqueness of solutions of the three-dimensional Navier–Stokes equations. However, we will ignore this circumstance for now.

fixed point which attracts all phase curves. This is the so-called laminar motion. Every other flow tends to turn into a laminar flow under the effect of viscosity. With the decrease of viscosity, the laminar flow may loose stability. Then a stable limit cycle appears (cf., chap. 6). With further decrease of viscosity, the cycle may lose its stability and give rise to a more complicated aperiodic motion which attracts its neighbors. It is expected that this motion will, in general, have the properties of exponential instability of phase curves on the attracting set. Although much theoretical and experimental work has been devoted to this question in the past (cf., for example, the survey of J. B. McLaughlin, P. C. Martin, *Transition to turbulence of a statically stressed fluid system*, Phys. Rev. A **12** (1975), 186–203), the above conjecture is far from being a theorem.

It should be noted, however, that the appearance of the attracting set with exponentially unstable trajectories is not necessarily connected with the loss of stability of the laminar flow; this set may occur far from the equilibrium position and even for the values of viscosity for which the laminar flow is still stable.

§ 15. Structurally Stable Systems Are Not Everywhere Dense

In this section, a domain in the function space of smooth dynamical systems of class C^1 is constructed, which is free from structurally stable systems.

A. Smale's Example

In 1965, Smale constructed a diffeomorphism of the three-dimensional torus in the neighborhood of which there is no structurally stable diffeomorphism.

Consequently, on some four-dimensional manifold, a vector field exists which cannot be made structurally stable by a small perturbation.

Later, fields with the same property were constructed on three-dimensional manifolds, as well (cf., S. Newhouse, *Nondensity of Axiom A(a), Global Analysis*, Proc. Simp. Pure Math. AMS, **14** (1971), 191–203).

In this section, we discuss Smale's construction.

B. Description of the Example

On T^3 we introduce the coordinates $(x, y, z \bmod 2\pi)$. We define a diffeomorphism $A: T^3 \to T^3$ in the neighborhoods of the torus $T^2: z = 0$ and of some interval of the z-axis (the form of the diffeomorphism A is immaterial in the remaining part of the three-dimensional torus).

In the neighborhood U of the torus T^2, the mapping A is defined by the formula

$$A(x, y; z) = \left(2x + y, x + y; \frac{z}{2} \right).$$

In the neighborhood of the point O with coordinates $(0, 0, \pi)$, the mapping A is given by the formula

$$A(x, y; \pi + u) = \left(\frac{x}{2}, \frac{y}{2}; \pi + 2u \right).$$

Therefore, O is a saddle point and the outgoing invariant manifold is a curve γ containing the interval $(\pi, \pi - \varepsilon)$ of the z-axis.

The curve γ is invariant under A and expands under the action of A. Consequently, iterating A, we obtain from the indicated interval one half of the invariant manifold, which either ends at a fixed point of A or has infinite length.

We require that the above curve enter the domain U indicated above and have infinite length there. It is easy to see that there are diffeomorphisms of the torus with these properties.

C. Stability Properties of the Diffeomorphisms A

1. The restriction of A to a sufficiently small neighborhood of T^2 is structurally stable.

◀ Indeed, this can be proved by using the same method as in the case of the Grobman–Hartman theorem. We replace the diffeomorphism $A: T^3 \to T^3$ by a mapping $A': T^2 \times \mathbb{R} \to T^2 \times \mathbb{R}$ defined everywhere by the same formula which defines A in the domain U. The diffeomorphism $\tilde{A}$ close to A can be replaced by a mapping $\tilde{A}': T^2 \times \mathbb{R} \to T^2 \times \mathbb{R}$ coordinated with $\tilde{A}$ in the neighborhood of the torus $T^2 \times O$ in such a way that the difference $\tilde{A}' - A'$ has compact support and is C^1-small. Now we can apply Anosov's theorem (more precisely, its proof). We obtain the topological equivalence of $\tilde{A}'$ with A' and, consequently, that of $\tilde{A}$ with A in the neighborhood of the torus T^2. ▶

2. From what has already been proved, it follows that $\tilde{A}$ has the invariant manifold $\tilde{T}^2$ close to T^2 and on it a countable everywhere dense set of periodic points. Through every point of the neighborhood $\tilde{U}$ of the torus $\tilde{T}^2$ there passes a uniquely determined smooth leaf of the two-dimensional contracting foliation of $\tilde{A}$, which depends continuously on the point (it consists of the points which approach each other under the iterates of the diffeomorphism).

3. The mapping $\tilde{A}$ has a fixed point $\tilde{O}$ close to the fixed point O of A.

◀ This follows from the implicit function theorem, since O is nondegenerate and $\tilde{A}$ is close to A.

The eigenvalues of the linearization of $\tilde{A}$ at $\tilde{O}$ are close to those of A at O.

By the Grobman–Harman theorem, the point $\tilde{O}$, just like O, is a saddle point and has a one-dimensional unstable invariant manifold $\tilde{\gamma}$ close to γ, as is easily seen. In particular, $\tilde{\gamma}$ enters the neighborhood $\tilde{U}$ of the torus $\tilde{T}^2$.

4. Far from $\tilde{U}$, by an arbitrarily small perturbation of A, the curve γ can be changed so that it "has a nose": locally it lies on one side of one of the leaves of the contractive foliation of A containing a point of γ, and the tangency of γ with this leaf is of the first order. We denote the mapping thus obtained by A_1.

D. Structural Instability

We prove that *the diffeomorphism A_1, together with all diffeomorphisms close to it is structurally unstable.*

◄ We imbed the mapping A_1 in a one-parameter family of diffeomorphisms A_s differing only slightly from each other in the neighborhood of the inverse image of the nose under A_1. Each of the mappings A_s sufficiently close to A_1 has the above properties of A_1: an invariant torus, two-dimensional contractive leaves, a saddle point with outgoing invariant manifold, and a nose on it. We assume that as s changes, the nose moves across the leaves of the contractive foliation.

Now we consider the contractive leaf containing the nose. This leaf may or may not contain a periodic point on the torus. Since periodic points are everywhere dense on the torus, by an arbitrarily small perturbation of A_1, in our family we can place a nose either on a leaf containing a periodic point or on a leaf not containing a periodic point.

On the other hand, the property that this leaf contains a periodic point is topologically invariant. Therefore, the topological type of A_1 changes under an arbitrarily small perturbation of A_1. Consequently, A_1 is structurally unstable.

Now let $\tilde{A}_1$ be any diffeomorphism sufficiently close to A_1. According to § 15C, the above construction for A_1 may be repeated for $\tilde{A}_1$. Consequently, $\tilde{A}_1$ is a structurally unstable diffeomorphism. ►

Chapter 4

Perturbation Theory

Most differential equations admit neither an exact analytic solution nor a complete qualitative description. Perturbation theory provides a most useful collection of methods for the study of equations close to equations of a specific form. These equations are called *unperturbed*, and their solutions are assumed to be known. Perturbation theory studies the effect of small changes in the differential equations on the behavior of solutions.

If the size of the perturbation is characterized by a small parameter ε, then the effect of perturbations over time of order 1 leads to a change of order ε of the solution. This change can be calculated approximately by solving a variational equation along the unperturbed solution. However, if we are interested in the behavior of a solution over a large time interval, say, of order $1/\varepsilon$, a much more complicated problem arises. This is the subject of the so-called *asymptotic methods of perturbation theory*. The most important of these methods is the *averaging* method, which is discussed in this chapter.

The averaging method has been used to determine the evolution of planetary orbits under the influence of the mutual perturbation of planets since the time of Lagrange and Laplace. Gauss formulated it in the following way: to determine evolution, one has to smear the mass of each planet over the orbit in proportion to the time spent in every part of the orbit and replace the attraction of planets by the attraction of the rings thus obtained.

Nevertheless, the problem of strict justification of the averaging method is still far from being solved.

§ 16. The Averaging Method

In this section, we describe the recipe of the averaging method in its simplest variant. The problems of the justification of this method are discussed in the following subsections.

A. Unperturbed and Perturbed Systems

We consider a smooth fiber bundle $\pi\colon M \to B$. A vector field v on the manifold M is said to be *vertical* if it is tangent to every fiber (Fig. 93). In applications, the fibers are usually tori.

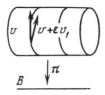

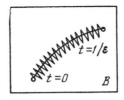

Figure 93.   Figure 94.

The functions on the base B of the fibering π determine first integrals of the equation $\dot{x} = v(x)$ on M. The vertical vector field v is said to be *unperturbed*. A *perturbed field* is, by definition, a field $v + \varepsilon v_1$ close to v. Consider the following perturbed differential equation:

$$\dot{x} = v(x) + \varepsilon v_1(x).$$

Every phase curve of the unperturbed equation is projected by π onto a point of the base. The motion on a phase curve of the unperturbed equation is projected on the base in the form of a slow motion whose velocity is of order ε. Noticeable displacement of the projection on the base takes place over time of order $1/\varepsilon$. The averaging method is intended to describe this slow motion on the base by means of a vector field on the base. In the averaging method, this slow motion is described as a combination of small oscillations and systematic evolution or drift (Fig. 94).

EXAMPLE

Consider the planetary system. The unperturbed equations only take into account the interaction of the Sun and the planets. In the unperturbed motion, the planets move along Keplerian orbits. The role of perturbation is played by the mutual attraction of planets. The role of ε is played by the ratio of the mass of the planets to that of the Sun; this is a quantity of the order $1/1000$. The characteristic unit of time is the period of revolution around the Sun, i.e., a quantity of the order of a year or decade, and the characteristic unit of length is the radius of a planetary orbit.

In this example, M is the phase space, the base B is the space of collections of Keplerian ellipses, and the fibers are tori of dimension equal to the number of planets (every collection of Keplerian ellipses determines a torus whose point is given by indicating the positions of the planets on the ellipses). Then a displacement on the base by a quantity of order 1 corresponds to a change in the orbit, say, a doubling of the radius. Time on the order of $1/\varepsilon$ is time on the order of thousands or tens of thousands of years.

Consequently, in this example, a systematic slow motion (drift) on the base at velocity ε could double the radius of the orbit of the Earth over a time period on the order of thousands of years. This would be fatal to our civilization, which owes its existence to the fact that this drift does not occur (at least not in the direction of the change of radii of orbits, a change in the eccentricities takes place and apparently has effects on ice ages).

B. The Procedure of Averaging

To describe the averaging method, we introduce some notation. We shall assume that the fibers of our bundle are n-dimensional tori. In the neighborhood of every point of the base, the fibering is assumed to be a direct product. We restrict ourselves to such a neighborhood and shall describe a point of the fiber space M by a pair (φ, I), where I is a point of the base and φ is a point of an n-dimensional torus F.

The notation I is chosen because the coordinates $(I_1, \ldots, I_k)$ of I determine first integrals of the unperturbed system on M. A point φ of the torus F is given by a collection of n angular coordinates $(\varphi_1, \ldots, \varphi_n)$ mod 2π.

In applications, by the nature of the problem, the coordinates φ_k are usually determined uniquely up to the choice of an origin on every torus and up to integer-valued unimodular linear transformations. We fix a coordinate system (φ, I).

Definition. The *unperturbed equation* of the averaging method is the equation

$$\begin{cases} \dot{\varphi} = \omega(I), \\ \dot{I} = 0, \end{cases}$$

where ω is a vertical vector field given by a frequency vector $(\omega_1(I), \ldots, \omega_n(I))$ depending on the point I of the base.

Definition. The *perturbed equation* of the averaging method is the system

$$\begin{cases} \dot{\varphi} = \omega(I) + \varepsilon f(I, \varphi_1 \varepsilon), \\ \dot{I} = \qquad \varepsilon g(I, \varphi_1 \varepsilon), \end{cases}$$

where f and g are 2π-periodic in φ and $\varepsilon \ll 1$ is a small parameter. The angular coordinates φ_i are called *fast variables* and the coordinates I_j on the base are called *slow variables*.

Definition. The *averaged equation* is the equation

$$\dot{J} = \varepsilon G(J),$$

where $G(J) = \oint g(J, \varphi, 0)\, d\varphi / \oint d\varphi$ is the mean value of the function g on a fiber.

The solutions of the averaged equation are called *averaged motions*.

EXAMPLE

Consider the perturbed equation

$$\dot{\varphi} = \omega, \qquad \dot{I} = \varepsilon(a + b\cos\varphi).$$

The averaged equation has the form

$$\dot{J} = \varepsilon a.$$

This means that, in passing to the averaged equation, we omit quantities of the same order as the remaining quantities on the right-hand side of the equation for I. Over times of order 1, both omitted and remaining quantities have the same effect (of order ε). However, their effects over times of order $1/\varepsilon$ are entirely different: the remaining terms lead to a systematic drift, and the *omitted* ones lead only to a small tremor.

◀ The solution of the perturbed equation yields (say, for $\varphi_0 = 0$)

$$I(t) = I_0 + \varepsilon a t + \frac{\varepsilon b \sin \omega t}{\omega}$$

which differs from the solution of the averaged equation $J(t) = I_0 + \varepsilon a t$ by an oscillating small term only. ▶

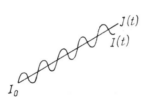

$$J(t)$$
$$I(t)$$
$$I_0$$

Figure 95.

C. Space and Time Averages

We consider a time interval T large compared to 1 but small compared to $1/\varepsilon$. Over the time T, the trajectory of the perturbed motion does not noticeably deviate from the initial fiber.

We calculate the displacement of the projection of the perturbed trajectory on the base over time T. This displacement is on the order of $\varepsilon T \ll 1$. The velocity of the displacement is equal to $\varepsilon g(I, \varphi, \varepsilon)$. In the first approximation, I can be assumed constant, ε equal to zero, and φ changing according to the unperturbed equation. Then for the displacement over time T, we obtain the approximate expression

$$\Delta I = \varepsilon T \left[\frac{1}{T} \int_0^T g(I, \varphi(t), 0)\, dt \right] + o(\varepsilon T).$$

Time $T \gg 1$ is large; therefore, the expression within the square brackets is close to the time average of g.

We introduce the slow time $\tau = \varepsilon t$. The variation of τ from 0 to 1 corresponds to the variation of t from 0 to $1/\varepsilon$. We will denote by a prime the velocity of motion with respect to the slow time. Then the preceding equation takes the form

$$\frac{\Delta I}{\Delta \tau} \approx \text{time average of } g, \qquad I' = \text{time average of } g.$$

We replace the time average by the space average and obtain the averaged equation

$$J' = G(J), \qquad G = \text{space average of } g.$$

Consequently, passage to the averaged equation corresponds to the replacement of time averages by space averages along the unperturbed motion.

D. Discussion

The use of the averaging method consists of replacing the perturbed equation by the much simpler averaged equation. The solutions of the averaged equation are studied on time intervals on the order of $1/\varepsilon$ (i.e., on intervals of the order of 1 of the slow time). Then conclusions are drawn on the behavior of the perturbed motion over time on the order of $1/\varepsilon$. (The conclusion is usually that the I component of the solution of the perturbed equation is close to the solution of the averaged equation over time $1/\varepsilon$.)

This conclusion does not follow from the preceding reasoning and requires a proof. Indeed, in deducing the averaged equation, we replaced time averages by space averages. The replacement is reasonable if the trajectory of the unperturbed motion is uniformly distributed over the n-dimensional torus, i.e., when the frequencies are incommensurable. In cases of resonance, the trajectory of the unperturbed motion is everywhere dense on a torus of dimension smaller than n. Therefore, the replacement of the time average by space average on the n-dimensional torus is not legitimate near resonances.

Indeed, there are examples which show that the difference between the projection of the perturbed trajectory on the base and the solution of the averaged equation may be on the order of 1 over time $1/\varepsilon$: the averaged drift and the projection of the true motion point in different directions.

Practically, the only case studied completely is that of single-frequency systems where the fibers are one-dimensional tori, i.e., circles.

§ 17. Averaging in Single-Frequency Systems

Here we formulate and prove a theorem which justifies the averaging method for single-frequency systems.

A. Formulation of the Theorem

We consider a phase space M which is the direct product of a domain B in the Euclidean space $\mathbb{R}^k$ and the circle S^1. The angular coordinate on the circle is denoted by $\varphi \bmod 2\pi$ and points of B are denoted by I.

The perturbed equation

$$\dot{\varphi} = \omega(I) + \varepsilon f(I, \varphi, \varepsilon),$$

$$\dot{I} = \varepsilon g(I, \varphi, \varepsilon)$$

with functions f and g 2π-periodic in φ yields the averaged equation

$$\dot{J} = \varepsilon G(J), \qquad G(J) = \frac{1}{2\pi} \int_0^{2\pi} g(J, \varphi, 0) \, d\varphi.$$

We consider an initial point I_0 in B and assume that the solution $J(t)$ of the averaged equation with initial condition $J(0) = I_0$ remains in B over time $t = T/\varepsilon$ (i.e., over slow time $\tau = T$ the solution of the equation $dJ/d\tau = G(J)$ with initial condition I_0 does not leave B).

Theorem. *Assume that the frequency ω does not vanish in the domain B. Then the distance between the value of the solution $J(t)$ of the averaged equation and the I component $I(t)$ of the solution of the perturbed equation with $I(0) = J(0)$ remains small for $t \in [0, T/\varepsilon]$ if ε is sufficiently small:*

$$|I(t) - J(t)| \leqslant C\varepsilon,$$

where the constant C is independent of ε.

B. The Main Construction

The basic idea of the proof of the theorem consists of trying to annihilate the perturbation by means of an appropriate change of variables. This idea has many applications (cf., for example; the preceding and following chapters) and is the basis of the whole formal apparatus of perturbation theory.

In place of I we choose a new coordinate $P = I + \varepsilon h(I, \varphi)$ such that the P component of the solution cease to oscillate. For this we want to annihilate the terms of order ε depending on φ on the right-hand side of the equation for P.

In other words, we try to construct a diffeomorphism $(I, \varphi) \mapsto (P, \varphi)$ of the manifold M so that the perturbed field turns into a field having almost constant projection on the base on every fiber (up to an error of order ε^2).

Differentiating $P = I + \varepsilon h(I, \varphi)$ with respect to time and collecting the terms of first order in ε, we obtain

$$\dot{P} = \varepsilon \left[g + \omega \frac{\partial h}{\partial \varphi} \right] + r,$$

where the argument ε of the function g is replaced by zero; the remainder r (as we shall verify below) is of second order in ε.

We try to choose h so that the terms of first order in ε are annihilated, i.e., the square brackets vanish. Formally, we obtain

$$h(I, \varphi) = -\frac{1}{\omega(I)} \int_{\varphi_0}^{\varphi} g(I, \psi, 0) \, d\psi.$$

(Here the condition $\omega \neq 0$ of the theorem has been used.) Actually, such a method of solving the equation $g + \omega \partial h / \partial \varphi = 0$ is not legitimate, since the function h has to be 2π-periodic in φ for the mapping $(I, \varphi) \mapsto (P, \varphi)$ to be defined on M.

The preceding formula defines the function h on the circle (and not on a covering line of it) only if the average value of g on the circle is equal to zero.

Consequently, the choice of h does not allow us to annihilate the whole perturbation g, but only its oscillating part

$$\tilde{g}(I, \varphi, 0) = g(I, \varphi, 0) - G(I).$$

The average of $\tilde{g}$ over the period is equal to zero, and we can define a periodic function h by the formula

$$h(I, \varphi) = -\frac{1}{\omega(I)} \int_0^{\varphi} \tilde{g}(I, \psi, 0) \, d\psi. \tag{1}$$

Now for P, we obtain the equation

$$\dot{P} = \varepsilon G(P) + \varepsilon R.$$

This equation differs from the averaged equation

$$\dot{J} = \varepsilon G(J)$$

by the small quantity εR of order ε^2. Therefore, the solutions diverge with velocity of order ε^2. Consequently, over time $1/\varepsilon$, they diverge to a distance of order ε. The difference between P and I is also of order ε. Therefore, the distance between $I(t)$ and $J(t)$ remains of order ε over a time period of order $1/\varepsilon$.

To prove this assertion, one needs (simple) estimates of the terms that were omitted above.

C. Estimates

1. Notation. Let $K \subset B$ be a convex compact domain containing the point I_0. We assume that $J(t)$ does not go out to the boundary of K over time T/ε. We shall denote the norms in the spaces C^0 and C^1 by $|\cdot|_0$ and $|\cdot|_1$ (the maximum of the modulus of the function and the maximum of the moduli of the function and its first derivative). Let c_1 be a constant with

$$|f|_1 \leqslant c_1, \qquad |g|_1 \leqslant c_1, \qquad |\omega^{-1}|_1 \leqslant c_1 \text{ for } I \text{ in } K.$$

2. We prove that *the mapping* $A: (I, \varphi) \mapsto (P, \varphi)$ *is a diffeomorphism of* $K \times S^1$ *for sufficiently small* ε.

◀ It follows from the definition of h (Eq. 1) that $h \in C^1$. Consequently, $|\varepsilon h|_1 < 1$ for sufficiently small ε. If two points were mapped onto one by A, then the difference of the values of εh at these points would be equal to the difference of the values of I; this contradicts the inequality $|\varepsilon h|_1 < 1$, since K is convex. It also follows from $|\varepsilon h|_1 < 1$ that A is a local diffeomorphism. Hence A is a diffeomorphism. ▶

3. Estimation of R. We have

$$R(P(I, \varphi, \varepsilon), \varphi, \varepsilon) = R_1 + R_2 + R_3 + R_4 + R_5,$$

$$R_1 = g(I, \varphi, 0) - g(P(I, \varphi, \varepsilon), \varphi, 0), \qquad R_2 = g(I, \varphi, \varepsilon) - g(I, \varphi, 0),$$

$$R_3 = h(I, \varphi) - h(P(I, \varphi, \varepsilon), \varphi), \qquad R_4 = \varepsilon g(I, \varphi, \varepsilon)\partial h/\partial I,$$

$$R_5 = \varepsilon f(I, \varphi, \varepsilon)\partial h/\partial \varphi.$$

We assume that I and $P(I, \varphi, \varepsilon)$ belong to K. Since

$$P = I + \varepsilon h(I, \varphi),$$

we obtain

$$|R_1| \leqslant \varepsilon|g|_1 |h|_0, \qquad |R_2| \leqslant \varepsilon|g|_1, \qquad |R_3| \leqslant \varepsilon|h|_1 |h|_0,$$

$$|R_4| \leqslant \varepsilon|h|_1 |g|_0, \qquad |R_5| \leqslant \varepsilon|h|_1 |f|_0.$$

The norms of f, g, and h appearing here can be estimated by c_1. Finally, if I and $P(I, \varphi, \varepsilon)$ belong to K, then

$$|R(P(I, \varphi, \varepsilon), \varphi, \varepsilon)| \leqslant c_2\varepsilon,$$

where $c_2(c_1) > 0$ is a constant independent of I, φ, and ε.

4. *Estimation of* $P(t) - J(t)$. Denoting by prime the derivative with respect to the slow time $\tau = \varepsilon t$, we see that P and J satisfy the relations

$$P' = G(P) + \varepsilon R(P, \varphi(t), \varepsilon),$$

$$J' = G(J).$$

Consequently, $P - J = Z$ satisfies the inequality

$$|Z|' \leqslant a|Z| + b,$$

where $a = |G|_1$ and $b = c_2\varepsilon$ as long as P, I, and J remain in K.

We write $|Z(0)| = c$. Solving the equation $z' = az + b$ with initial condition c, we obtain the estimate

$$|Z(\tau)| \leqslant (c + b\tau)e^{a\tau}$$

as long as P, I, and J remain in K.

5. *Completion of the Proof of the Theorem of* § 17A.
◀ Denote by c_3 the quantity $|h|_0$. Then we have

$$|P(I, \varphi, \varepsilon) - I| \leqslant c_3\varepsilon.$$

At the same time, the estimate proved above yields

$$|P(t) - J(t)| \leqslant c_4\varepsilon, \qquad c_4 = (c_3 + c_2 T)e^{aT}$$

for $\varepsilon t \leqslant T$ as long as $I(t)$, $P(t) = P(I(t), \varphi(t), \varepsilon)$, and $J(t)$ remain in K.

We denote by ρ the distance from the trajectory of the averaged motion $\{J(t), \varepsilon t \leqslant T\}$ to the boundary of K. If $(c_3 + c_4)\varepsilon < \rho$, then, according to the preceding estimates, $I(t)$, $P(t)$, and $J(t)$ cannot go out to the boundary of K for $\varepsilon t \leqslant T$. Over all this time interval we obtain

$$|I(t) - J(t)| \leqslant |I(t) - P(t)| + |P(t) - J(t)| \leqslant c_3\varepsilon + c_4\varepsilon. \blacktriangleright$$

D. Example

The *van der Pol equation* is, by definition, the equation

$$\ddot{x} = -x + \varepsilon(1 - x^2)\dot{x}.$$

This is the equation of a pendulum, in which a nonlinear "friction" is included, positive for large amplitudes and negative for small ones.

The unperturbed equation $\ddot{x} = -x$ can be written in the standard form $\dot{\varphi} = -1$, $\dot{I} = 0$, where

$$\varphi = \arg(x + i\dot{x}), \qquad 2I = x^2 + \dot{x}^2.$$

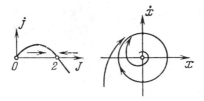

Figure 96.

The equation for I in the perturbed motion has the form

$$\dot{I} = \varepsilon(1 - x^2)\dot{x}^2 = 2\varepsilon I(1 - 2I\cos^2 \varphi)\sin^2 \varphi.$$

The averaged equation is therefore

$$\dot{J} = \varepsilon\left(J - \frac{J^2}{2}\right).$$

This equation has repelling equilibrium position $J = 0$ and attracting equilibrium position $J = 2$.

The equilibria of the equation for J correspond to cycles of the perturbed system. The theorem proved above enables us to assert that the variation of I in the perturbed system is close to the variation of J in the averaged system over time of order $1/\varepsilon$. On the other hand, if the averaged system has a nondegenerate (for example, stable in the first approximation) equilibrium position, then (for sufficiently small ε) the perturbed system will have a nondegenerate (for example, stable in the first approximation) cycle; this follows easily from the implicit function theorem.

In particular, for small ε, the van der Pol equation has a stable limit cycle close to the circle $x^2 + \dot{x}^2 = 4$ (Fig. 96).

§ 18. Averaging in Systems with Several Frequencies

The case of several frequencies is studied much less than that of a single frequency. In this section, we give a survey of the basic results in this area.

A. Resonance Surfaces

We consider the usual perturbed system of the averaging method:

$$\begin{cases} \dot{\varphi} = \omega(I) + \varepsilon f(I, \varphi, \varepsilon), & \varphi \in T^n, \quad \varepsilon \ll 1, \quad \omega \neq 0, \\ \dot{I} = \varepsilon g(I, \varphi, \varepsilon), & I \in B \subset \mathbb{R}^k. \end{cases}$$

The frequency vector $\omega = (\omega_1, \ldots, \omega_n)$ is said to be a *resonance vector* if there exists an integer-valued nonzero vector $m = (m_1, \ldots, m_n)$ for which $(m, \omega) = 0$.

The integer-valued vector m is called the *resonance index*.

A point I in the base B is called a *resonance point* if the vector $\omega(I)$ is a resonance vector. All resonance points I corresponding to a resonance with index m form a hypersurface

$$\Gamma_m = \{I \in B \colon (m, \omega(I)) = 0\}$$

in the base B of our fibering.

This hypersurface is called the *resonance hypersurface*.

In the general case, both the resonance and nonresonance points are everywhere dense in B (if the number of frequencies $n > 1$).

EXAMPLE

1. Consider the unperturbed system

$$\dot\varphi_1 = I_1, \qquad \dot\varphi_2 = I_2, \qquad \dot I = 0$$

with two frequencies. Here B is the plane with coordinates I_1 and I_2 (without zero, since we assume that $\omega \neq 0$); the resonance surfaces are all straight lines passing through 0 with rational slope with respect to the I_1-axis.

In the general case of a system with two frequencies, just as in this example, the resonance surfaces in general form a family of nonintersecting hypersurfaces (Fig. 97; "in general" $\equiv$ if rank $(\partial\omega/\partial I)$ is maximal). In this case, during the motion of the point I on the base, the point intersects the resonance surfaces transversally, in general.

The resonance surfaces are distributed in a completely different way when the number of frequencies is three or more.

EXAMPLE

2. Consider the unperturbed system

$$\dot\varphi_1 = I_1, \qquad \dot\varphi_2 = I_2, \qquad \dot\varphi_3 = 1, \qquad \dot I = 0$$

with three frequencies. Here B is the plane with coordinates I_1 and I_2, and the resonance surfaces are all straight lines given by rational equations.

In this case, during the motion of the point I in the plane, the point I, even if it intersects all resonance curves transversally, will always intersect many of them at small angles, because for any linear element there exists a linear element of a resonance curve arbitrarily close to it (Fig. 98).

Figure 97.

Figure 98.

Remark. What has been said above may become clearer if we consider the following mapping of the base into the projective $(n - 1)$-space:

$$\Omega: B \to \mathbb{R}P^{n-1}, \qquad \Omega(I) = (\omega_1(I): \cdots : \omega_n(I)).$$

The resonance surfaces are pre-images of rational hyperplanes in $\mathbb{R}P^{n-1}$. In the case of two frequencies, n is 2; rational points on the projective line correspond to the resonances.

If the number of frequencies n exceeds 2, then the rational hyperplanes form a connected, everywhere dense set, so that the neighborhood of any point can be reached along resonances from the neighborhood of any other point.

According to the above discussion, the basic effect in systems with two frequencies is passage through resonances. In the case of a larger number of frequencies, one also has to take into account the tangencies to resonances as well.

B. The Effect of a Single Resonance

In order to understand the possible effect of a single resonance, we consider some simple examples.

EXAMPLE

1. Consider the perturbed system:

$$\dot{\varphi}_1 = I_1, \qquad \dot{\varphi}_2 = 1, \qquad \dot{I}_1 = \varepsilon, \qquad \dot{I}_2 = \varepsilon \cos \varphi_1.$$

We consider the resonance $\omega_1 = 0$. The averaged motion intersects the resonance curve $I_1 = 0$ with nonzero velocity. The variation of I_2 in the exact solution over time from $-\infty$ to $+\infty$ is given by the Fresnel integral

$$\Delta I_2 = \varepsilon \int_{-\infty}^{\infty} \cos\left(\varphi_0 + \varepsilon \frac{t^2}{2}\right) dt = c(\varphi_0)\sqrt{\varepsilon}.$$

as is easily seen. In the averaged system, J_2 does not change over time.

We note that the major contribution to the integral is given by the neighborhood of the resonance of width of order $\sqrt{\varepsilon}$; the integral itself is of order $\sqrt{\varepsilon}$ and depends on the initial phase φ_0.

Consequently, in this simple example, the intersection of the resonance leads to the divergence to a distance of order $\sqrt{\varepsilon}$, of the solutions of the perturbed equation having a common initial value I. This divergence takes place in a neighborhood of width on the order of $\sqrt{\varepsilon}$ of the resonance surface.

The occurrence of quantities of order $\sqrt{\varepsilon}$ is characteristic of all problems connected with passage through a resonance.

While in Example 1 passage through a resonance only leads to a small scattering of the projections of trajectories of the perturbed system on the base with respect to trajectories of the averaged system, in the following example the perturbed and averaged motions are completely different.

EXAMPLE

2. Consider the following perturbed system:

$$\dot{\varphi}_1 = I_1, \qquad \dot{\varphi}_2 = I_2, \qquad \dot{I}_1 = \varepsilon, \qquad \dot{I}_2 = \varepsilon\cos(\varphi_1 - \varphi_2).$$

The averaged system is

$$\dot{J}_1 = \varepsilon, \qquad \dot{J}_2 = 0.$$

The averaged system with the initial conditions

$$J_1(0) = 1, \qquad J_2(0) = 1,$$

gives

$$J_1(t) = 2, \qquad J_2(t) = 1 \qquad \text{for } t = 1/\varepsilon.$$

The perturbed motion with the initial conditions

$$I_1(0) = 1, \qquad I_2(0) = 1, \qquad \varphi_1(0) = 0, \qquad \varphi_2(0) = 0$$

leads to $I_1(t) = 2, I_2(t) = 2$ for $t = 1/\varepsilon$.

Consequently, the projection of the perturbed motion on the base moves systematically in a direction quite different from that of the trajectory of the averaged motion. Over time $t = 1/\varepsilon$, these two trajectories on the base diverge to a great distance (of order 1).

The reason why the averaged equation is not suitable for the descrip-

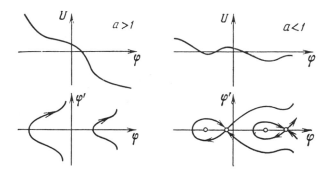

Figure 99.

tion of the perturbed motion under consideration lies in the fact that the perturbed trajectory remains on the resonance surface all the time, where averaging is evidently not applicable, since the time average is not close to the space average over the whole n-dimensional torus.

The capture of a part of trajectories by resonances is characteristic of systems with more than one frequency.

EXAMPLE

3 (A. I. Neistadt). Consider the system*

$$\dot{\varphi}_1 = I, \qquad \dot{\varphi}_2 = 1, \qquad \dot{I} = \varepsilon(a + \sin\varphi_1 - I).$$

For the study of this system, we consider the equation $\ddot{\varphi} = \varepsilon(a + \sin\varphi - \dot{\varphi})$ of a pendulum with torque and friction to which it can be reduced easily. We introduce the slow time $\tau = \sqrt{\varepsilon}t$ ($\tau \sim 1/\sqrt{\varepsilon}$ corresponds to the interval $t \sim 1/\varepsilon$). Denoting by a prime the derivation with respect to slow time, we obtain the equation

$$\varphi'' = a + \sin\varphi - \sqrt{\varepsilon}\varphi'.$$

The phase portraits are illustrated in Fig. 99 for $\varepsilon = 0$ (U is the potential energy).

We shall assume that $a > 0$. Depending on the value of the torque a, two cases are possible. If $a > 1$ (the external torque is great compared to gravitational torque), then the term $\sin\varphi$ does not play an essential role: I is monotone. To the passage through the resonance $I = 0$ there corresponds a change in the direction of revolution of the pendulum.

---

*This system is obtained from a single-frequency system by adding the trivial equation $\dot{\varphi}_2 = 1$; the resonance in the system thus obtained corresponds to the vanishing of the frequency in the single-frequency system.

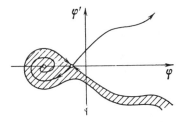

Figure 100.

If $a < 1$, then the pendulum may oscillate (the loop inside the separatrix in the phase picture). This oscillating behavior corresponds to trajectories constantly remaining near resonance.

The effect of the small friction consists, in essence, of the destruction of the loop of the separatrix. In its place, a narrow (of width on the order of $\sqrt{\varepsilon}$) strip appears in the plane (φ, φ') along the unbounded part of the separatrix consisting of the phase points captured by an attracting equilibrium position; the entire domain inside the separatrix is also captured (Fig. 100).

Returning to the initial system, we see that for $a < 1$ there is capture into resonance. The resonance captures a small portion of all the trajectories. [The measure of the set of initial conditions (I, φ) captured over time $1/\varepsilon$ is of order $\sqrt{\varepsilon}$.] For these captured initial conditions, the difference between the variation of the slow variable I and the variation of the solution J of the averaged equation reaches order 1 over time $1/\varepsilon$.

For the remaining initial conditions (i.e., for all initial conditions outside a set of measure of order $\sqrt{\varepsilon}$), the difference between I and J remains small over time $1/\varepsilon$ (it is of order $\sqrt{\varepsilon} \ln \varepsilon$, as can be calculated).

On the other hand, if $a > 1$, then capture into resonance does not occur at all.

C. Passage through Resonances in the Case of Two Frequencies

We consider the following system with two frequencies $\omega_1(I)$ and $\omega_2(I)$:

$$\dot{\varphi}_1 = \omega_1 + \varepsilon f_2, \qquad \dot{\varphi}_2 = \omega_2 + \varepsilon f_2$$
$$\dot{I}_1 = \varepsilon g_1, \qquad \dot{I}_2 = \varepsilon g_2.$$

Definition. The system *satisfies condition A* if the rate of change of the ratio ω_1/ω_2 of the frequencies along the trajectories of the perturbed system is everywhere different from zero:

$$\omega_2 \frac{\partial \omega_1}{\partial I} g \neq \omega_1 \frac{\partial \omega_2}{\partial I} g.$$

The system *satisfies condition $\bar{A}$* if the rate of change of the ratio ω_1/ω_2 of the frequencies is different from zero everywhere along trajectories of the averaged system:

$$\omega_2 \frac{\partial \omega_1}{\partial I} G \neq \omega_1 \frac{\partial \omega_2}{\partial I} G.$$

We will assume that all systems under consideration are analytic.

Theorem. *If the system satisfies condition A then the difference between the slow motion $I(t)$ in the perturbed system and $J(t)$ in the averaged one remains small over time $1/\varepsilon$:*

$$|I(t) - J(t)| \leqslant c\sqrt{\varepsilon} \qquad \text{if } I(0) = J(0), 0 \leqslant t \leqslant \frac{1}{\varepsilon}.$$

◀ The proof is based on singling out a finite number of resonances with small indices (the number of resonances is large for small ε). Outside small neighborhoods of the chosen resonance surfaces, the usual changes of variables are used (cf., § 17).

Passage through the neighborhoods of the resonances chosen leads to a divergence of order $\sqrt{\varepsilon}$ (as in the examples above).

Combining results for the divergence near the resonances chosen and the drift in the segments between them, we obtain the estimate above. ▶

For details, cf., V. I. Arnold, *Conditions for the applicability and estimation of the error of the averaging method for systems which go through resonances in the process of evolution*, Sov. Math. Dokl. **161**, 1 (1965); A. I. Neistadt, *On passage through resonances in the problem with two frequencies*, Sov. Math. Dokl. **22**, 2, (1975); A. I. Neistadt's dissertation "On some resonance problems in nonlinear systems", Moscow University, 1975 contains a proof of the above estimate with $\sqrt{\varepsilon}$ replacing the original estimate $\sqrt{\varepsilon} \ln^2 \varepsilon$ in the first reference.

Theorem (A. I. Neistadt). *If the system satisfies condition $\bar{A}$ and some other condition B (satisfied almost always), then for all initial points (I_0, φ_0) outside a set of measure not exceeding $c_1\sqrt{\varepsilon}$, the difference between the slow motion $I(t)$ in the perturbed system and the motion $J(t)$ in the averaged system remains small over time $1/\varepsilon$:*

$$|I(t) - J(t)| \leqslant c_2\sqrt{\varepsilon}|\ln \varepsilon| \qquad \text{if } I(0) = J(0).$$

◀ The proof is based on the choice of a finite number of resonances with small indices. Outside small neighborhoods of the resonance surfaces chosen, the usual changes of variables are used.

In the study of the resonances chosen, one averages over circles which are trajectories of the unperturbed motion at the resonance.

To do this, we fix the index (m_1, m_2) of the resonance, where m_1 and m_2 are relatively prime, and choose new angular coordinates (γ, δ) on the torus in place of the angular coordinates (φ_1, φ_2), where $\gamma = m_1\varphi_1 + m_2\varphi_2$. The rate of change of the angular coordinate γ in the unperturbed motion vanishes at resonance, since $m_1\omega_1 + m_2\omega_2 = 0$.

For the base, we also introduce the special coordinate $\rho = m_1\omega_1 + m_2\omega_2$. Now the equation of the resonance surface has the form $\rho = 0$, i.e., ρ characterizes the deviation from resonance. A point of the resonance surface will be denoted by σ. In the neighborhood of this surface, a point of the base can be characterized by the distance ρ from resonance and the projection σ on the resonance surface.

In these coordinates, the perturbed system takes the form

$$\dot{\gamma} = \rho + \varepsilon F_1, \qquad \dot{\delta} = \alpha(I) + \varepsilon F_2, \qquad \dot{\rho} = \varepsilon F_3, \qquad \dot{\sigma} = \varepsilon F_4,$$

where the functions F_k have period 2π with respect to γ and δ.

Averaging with respect to trajectories of the resonance motion reduces to averaging with respect to δ. The averaged system assumes the form

$$\dot{\gamma} = \rho + \varepsilon G_1, \qquad \dot{\rho} = \varepsilon G_3, \qquad \dot{\sigma} = \varepsilon G_4.$$

The functions G_k have period 2π with respect to γ. They also depend on ρ and ε.

We introduce the slow time $\tau = \sqrt{\varepsilon} t$ and the normalized distance $r = \rho/\sqrt{\varepsilon}$ from resonance. Denoting by a prime derivatives with respect to τ, we write the averaged equation in the form

$$\gamma' = r + \sqrt{\varepsilon} G_1, \qquad r' = G_3, \qquad \sigma' = \sqrt{\varepsilon} G_4.$$

The arguments of G_k are γ, $\sqrt{\varepsilon} r$, σ, and ε.

Substituting $\varepsilon = 0$ in this equation, as a first approximation we obtain the equation

$$\gamma' = r, \qquad r' = u(\gamma, \sigma), \qquad \sigma' = 0.$$

This means that we obtain as a first approximation the equation of a pendulum with torque depending on the parameter σ. The Hamiltonian character of this equation is a surprising fact, discovered through calculation and by no means obvious beforehand.

We consider the phase portrait of the equation of the first approximation in the plane (γ, r). It looks similar to the portrait in Example 3 of § 18B for $a < 1$ or $a > 1$, depending on whether or not the function u changes sign.

It turns out that loops occur in the separatrix only for a small number of resonances with indices which are not too large (here condition $\bar{A}$ is used). Indeed, condition $\bar{A}$ implies that the average value of the function u with respect to γ is different from zero. For resonances with large indices in the equation of the first approximation, we obtain a function u which differs little from its average value (since in this case the average with respect to δ is close to the average on the torus). Therefore, it is everywhere different from zero. This corresponds to a pendulum with external torque that is large compared to the gravitational torque. In this case, the equation of the first approximation has neither an equilibrium position nor a domain of oscillation.

When passing from the equation of the first approximation to the complete equation, a loop of the separatrix changes to a zone of capture to resonance, as in Example 3 of § 18B*. The measure of the set of captured points of the phase space can be estimated by a quantity of order $\sqrt{\varepsilon}$ if all equilibrium positions of the equation of the first approximation are simple (i.e., if the zeros of u are simple: $u = 0 \Rightarrow \partial u/\partial \gamma \neq 0$). This restriction concerning simplicity is in fact condition B in Neishtadt's theorem. We note that the condition is imposed on equations of the first approximation corresponding to a

---

*In contrast with Example 3 of § 18B, in the general case, the "captured" trajectories are not bound to remain close to resonance forever.

finite number of resonances (since under condition $\overline{A}$ the equations of the first approximation have equilibrium positions only for a finite number of resonances).

The proof of the theorem is completed by combining the estimates of the variation of I in the nonresonance regions and near resonances—in the uncapturable portion of the phase space. For details, cf., Neistadt's dissertation cited above.

Remark. For systems with two frequencies, the case where condition $\overline{A}$ is violated has not been studied yet. In this case the ratio of the frequencies of the fast motion does not vary monotonically in the averaged motion. Such a behavior is not possible in the case of a one-dimensional base: however, if the number of slow variables I is two or larger, then the reversal of evolution of the ratio of the frequencies is a generic phenomenon not removable by a perturbation of the system.

D. Systems with Several Frequencies

The case where the number of frequencies is larger than two has been studied much less than the case of two frequencies. For generic systems, the frequencies of the fast motion are incommensurable for almost all values of the slow variables. Therefore, it is natural to expect that for the majority of initial conditions, the averaging method describes the evolution of slow variables on time intervals of order $1/\varepsilon$ accurately.

The first general theorems in this direction are due to Anosov [D. V. Anosov, *Averaging in systems of ordinary differential equations with rapidly oscillating solutions*, Izv. A. N. USSR, Ser. Math. **24**, 5 (1960), 721–742] and Casuga [T. Casuga, *On the adiabatic theorem for the hamiltonian system of differential equations in the classical mechanics, I, II, III,* Proc. Japan Acad. **37**, 7 (1961)]. Anosov's theorem asserts that for any positive number ρ, the measure of the set of initial conditions (in a compactum in the phase space) for which

$$\max_{0 \le t \le 1/\varepsilon} |I(t) - J(t)| > \rho \qquad \text{with} \qquad I(0) = J(0)$$

converges to zero as ε converges to zero. (Here, as usual, I is the projection of the perturbed motion and J is the averaged motion; it is assumed that the frequencies are independent in the sense that the rank of the derivative $\partial \omega / \partial I$ of the frequencies with respect to the slow variables is equal to the number of fast variables.)

This theorem is in fact proved under more general assumptions: the quasi-periodicity of fast motions is not assumed, but the ergodicity of fast motion is assumed on almost all tori; as usual, it is supposed that the solution J of the averaged equation is continued over time $1/\varepsilon$.

The set of small measure (provided ε is small), where large deviations are possible from the averaged motion over time $1/\varepsilon$, corresponds to all trajectories captured into resonance or wandering along resonance surfaces passing from one to another. This is also possible if the number of frequencies is larger than two.

It is of interest to estimate the measure of this set accurately. For example, for systems with two frequencies Neistadt's results (cf., § 18C) imply the estimate $|I(t) - J(t)| \leqslant c_2\sqrt{\varepsilon}|\ln \varepsilon|$ outside a set of measure not greater than $c_1\sqrt{\varepsilon}$ (under insignificant restrictions on the system).

We assume that the frequencies are independent, i.e., the rank of $\partial\omega/\partial I$ is equal to the number of frequencies.

Theorem (A. I. Neistadt). *For a system with independent frequencies outside a set of small measure $\varkappa$, the error*

$$\max_{0 \leqslant t \leqslant 1/\varepsilon} |I(t) - J(t)| \qquad \text{with} \qquad I(0) = J(0)$$

of the averaging method is estimated from above by $c_3\sqrt{\varepsilon}/\varkappa$.

An equivalent formulation: denote by $E(\varepsilon, \rho)$ the set of initial conditions within a fixed compactum for which the error attains ρ or is larger than ρ for the indicated value of ε.

Then we have

$$\text{mes } E(\varepsilon, \rho) \leqslant \frac{c_4\sqrt{\varepsilon}}{\rho}.$$

For the proof, cf., A. I. Neistadt, *On averaging in systems with several frequencies, II*, DAN USSR **226**, 6 (1976) 1295–1298. The proof uses the following idea in Casuga's work cited above: the change of variables in the averaging method is modified (smoothed) in such a way that it is given by smooth functions not only outside neighborhoods of resonances, but everywhere.

Neistadt's result can be interpreted as an indication of the statistical independence of the increments of the deviation of I from J on subsequent time intervals of length 1. Indeed, the increment of $I-J$ over time T of order 1 is of order ε, and the number of intervals of length T in the interval $1/\varepsilon$ is of order $1/\varepsilon$. If the increments were independent on every interval of length T, then according to laws of probability theory, the expected value of the increment over time $1/\varepsilon$ would turn out to be proportional to the product of the increment over time T with the square root of the number of trials, i.e., it would turn out to be on the order of $\varepsilon\sqrt{1/\varepsilon} = \sqrt{\varepsilon}$.

Neistadt's theorem gives the same order for the increment but not for all initial conditions: we have to exclude the set of initial conditions of measure of order $\sqrt{\varepsilon}$ on which capture to resonance and large deviations not corresponding to a scheme with independent increments are observed.

The idea of the independence of the increments of the deviation of I from J can apparently be justified much more completely when the fast motion is not quasi-periodic but is an Anosov system. This is indicated, in particular, by the central limit theorem for functions on the phase space [Ja. G. Sinai, *The central limit theorem for geodesic flows on manifolds of negative constant curvature*, Sov. Math. Dokl. **133**, 6 (1960), 1303–1306; M. E. Ratner, *The central limit theorem for Anosov flows on three-dimensional manifolds*, Sov. Math. Dokl. **186**, 3 (1969), 519–521]. This theorem justifies the probabilistic approach when both the slow and fast motions are independent of the slow variables:

$$\dot{I} = \varepsilon g(\varphi), \qquad \dot{\varphi} = \omega(\varphi).$$

The probabilistic arguments become especially interesting when we study the behavior of the system over times large compared to $1/\varepsilon$ (say, of order $1/\varepsilon\sqrt{\varepsilon}$ or $1/\varepsilon^2$). If over time $1/\varepsilon$ a $\sqrt{\varepsilon}$th fraction of all trajectories is captured to resonance, and if on the subsequent time intervals of length $1/\varepsilon$ all new trajectories will be captured in the same way, then over time on the order of $1/\varepsilon\sqrt{\varepsilon}$, the majority of trajectories turns out to be captured by resonances. After time $1/\varepsilon^2$, only resonance motions will be observed. However, the independence of captures on the different intervals of length $1/\varepsilon$ is a strong additional assumption; along with capture to resonance, a reverse process also takes place.

The assumption in Neistadt's theorem, independence of frequencies, essentially restricts its area of application. The condition

$$\text{rank } \partial\omega/\partial I = \text{number of frequencies}$$

can be replaced by the following condition of independence of the ratios of frequencies:

$$\text{rank of the mapping } I \mapsto (\omega_1(I); \cdots ; \omega_n(I)) \text{ equals } n - 1.$$

However, in the case where the number of slow variables is small (smaller than the number of frequencies minus one), even this condition cannot be satisfied.

An extension of Neistadt's theorem to the case where the number of slow variables is essentially smaller than that of frequencies requires a study of Diophantine approximations on submanifolds of Euclidean spaces.

For mappings

$$\omega : \mathbb{R}^k \to \mathbb{R}^n, \qquad k < n$$

satisfying certain nondegeneracy conditions (the nonvanishing of some determinants), we may expect the same lower estimate

$$|(m, \omega(I))| \geqslant c|m|^{-\nu}, \qquad m \in \mathbb{Z}^n \backslash 0$$

for almost all I in $\mathbb{R}^k$, which holds for almost all points of $\mathbb{R}^n$.

Results of this kind have been obtained for special curves ($\omega_s = I^s$); cf., V. G. Sprindžuk, Mahler's Problem in Metric Number Theory, Minsk, 1967; concerning the general case, cf., A. S. Pjartli, *Diophantine approximations on submanifolds of a Euclidean space*, Funct. Anal. Appl. **3**, 4 (1969), 59–62.

We note that these publications do not solve either the above problem of generalizing Neistadt's theorem or the arithmetic problem of giving a sharp estimate for v (which, by the way, does not have a great significance for our problem, in which the variation of v will only change the necessary smoothness of the equations).

In generic systems with arbitrary numbers of fast and slow variables depending on sufficiently many parameters for almost every (in the sense of Lebesgue measure) value of the parameter one has:

(i) the measure of the set of the initial conditions, for which the deviation is greater than ρ for some t in $(0, 1/\varepsilon)$, does not exceed $C\sqrt{\varepsilon}/\rho$;

(ii) the integral of the deviation over the set of other initial conditions (the phase space is supposed to be compact) does not exceed $C\sqrt{\varepsilon}$ (the deviation equals $\max |I(t) - J(t)|$ for $t \in [0, 1/\varepsilon]$.)

For mappings ω outside some exceptional set of codimension N in the space of mappings we may substitute $\varepsilon^{1/(p+1)}$ for $\sqrt{\varepsilon}$ in the previous estimate, where

$$n \leqslant C_{k+p}^p - k - N + 1$$

(here as above n is the number of fast variables and k is the number of slow variables). The constant C in the estimate depends on the mapping, but this time the estimate holds for all systems, not for almost all, and we do not need to consider parametrized families—the result holds for individual systems.

Both theorems formulated above are proved by V. I. Bachtin, *On the averaging in systems with several frequencies*, Uspekhi Math. Nauk **40**, 5 (1985), 304–305, Funct. Anal. Appl. **20**, 2 (1986), 1–7; see also V. I. Arnold, V. V. Kozlov, A. I. Neistadt, *The Mathematical Aspects of the Classical and Celestial Mechanics*, Itogi Nauki i Techniki, Sovremennye Problemy Matematiki, Fundamentalnye Napravleniia. Dynamicheskie Systemy, Vol. 3. pp. 3–304, see p. 181, Mosc. VINITI, 1985 (Springer translation, 1988).

The proofs depend on the estimate

$$|(m, \omega(I))| + |\partial(m, \omega(I))/\partial I| \geqslant C_I |m|^v, \qquad v > n + 1, \quad \forall m \in \mathbb{Z}^n \setminus \{0\},$$

holding generically for almost all values of I. This estimate implies that the deviation, averaged over the initial conditions, does not exceed (generically) a value of order $\varepsilon^{1/(p+1)}$ if $n \leqslant C_{k+p}^p$.

Let us note that these estimates provide new information, even for those values of the dimensions (n and k) where the Neistadt theorem is applicable, in the case where the Jacobian vanishes (but not identically) on some surface. Indeed, to apply the Neistadt theorem, we have to consider time intervals of lengths C/ε smaller than $1/\varepsilon$, where C depends on the initial point (so as to prevent the averaged orbit from the intersection with the surface of zero Jacobian). In the theorem of Bachtin the set swept by the orbits intersecting the zero surface of the Jacobian, is included in the exceptional set of small measure controlled by the estimate.

§ 19. Averaging in Hamiltonian Systems

In this section, we briefly describe the characteristic properties of averaging in the case where both the unperturbed and perturbed systems are Hamiltonian.

A. Calculation of the Averaged System

We assume that in the unperturbed system, action-angle variables are intro-
duced, i.e., canonically conjugate* variables $(I_1, \ldots, I_n; \varphi_1, \ldots, \varphi_n \bmod 2\pi)$
such that the unperturbed Hamiltonian H_0 depends only on the action
variables I. Hamilton's canonical equations take the form

$$\dot{\varphi} = \frac{\partial H}{\partial I}, \qquad \dot{I} = -\frac{\partial H}{\partial \varphi},$$

i.e., for $H = H_0(I)$, we have

$$\dot{\varphi} = \omega(I), \qquad \dot{I} = 0,$$

where the frequency vector $\omega(I)$ is equal to $\partial H_0/\partial I$.

The perturbed system is given by the Hamiltonian $H = H_0(I) +$
$\varepsilon H_1(I, \varphi, \varepsilon)$, where the function H_1 has period 2π with respect to the angular
variables φ. Consequently, the equations of the perturbed motion have the
form

$$\dot{\varphi} = \omega(I) + \varepsilon\frac{\partial H_1}{\partial I}, \qquad \dot{I} = -\varepsilon\frac{\partial H_1}{\partial \varphi}.$$

Theorem. *In a Hamiltonian system with n degrees of freedom and n frequencies,
evolution of slow variables does not occur in the sense that the averaged
system has the form $\dot{J} = 0$.*

◀ When calculating the integral of $\partial H_1/\partial \varphi_s$ over the n-dimensional torus,
we may first integrate with respect to the variable φ_s. This one-dimensional
integral is equal to the increment of the periodic function H_1 on its period,
i.e., it is zero. ▶

This simple theorem shows that in a Hamiltonian system, in general,
the evolution of slow variables differs drastically from the situation of non-
Hamiltonian systems.

B. Kolmogorov's Theorem

We assume that the frequencies are independent in the sense that the derivative $\partial\omega/\partial I$ of the frequencies with respect to the action variables is nondegenerate. In this case, as has been established by Kolmogorov [A. N. Kolmogorov, *On the preservation of quasi-periodic motions under a small variation
of Hamilton's function*, Sov. Math. Dokl. **98**, 4 (1954), 527–530], the majority

---

*The coordinates (I, φ) are said to be *canonically conjugate* if the symplectic structure of
the phase space can be written in the form $\omega = \sum dI_k \wedge d\varphi_k$.

of invariant tori $I = $ const are only slightly deformed without disappearing under a small Hamiltonian perturbation: for the majority of initial conditions, the phase curves fill out the invariant tori densely in both the perturbed and unperturbed systems.

Let the Jacobian of the mapping of the $(n - 1)$-dimensional surface $H_0(I) = h$ into the $(n - 1)$-dimensional projective space, given by the formula $I \mapsto (\partial H_0/\partial I_1 : \cdots : \partial H_0/\partial I_n)$, be different from zero. Then the invariant tori of the perturbed system fill almost completely the entire $(2n - 1)$-dimensional level manifold $H(I, \varphi) = h$ of the Hamiltonian function H (the complementary set is of small measure).

In particular, if the number of frequencies n is equal to 2, then these two-dimensional tori divide the three-dimensional level manifold. Therefore, even for those phase curves which do not lie on the tori, the action variables vary little over an infinite time interval: a phase curve starting in a gap between two invariant tori cannot get out of it.

On the other hand, if the number of frequencies is larger than two, then the tori do not divide the level manifold of the Hamiltonian function. Then some phase curves (constituting a set of small measure) can go far from the initial values of the action variables, wandering near resonance surfaces between the invariant tori.

There exist examples [V. I. Arnold, *On the instability of dynamical systems with many degrees of freedom*, Sov. Math. Dokl. **156**, 1 (1964), 9–12] in which such an exit actually occurs. The average velocity of exit is exponentially small (of order $e^{-1/\sqrt{\varepsilon}}$) in these examples.

C. Nehoroshev's Theorem

It turns out that the average velocity of the deviation of the action variables from their initial values is so small in every generic Hamiltonian system that it cannot be detected by any approximation method of perturbation theory (i.e., it does not appear in the form of noticeable deviation in time of order $1/\varepsilon^N$ for any N, where ε is the parameter of the perturbation).

More precisely, Nehoroshev [N. N. Nehoroshev, *On the behavior of Hamiltonian systems close to integrable ones*, Funct. Anal. Appl., **5**, 4 (1971), 82–83; N. N. Nehoroshev, *Exponential estimate of the time of stability of Hamiltonian systems close to integrable ones*, I, UMN, **32**, 6, (1977), 5–66; II Trudy seminara im. I. G. Petrovskogo, V. **5** (1979), 5–50; N. N. Nehoroshev, *Stable estimates from above for smooth mappings and smooth function gradients*, Math. Sbornik **90**, 3 (1973), 432–472. Cf., also his dissertation, MGU, 1973] proved that for almost every unperturbed Hamiltonian function $H_0(I)$, there exist positive numbers a and b such that the average rate of change of the action variables I does not exceed ε^b in the perturbed system over time $T = e^{1/\varepsilon^a}$.

We note that T increases more rapidly then any power of $1/\varepsilon$ as $\varepsilon \to 0$, so that the variation of I over time $1/\varepsilon^N$ is small for any N.

The constants a and b depend on geometric properties of the unperturbed

Hamiltonian H_0. For example, if H_0 is strictly convex (the matrix $\partial^2 H_0/\partial I^2$ is positive definite), then we may set $a = 2/(6n^2 - 3n + 14)$, $b = 3a/2$, where n is the number of frequencies.

The theorem is proved for almost all H_0 in the sense that only those functions H_0 are excluded which satisfy infinitely many explicitly given algebraic equations for the Taylor coefficients. Nehoroshev calls the exceptional functions *nonsteep*.* For nonsteep H_0, exit is possible in time on the order of $1/\varepsilon$. In the examples of exponentially slow exit (cf., § 19B), H_0 is steep.

The proof of Nehoroshev's theorem is based on the following simple property of averaging in a Hamiltonian system.

Assume that for some values of the slow variables I, the resonance $(m, \omega) = 0$ takes place in a Hamiltonian system with n frequencies. Then near the corresponding resonance surface it is natural to perform averaging not over n-dimensional tori, but over the resonance tori of smaller dimension. The dimension of the resonance torus is $n - 1$ if the resonance is simple, i.e., if the direction of the integer-valued vector m is uniquely determined. If the equation $(m, \omega) = 0$ for m has k rationally independent solutions, then the trajectories of the slow motion fill the resonance tori of dimension $n - k$ densely; we have to average over these tori.

Theorem. *Upon averaging over resonance tori corresponding to the resonance* $(m, \omega) = 0$, *the direction of the evolution of the action variables I in the averaged system lies in the plane spanned by the resonance vectors m.*[†] *(In the case of a simple resonance, the direction of the evolution is uniquely determined: it is the direction of the straight line spanned by the vector m.)*

◀ For the sake of simplicity, we consider the case of simple resonance. We denote by γ the angular coordinate which does not change in resonance: $\gamma = (m, \varphi)$. In order to average the perturbed system, it is sufficient to average the Hamiltonian with respect to the fast variables. As a result, we obtain the averaged Hamiltonian $H_0 + \varepsilon \overline{H}_1$, where $\overline{H}_1$ depends on the action variables and one angular variable γ.

The equations of the averaged motion now give

$$\dot{I} = \varepsilon \frac{\partial \overline{H}_1}{\partial \varphi}$$

---

*The nonsteepness condition, equivalent to that of Nekhoroshev for all analytic systems (with the exception of codimension infinity), admits a very simple formulation: The critical points of all the restrictions of the unperturbed Hamiltonian function to the affine subspaces of all dimensions of the space of action variables should be of finite multiplicity (that is, isolated over complex numbers)—this remark is due to Ju. S. Iliashenko, 1985; it is assumed that the unperturbed Hamiltonian function has no critical points on any part of the action variable spaces which we consider.

[†]We note that in the space of action variables, the affine structure is uniquely determined and so is the identification of vectors of the space dual to the space of frequencies with vectors of the space of action variables.

On the other hand, $\partial \overline{H}_1/\partial \varphi = (\partial \overline{H}_1/\partial \gamma)(\partial \gamma/\partial \varphi)$ has the direction of the vector $\partial \gamma/\partial \varphi = m$. ▶

Nehoroshev's theorem can be deduced from the theorem just proved with the help of the following arguments. Fast evolution (with velocity of order ε) is possible only at resonance and in directions generated by resonance vectors. However, conditions of steepness imposed on H_0 (for example, strict convexity of H_0 will suffice) guarantee that such evolution takes place in a direction leading out of the resonance surface. Consequently, resonance is violated and evolution lasts only a short time; as a consequence, an exponentially small upper estimate of the average velocity is indeed obtained.

On the other hand, if the steepness conditions are violated, then a curve exists on the resonance surface, whose tangent at every point belongs to the plane spanned by the resonance vectors. Along such a curve, evolution may have an average velocity of order ε. This leads to deviation of the action variables from their initial values over time on the order of $1/\varepsilon$.

§ 20. Adiabatic Invariants

Here we give a review of the basic results of the theory of adiabatic invariants in Hamiltonian systems with slowly varying parameters.

A. The Notion of the Adiabatic Invariant

In considering Hamiltonian systems with slowly varying parameters, a peculiar phenomenon occurs: quantities which are generally independent become functions of one another asymptotically (as the rate of change of the parameters converges to zero).

For example, consider a pendulum of variable length. The length of the pendulum and the amplitude of oscillations are generally independent: if we change the length of the oscillating pendulum, then upon restoring the original length of the pendulum, the amplitude of oscillations can, in general, change in an arbitrary way, depending on the specific way in which the length has been changed.

Nevertheless, it turns out that if the change in the length of the pendulum occurs sufficiently slowly, then the amplitude of oscillations remains almost unchanged upon restoring the length. Moreover, the ratio of the energy of the pendulum to the frequency remains almost unchanged during the entire process, although both the energy and the frequency vary as the length varies.

Quantities asymptotically preserved under sufficiently slow variation of the parameters of a Hamiltonian system are called *adiabatic invariants*.

More precisely, we consider the Hamiltonian system of differential equations $\dot{x} = v(x, \lambda)$, where λ is a parameter.

A function I of the phase point x and the parameter λ is called an *adiabatic invariant* if for any smooth (i.e., sufficiently often differentiable) function $\lambda(\tau)$ of the slow time $\tau = \varepsilon t$, the variation of $I(x(t), \lambda(\varepsilon t))$ along a solution of the equation $\dot{x} = v(x, \lambda(\varepsilon t))$ remains small along the time interval $0 \leqslant t \leqslant 1/\varepsilon$, provided that ε is sufficiently small.

B. Construction of an Adiabatic Invariant of a System with One Degree of Freedom

We assume that the Hamiltonian $H(p, q; \lambda)$ has closed phase curves $H(p, q; \lambda) = h$ for every value of the parameter λ (say, surrounding an equilibrium position in which the frequency of small oscillations is different from zero).

We denote by $I(p, q; \lambda)$ the area bounded by the phase curve passing through the point with coordinates (p, q) for a fixed value of λ divided by 2π (according to tradition). The quantity I is called the *action variable*.

EXAMPLE

For a pendulum, we have $H = ap^2/2 + bq^2/2$. The phase curve $H = h$ is an ellipse with area $\pi\sqrt{2h/a}\sqrt{2h/b} = 2\pi h/\sqrt{ab}$. The frequency of the oscillations is $\omega = \sqrt{ab}$. Consequently, for a pendulum,

$$I = \frac{H}{\omega}.$$

Here the role of λ is played by the pair (a, b).

Theorem. *The action variable I is an adiabatic invariant of a Hamiltonian system with one degree of freedom.*

C. Proof of the Adiabatic Invariance of the Action Variable

The proof of this theorem is based on the averaging method. Let φ be the angular coordinate on closed phase curves chosen in such a way that it varies in proportion to the time of the motion on the curve and increases by 2π during every revolution (of course, the angular coordinate φ and the action variable I both depend on not only the phase coordinates (p, q), but also on the parameter λ).

For every value of λ, the equation of our system can be written in the form of the standard unperturbed system of the averaging method:

$$\dot{\varphi} = \omega(I, \lambda(\tau)), \qquad \dot{I} = 0, \qquad \dot{\tau} = 0.$$

If λ is slowly varying, then, instead of the unperturbed system, we obtain the perturbed system

$$\dot{\varphi} = \omega + \varepsilon f, \qquad \dot{I} = \varepsilon g, \qquad \dot{\tau} = \varepsilon,$$

where the functions f and g are 2π-periodic in φ.

Now compose the averaged system.

Lemma. *The action variable is a first integral of the averaged system (that is, the average of g with respect to φ is equal to zero).*

◀ We consider the domain bounded by the closed phase curve $I = I_0$ for the initial value of the parameter. According to the theorem on averaging, the image of this domain at any time t in the interval $[0, 1/\varepsilon]$ is, up to an error of order ε, the domain bounded by some closed phase curve $I = I_t$ for the value $\lambda = \lambda(\varepsilon t)$ of the parameter.

On the other hand, the equations of motion are Hamiltonian (even if nonautonomous). By Liouville's theorem, the area of the image is equal to the area of the preimage. This implies that $I_t = I_0$. ▶

Corollary. *The ratio of the energy of a pendulum to the frequency is an adiabatic invariant.*

Problem. A little ball moves horizontally between two vertical absolutely elastic walls whose distance apart varies slowly. Prove that the product of the velocity of the ball with the distance between the walls is an adiabatic invariant.

Problem. A charged particle moves in a magnetic field which varies slowly in the course of a Larmour loop of the particle around a magnetic line of force. Prove that an adiabatic invariant is the ratio $v_{\perp}^2/H$ of the square of the component of the velocity of the particle normal to the line of force to the magnetic field strength (cf., for example, L. A. Arcimovich, Controllable Thermonuclear Reactions, Moscow, Fizmatgiz, 1961).

D. Adiabatic Invariants of Hamiltonian Systems with Several Frequencies

We consider a Hamiltonian system $\dot{p} = -H_q$, $\dot{q} = H_p$ with several frequencies, depending on a parameter λ and admitting, for fixed λ, the action-angle variables $\dot{\varphi} = \omega(I, \lambda)$, $\dot{I} = 0$ (where $\omega = \partial H_0/\partial I$) with the Hamiltonian $H_0(I, \lambda)$ depending on n action variables in a nondegenerate way, such that

$$\det\left(\frac{\partial \omega}{\partial I}\right) = \det\left(\frac{\partial^2 H_0}{\partial I^2}\right) \neq 0.$$

We assume, as above, that the parameter λ begins to vary slowly. The variation of p and q can be described by Hamilton's equations with a varying function H, and the behavior of the variables I is described by the perturbed system (we assume that $\lambda = \varepsilon t$, where ε is a small parameter).

Lemma. *The perturbed system is Hamiltonian with the single-valued Hamiltonian function* $H = H_0(I, \lambda) + \varepsilon H_1(I, \varphi, \lambda, \varepsilon)$.

The proof of this lemma requires either some knowledge of symplectic geometry or Hamiltonian formalism (cf., for example, V. I. Arnold, Mathematical Methods of Classical Mechanics, New York, Springer-Verlag, 1978), or else cumbersome calculations, which we omit.

Corollary. *The action variables I are first integrals of the averaged system.*

◀ Indeed, the averaged function—the right-hand side of the equation $\dot{I} = -\varepsilon \partial H_1/\partial \varphi$—is the derivative of a periodic function and, therefore, has a zero average value (cf., the theorem in § 19A). ▶

Combining this corollary with Neistadt's theorem (cf., § 18D), we arrive at the following conclusion.

The variation of the action variables I in a Hamiltonian system with several frequencies and slowly varying parameters remains smaller than ρ over time $1/\varepsilon$ if we neglect a set of initial conditions of measure not greater than $c\sqrt{\varepsilon}/\rho$ in the original phase space. (Here the phase space is assumed to be compact and the derivative $\partial \omega/\partial I$ is assumed to be nondegenerate.)

Definition. A function F of the phase point of a Hamiltonian system and of a parameter is called an *almost adiabatic invariant* if for every $\rho > 0$ the measure of the set of initial conditions in the compact phase space for which the variation of F along a solution of Hamilton's equation with slowly varying parameter exceeds ρ in time $1/\varepsilon$ converges to zero as ε converges to zero.

Consequently, *the action variables $(I_1, \ldots, I_n)$ are almost adiabatic invariants of any nondegenerate Hamiltonian system with several frequencies.*

E. Behavior of Adiabatic Invariants for $t \gg 1/\varepsilon$

Although an adiabatic invariant varies slowly over time $1/\varepsilon$, there is no reason to assume that the variation remains small over larger time intervals (say, on the order of $1/\varepsilon^2$) or, a fortiori, over an infinite time interval.

EXAMPLE

We consider a pendulum with a slow (periodically varying) parameter:

$$\ddot{x} = -\omega^2(1 + a\cos\varepsilon t)x.$$

For arbitrarily small ε (i.e., for an arbitrarily slow variation of the parameter), parametric resonance is possible in which the equilibrium position $x = 0$ becomes unstable. It is clear that in parametric resonance an adiabatic invariant of a linear pendulum changes unboundedly (over an infinite time interval).

It turns out that this behavior of the adiabatic invariant in a system with a *periodic* slowly varying parameter is connected with the linearity of the system, more precisely, with the independence of the period of oscillations from the amplitude. If, in a Hamiltonian system with periodic slowly varying parameter, the derivative of the frequency of the fast motion with respect to the action variable is different from zero, then the action variable varies little over an infinite time interval [cf., V. I. Arnold, *On the behavior of an adiabatic invariant in a periodic slow variation of the Hamiltonian function*, Sov. Math. Dokl., **142**, 4(1962), 758–761].

The proof is based on the fact that invariant tori exist in this situation (cf., Kolmogorov's theorem, § 19B).

Another interesting case is the one in which the parameter varies in such a way that it has definite limits as $t \to -\infty$ or $t \to +\infty$. In this case, it makes sense to speak about the values of an adiabatic invariant at minus infinity and plus infinity and its increment

$$\Delta I = I(+\infty) - I(-\infty)$$

over the entire time.

For the linear equation

$$\ddot{x} = -\omega^2(\varepsilon t)x, \qquad \omega(-\infty) = \omega_-, \qquad \omega(+\infty) = \omega_+,$$

one can prove that the increment of an adiabatic invariant over infinite time is exponentially small in ε (under the assumption that ω is an analytic function which must not change sign and must behave reasonably at infinity). Moreover, the principal term of the asymptotics of the increment of an adiabatic invariant can be given explicitly as $\varepsilon \to 0$ [cf., A. M. Dyhne, *Quantum passages in adiabatic approach*, JETP **38**, 2 (1960), 570–578 and A. A. Sludskin, JETP **45**, 4 (1963), 978–988]. Analogous results have been obtained for multidimensional systems as well. The accurate formulations and proofs can be found in M. V. Fedorjuk, *Adiabatic invariant of a system of linear oscillators and scattering theory*, Differential Equations **12**, 6 (1976), 1012–1018 (in which references to earlier work by physicists have been omitted, however). See also M. Levi, *Adiabatic invariants of linear Hamiltonian systems with periodic coefficients*, J. Diff. Equat. **42**, 1 (1981), 47–71.

The problem of the increment of an adiabatic invariant of a one-dimensional nonlinear system has also been studied by physicists: they have proved the smallness of the increment compared to ε^N, i.e., the absence of variations of an adiabatic invariant in all orders of perturbation theory [A. Lenard, Ann. Phys. **6** (1959), 261–276]. Neistadt obtained an exponential estimate in the analytic case. A. I. Neistadt, *On the degree of precision of the adiabatic invariant conservation*, Prikl. Mat. Mekh. **45**, 1 (1981), 80–87.

Some exponentially small deviation is unavoidable in all the versions of the theory of averaging for the following reason: the size of the projection on the base space of *any* closed curve, close to the fiber of our fibering, grows under the perturbed phase flow action (exponentially slowly, that is, with average speed of order $\exp(-C/\varepsilon)$): see A. I. Neistadt, *On the separation of motions in the systems with fast rotation of the phase*, Prikl. Mat. Mekh. **48**, 2 (1984), 197–204.

The adiabatic invariant changes its value with a jump when the slow motion forces the fast orbit to intersect the separatrix of the instantaneous phase portrait. The asymptotics of these jumps have been studied by A. I. Neistadt and, according to Wisdom (*The origin of the Kirkwood gaps: a mapping for asteriodal motion near 3/1 commensurability*, Astrophys. J. **87** (1982), 577–593; Icarus **56** (1983), 51–74; **63** (1985), 272–289), those jumps imply the resonance gaps in the asteroid distribution (namely, those asteroids whose mean motion is commensurable with that of Jupiter change their eccentricity abruptly, which leads to collision with Mars).

The theorems of Neistadt on asymptotics of the separatrix intersections appear as: *On the separatrix passage in resonance problems with slowly varying parameter*, Prikl. Mat. Mekh. **39**, 4 (1975), 621–632; Phisika plasmy **12** (1986), 992–1001; see also his paper *Jumps of adiabatic invariants and the origin of the Kirkwood gap 3 : 1*, Prikl. Mat. Mekh. **51**, 5 (1987), 750–757. See also the paper by A. V. Timofeev in JETP **75**, 4 (1978), 1203–1308; and J. L. Tennyson, J. R. Cary, D. F. Escande, Phys. Rev. Lett. **56**, 20 (1986), 2117–2120.

For nonlinear systems with several degrees of freedom, the adiabatic invariance of the action variables does not hold, in spite of assertions in the physical literature: they are only almost adiabatic invariants, i.e., change little for the majority of initial conditions.

§ 21. Averaging in Seifert's Foliation

In the study of the neighborhood of a closed phase curve, one encounters the case where the nearby phase curves also close in first approximations; but, before closing, they revolve several times along the initial closed phase curve (this is the so-called case of resonance). The study of the behavior of a system close to a resonant or approximately resonant motion leads to a peculiar variant of the averaging method; averaging in Seifert's foliation.

A. Seifert's Foliation

Seifert's foliation is a partition of the direct product $\mathbb{R}^2 \times S^1$ into circles. It is constructed in the following way. In Euclidean three-dimensional space, we consider a cylinder with horizontal bases and vertical axis. We foliate

$p = 1$
$q = 2$

Figure 101.

the interior of the cylinder by vertical segments and identify the upper and lower bases of the cylinder after having rotated the upper base by an angle of $2\pi p/q$ [we glue the point $(z, 0)$ of the lower base to the point $(Az, 1)$ of the upper base, where A is the rotation by angle $2\pi p/q$ and p and q are relatively prime numbers].

Definition. The *Seifert foliation* of type (p, q) is the three-dimensional manifold $\mathbb{R}^2 \times S^1$, together with its partition into circles obtained from the partition of the interior of the cylinder into segments parallel to the axis upon pasting the bases after a rotation by angle $2\pi p/q$.

Therefore, each of the circles of the Seifert foliation is obtained by pasting q segments except for one central circle obtained from the axis of the cylinder.

We consider the q-sheeted covering of the space $\mathbb{R}^2 \times S^1$ of Seifert's foliation of type (p, q). The covering space itself is diffeomorphic to $\mathbb{R}^2 \times S^1$. Seifert's foliation in the original manifold induces a partition into circles on the covering manifold. This partition can be considered as Seifert's foliation of type $(p, 1)$. (The pasting is now performed with a rotation by an angle $2\pi p$.)

The Seifert foliation of type $(p, 1)$ is already a fibration into circles and, moreover, a direct product. In the covering, every circle of the original Seifert foliation is covered diffeomorphically by q circles, except for one central circle covered q-fold (Fig. 101).

B. Definition of Averaging in Seifert's Foliation

Let a vector field be given in the space $\mathbb{R}^2 \times S^1$ of Seifert's foliation. Then the covering foliation contains a vector field, too. Every vector of the field can be projected on the base $\mathbb{R}^2$ of the covering foliation. We average the vector thus obtained on the base along a leaf of the covering foliation. At every point of the base we obtain a well-defined vector. Consequently, we have defined a vector field on the base. The above operation of constructing (from a field in the Seifert foliation space) a field on the plane is called *averaging the original field along Seifert's foliation*.

In other words, averaging on Seifert's foliation of type (p, q) is defined as the usual averaging in its q-sheeted covering foliation.

C. Properties of the Averaged Field

Upon averaging in an ordinary fibration, any vector field can be obtained on the base. Upon averaging along Seifert's foliation, a vector field with specific properties is obtained on the base: for example, at the central point the vector of the averaged field vanishes if $q > 1$.

Theorem. *As a result of averaging along Seifert's foliation of type (p, q), we obtain a field invariant with respect to rotation of the plane by an angle $2\pi/q$.*

◀ We realize the base as one of the bases of the initial cylinder. Then the averaging along Seifert's foliation turns into averaging on q intervals parallel to the axis of a cylinder. Upon rotation by an angle $2\pi/q$, these q intervals turn into each other. It is easy to see that averaging commutes with rotation by angle $2\pi/q$. (After the rotation, we have to average along the same intervals but in a different order.) ▶

D. An Example

We consider the differential equation

$$\dot{z} = i\omega z + \varepsilon f(z, t), \qquad \text{where } z \in \mathbb{C},$$

where f is a complex-valued (not necessarily holomorphic) function having period 2π in the real time variable t, and ε is a small parameter. The equation corresponding to $\varepsilon = 0$ will be called *unperturbed*.

We assume that the frequency ω of the unperturbed motion is rational or close to a rational number p/q.

The integral curves of the unperturbed equation with $\omega = p/q$ form Seifert's foliation of type p/q in $\mathbb{C} \times S^1 = \{z, t \bmod 2\pi\}$.

Upon averaging along this foliation, we obtain the averaged equation

$$\dot{z} = \varepsilon F(z),$$

where the vector field F turns into itself under rotation of the z-plane by an angle $2\pi/q$.

E. Taylor Coefficients of a Symmetric Field

We shall define a vector field on the complex z-plane by a complex (not necessarily holomorphic) function F. The Taylor series of F in x and y (where $z = x + iy$) can be written in the form of a Taylor series in the variables z and $\bar{z}$. We write this series in the form

$$\sum F_{k,l} z^k \bar{z}^l.$$

Proposition. *If the field F is invariant under rotation by an angle $2\pi/q$, then only those coefficients $F_{k,l}$ are different from zero for which $k - l$ is congruent with 1 modulo q.*

◀ As a result of the uniqueness of the Taylor series, every term of the series defines a vector field invariant under rotation. Under rotation of z by the angle $2\pi/q$, the vector $z^k \bar{z}^l$ turns by the angle $(k - l)2\pi/q$. This rotation is a rotation by the angle $2\pi/q$ if and only if $k - l$ is congruent with 1 modulo q. ▶

We consider the quadrant of the lattice of integral nonnegative points (k, l). Among them, we distinguish those for which $k - l$ is congruent with 1 modulo q. The point $(1, 0)$ and all integral points on the line issued from this point and parallel to the bisector of the quadrant are always distinguished. These points correspond to fields $z\Phi(|z|^2)$ invariant under rotation by any angle.

The point $(0, q - 1)$ is always among the distinguished points. This point corresponds to the field $\bar{z}^{q-1}$ invariant with respect to rotation by the angle $2\pi/q$. All distinguished points form a series of rays parallel to the bisector issued from the points $(0, mq - 1)$ and $(mq + 1, 0)$ on the sides of the quadrant.

F. The Case of Symmetry of Order 3

We consider vector fields invariant under the symmetry group of order 3 (i.e., the case $q = 3$).

The monomials of smallest degree in the Taylor series of a field symmetric under rotation by 120° are given by distinguished points of the (k, l)-plane with the smallest $k + l$. The first two monomials are z and $\bar{z}^2$. Consequently, every field on the plane invariant under rotation by the angle $2\pi/3$ has the form

$$F(z) = az + b\bar{z}^2 + O(|z|^3).$$

Omitting the last term, we obtain the following simplest differential equation with symmetry of order 3:

$$\dot{z} = az + b\bar{z}^2.$$

Here the coefficients a and b and the phase coordinate z are complex.

We assume that $a \neq 0$ and $b \neq 0$. Multiplying z by a suitable number and changing the unit of time, one obtains $b = 1$ and $|a| = 1$. The variation of the phase portrait is shown for $a = e^{i\varphi}$, $b = 1$ in Fig. 102. For any a, there are four equilibrium positions at the vertices of an equilateral triangle and at its center. For purely imaginary a, the system is Hamiltonian. In order to study the system for arbitrary a, it is sufficient to notice that it

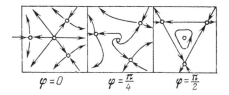

Figure 102.

can always be obtained from this Hamiltonian system by formal multi-plication of z and t by complex numbers (i.e. by rotation and dilation of the z-plane and rotation of the Hamiltonian field by a constant angle).

G. Effect of the Omitted Terms

Now we try to take into account the omitted terms $O(|z|^3)$. We assume that $|a|$ is small (this corresponds to the assumption that in the original system of differential equations a resonance of third order almost takes place). Then the radius of the triangle of singular points is also small (of order $|a|$). We consider our symmetric vector field in the neighborhood of $z = 0$, which is large compared to $|a|$, but small compared to 1.

In the same neighborhood, the omitted terms $O(|z|^3)$ are small compared to the terms we kept. It is easy to show that their effect does not essentially change the form of the phase portrait if it is structurally stable. In our case, the phase portrait is structurally unstable only for purely imaginary a, when the system is Hamiltonian. The Hamiltonian character is not preserved if we take into account the omitted terms.

For every ray in the complex a-plane not coinciding with the imaginary axis and for sufficiently small $|a| \neq 0$, the shape of the phase portrait of the complete system (under the condition $b \neq 0$) in a neighborhood of the origin small compared to 1 and large compared to $|a|$ is indicated in Fig. 102, $\varphi \neq \pm \pi/2$.

A study of the change of the phase portrait when a intersects the imaginary axis constitutes a special problem, to which we shall return in Chap. 6. In the generic case, the change is determined by just one term of the Taylor series: all happens in the same way as for the equation

$$\dot{z} = az + \bar{z}^2 + cz|z|^2,$$

where $\mathrm{Re}\, c \neq 0$.

H. Application to the Original Equation

The above analysis of the averaged equation yields significant information on the original system for the case when the parameter ε is sufficiently

Figure 103.

small. Without going into proofs, we only discuss a translation of our results into the language of phase curves of the original equation.

The three equilibrium positions at the vertices of the equilateral triangle correspond to one closed integral curve of the original equation. If the difference between the frequency ω of the unperturbed motion and the resonance frequency $p/q = \frac{1}{3}$ converges to zero, this closed curve coalesces with the initial closed curve, going three times around it.

The stability of equilibrium positions of the averaged system is interpreted as the stability of periodic solutions of the perturbed system, etc. An essential difference arises at only one point, namely, in the case where the averaged system has a separatrix going from saddle into saddle.

In the perturbed system, a closed curve corresponds to saddles. Stable and unstable invariant manifolds of the closed curve correspond to incoming and outgoing separatrices. However, while in the averaged system the separatrices merge upon intersection, in the perturbed system this is not so, in general. In order to see how the invariant manifolds of the perturbed system intersect in the three-dimensional space, we consider the section $t = 0$ of three-dimensional space.

The plane $t = 0$ is intersected by our solution at three points which are fixed points of the third iterate of the Poincaré mapping. Each of the three fixed points has an incoming and an outgoing invariant manifold (curve). These curves are not bound to coincide upon intersection (unlike phase curves of an equation on the plane, which have to coincide forever, once intersecting).

Under iterations by the Poincaré mapping, the intersecting arcs of the invariant manifolds form a complicated network called the homoclinic picture* (Fig. 103).

I. Resonances of Higher Order

For resonances of order q higher than 3, we obtain for the averaged system of the first nontrivial approximation

$$\dot{z} = az + zA(|z|^2) + \bar{z}^{q-1}$$

---

*A fixed point of a diffeomorphism of the plane is said to be homoclinic if its attracting and repelling invariant curves intersect without coinciding.

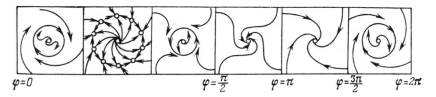

$\varphi=0$ $\varphi=\frac{\pi}{2}$ $\varphi=\pi$ $\varphi=\frac{3\pi}{2}$ $\varphi=2\pi$

Figure 104.

in the same way. In particular, for a resonance of order 4, we obtain the system

$$\dot{z} = az + Az|z|^2 + \bar{z}^3.$$

These systems, as well as the system corresponding to a resonance of order 2, are considered in detail in Chap. 6.

In Fig. 104, we illustrate the change of phase portraits in an averaged system corresponding to the resonance of order 5:

$$\dot{z} = az + Az|z|^2 + \bar{z}^4$$

for $\operatorname{Re} A < 0$, $\operatorname{Im} A < 0$, $a = \varepsilon e^{i\varphi}$, $\varepsilon \ll 1$.

Chapter 5

Normal Forms

A fruitful technique for many differential equations consists of transforming (but not solving) them to a simpler form. Poincaré's theory of normal forms produces such simple forms to which a differential equation can be reduced in the neighborhood of an equilibrium position or a periodic motion.

The reduction to normal forms is realized by means of power series in the deviation from the equilibrium position or periodic motion. The series are not always convergent. Nevertheless, even in cases where the series are divergent, the method of normal forms turns out to be a powerful device in the study of differential equations: a few initial terms of the series often give significant information on the behavior of solutions, which is sufficient for the construction of the phase portrait. The method of normal forms is also a basic tool in the study of bifurcations, where it is applied to families of equations depending on parameters.

In this chapter, the simplest basic aspects of the method of normal forms are presented.

§ 22. Formal Reduction to Linear Normal Forms

According to Poincaré's theorem, in the class of formal power series, a "nonresonant" vector field can be reduced to its linear form at a singular point by a formal diffeomorphism. We are going to formulate this condition of nonresonance.

A. Resonances

Instead of a vector field, we consider a formal vector-valued power series $v(x) = Ax + \cdots$ in n variables with complex coefficients. We assume that the eigenvalues of A are distinct.

Definition. The n-tuple $\lambda = (\lambda_1, \ldots, \lambda_n)$ of eigenvalues is said to be *resonant* if among the eigenvalues there exists an integral relation of the form

$$\lambda_s = (m, \lambda),$$

where $m = (m_1, \ldots, m_n)$, $m_k \geqslant 0$, $\sum m_k \geqslant 2$. Such a relation is called a *resonance*. The number $|m| = \sum m_k$ is called the *order* of the resonance.

EXAMPLE

The relation $\lambda_1 = 2\lambda_2$ is a resonance of order 2; the relation $2\lambda_1 = 3\lambda_2$ is not a resonance; the relation $\lambda_1 + \lambda_2 = 0$ is a resonance of order 3 (more precisely, this relation implies the resonance $\lambda_1 = 2\lambda_1 + \lambda_2$).

B. Poincaré's Theorem

The following theorem is the fundamental result of Poincaré's dissertation.

Theorem. *If the eigenvalues of the matrix A are nonresonant, then the equation*

$$\dot{x} = Ax + \cdots$$

can be reduced to a linear equation

$$\dot{y} = Ay$$

by a formal change of variable $x = y + \cdots$ (the dots denote series starting with terms of degree greater than one).

The proof of Poincaré's theorem consists of the successive annihilation of the terms of degree 2, 3, etc. on the right-hand side. Every step is based on the solution of a linear homological equation; we begin with deducing this equation.

C. Deduction of the Homological Equation

Let h be a vector-valued polynomial* in y of order $r \geqslant 2$ and let $h(0) = h'(0) = 0$.

Lemma. *The differential equation $\dot{y} = Ay$ is transformed into*

$$\dot{x} = Ax + v(x) + \cdots$$

by the change of variables $x = y + h(y)$, where $v(x) = (\partial h/\partial x)Ax - Ah(x)$ and the dots denote terms of degree greater than r.

---

*That is, let h be a vector field whose components are polynomials. A vector-valued polynomial is the sum of *vector-valued monomials*; the latter are fields, one component of which is a monomial and the remaining components zeros. The order of a polynomial is the degree of the lowest term.

$$\blacktriangleleft \dot{x} = \left(E + \frac{\partial h}{\partial y}\right) Ay = \left(E + \frac{\partial h}{\partial y}\right) A(x - h(x) + \cdots)$$

$$= Ax + \left[\frac{\partial h}{\partial x} Ax - Ah(x)\right] + \cdots . \blacktriangleright$$

Remark. The square brackets contain the Poisson bracket of the vector fields Ax and $h(x)$.

We shall denote by L_A the operator converting every field into the Poisson bracket of the linear field Ax with the given field:

$$L_A h = \frac{\partial h}{\partial x} Ax - Ah(x).$$

Definition. *The homological equation associated with the linear operator A* is the equation

$$L_A h = v,$$

where h is the unknown and v is the known vector field.

D. Solution of the Homological Equation

The linear operator L_A acts in the space of formal vector fields. It leaves the spaces of homogeneous vector-valued polynomials of any degree invariant.

We calculate the eigenvalues and eigenvectors of L_A. Denote by e_i an eigenvector of A with eigenvalue λ_i. We denote by $(x_1, \ldots, x_n)$ coordinates with respect to the basis $(e_1, \ldots, e_n)$. As usual, x^m will stand for $x_1^{m_1} \cdots x_n^{m_n}$.

Lemma. *If the operator A is diagonal, then the operator L_A is also diagonal on the space of homogeneous vector-valued polynomials. The eigenvectors of L_A are the vector-valued monomials $x^m e_s$. The eigenvalues of L_A depend linearly on the eigenvalues of A, namely,*

$$L_A x^m e_s = [(m, \lambda) - \lambda_s] x^m e_s.$$

$\blacktriangleleft$ Let $h = x^m e_s$. Then the only nonzero component of the vector $(\partial h/\partial x) Ax$ is the sth one, which is equal to

$$\frac{\partial x^m}{\partial x} Ax = \sum \frac{m_i}{x_i} x^m \lambda_i x_i = (m, \lambda) x^m.$$

On the other hand, $Ah(x) = \lambda_s h(x)$. $\blacktriangleright$

If all eigenvalues of L_A are different from zero, then it is invertible.

Corollary. *If the collection of eigenvalues of A is nonresonant, then the homological equation $L_A h = v$ is solvable in the class of formal power series h for every formal vector field v without a constant term and a linear part at zero.*

If there are no resonances of order k, then the homological equation $L_A h = v$ is solvable for every homogeneous vector-valued polynomial v of degree k in the class of homogeneous vector-valued polynomials of degree k (here $k \geqslant 2$).

Remark. If A is not diagonal (has Jordan blocks), then L_A also has Jordan blocks. Nevertheless, as is easily seen, the eigenvalues are given by the same formula as in the diagonal case. Therefore, for nonresonant (possibly multiple) eigenvalues, L_A is invertible on the space of homogeneous vector-valued polynomials. Hence, the corollary above holds in the case of multiple eigenvalues as well.

E. Proof of Poincaré's Theorem

◀ Let the original equation have the form $\dot{x} = Ax + v_r(x) + \cdots$, where the v_r are terms of degree r ($r \geqslant 2$).

We solve the homological equation $L_A h_r = v_r$ (on the basis of the corollary in § 22D). Substituting $x = y + h_r(y)$ transforms the original equation into $\dot{y} = Ay + w_{r+1}(y) + \cdots$ (we use the lemma in § 22C). Consequently, we have annihilated the terms of degree r on the right-hand side of the original equation.

Successively killing the terms of degree 2, 3, ..., we construct a sequence of substitutions. The product of these substitutions stabilizes in the class of formal series, i.e., terms of any fixed degree do not change from a certain step. The limit substitution converts our formal equation into $\dot{y} = Ay$. ▶

Remark 1. Although convergence of series has not been proved, in the nonresonant case the perturbation can be moved arbitrarily far by a convergent substitution: we have proved that for any N the original equation can be reduced to $\dot{y} = Ay + o(|y|^N)$ by a true (even polynomial) change of variable.

Remark 2. If the perturbation $v = v_r + v_{r+1} + \cdots$ has order r, then, solving the homological equation $L_A h = v$ after the substitution $x = y + h$, we obtain an equation with perturbation of order $2r - 1$, a fact related to the hyperconvergence of the approximations obtained by iterating the procedure (cf., § 12).

Remark 3. The proof of Poincaré's theorem remains valid in the case of multiple eigenvalues (cf., the remark at the end of § 22D) as long as they are nonresonant.

Remark 4. If the original equation is real but the eigenvalues are not, then an eigenbasis can be chosen from complex conjugate vectors. In this case all substitutions in Poincaré's theorem can be chosen real, i.e., converting complex conjugate vectors into complex conjugate vectors.

§ 23. The Case of Resonance

In the case of resonance, the Poincaré's–Dulac theorem asserts that all nonresonant terms in the equation can be annihilated by a formal change of variables.

A. Resonant Monomials

Let the n-tuple $\lambda = (\lambda_1, \ldots, \lambda_n)$ of the eigenvalues of the operator A be resonant. Let e_s be a vector in the eigenbasis; let x_i be coordinates with respect to the basis e_i; and let $x^m = x_1^{m_1} \cdots x_n^{m_n}$ be a monomial in terms of the coordinates x_i.

Definition. The vector-valued monomial $x^m e_s$ is said to be *resonant* if $\lambda_s = (m, \lambda), |m| \geq 2$.

EXAMPLE

For the resonance $\lambda_1 = 2\lambda_2$, the unique resonant monomial is $x_2^2 e_1$. For the resonance $\lambda_1 + \lambda_2 = 0$, all monomials $(x_1 x_2)^k x_s e_s$ are resonant.

B. The Poincaré–Dulac Theorem

We consider the following differential equation $\dot{x} = Ax + \cdots$ given by the formal series $v(x) = Ax + \cdots$.

Theorem. *The equation above can be reduced to the canonical form*

$$\dot{y} = Ay + w(y)$$

by means of a formal change of variables $x = y + \cdots$, where all monomials in the series w are resonant.

◀ We begin killing the nonlinear terms of the series v. After several steps, we may encounter an unsolvable homological equation

$$L_A h = v$$

for a homogeneous vector-valued polynomial h of degree r equal to the degree of resonance. In this case, we cannot annihilate all terms of degree r of the perturbation v by an appropriate substitution. Instead, we annihilate only those which can be annihilated. In other words, we represent v and h in the form of vector-valued polynomials

$$v = \sum v_{m,s} x^m e_s, \qquad h = \sum h_{m,s} x^m e_s$$

and set

$$h_{m,s} = \frac{v_{m,s}}{(m, \lambda) - \lambda_s}$$

for those m and s for which the denominator is different from zero. In this way we define the field h.

We perform the usual substitution $x = y + h(y)$ of Poincaré's theorem. In the original equation, all terms of degree r will vanish except the resonant ones, which will remain unchanged. The equation will take the form

$$\dot{y} = Ay + w_r(y) + \cdots,$$

where w_r consists of only resonant terms.

The subsequent steps are performed in the same way. The remaining resonant terms w_r do not affect the homological equation which we solve and do not change in the subsequent substitutions. Indeed, upon the substitution $y = z + g_s(z)$, the equation

$$\dot{y} = Ay + w_2(y) + \cdots + w_s(y) + \cdots$$

is converted into

$$\dot{z} = Az + w_2(z) + \cdots + w_{s-1}(z) + [w_s(z) - (L_A g_s)(z)] + \cdots;$$

the Poisson bracket of w_2 and g_s has degree $s + 1$.

Consequently, all nonresonant terms of degree s are annihilated by the choice of g_s, and the proof is completed in the same way as in the nonresonance case. ▶

C. Examples

In practice, the Poincaré–Dulac theorem is usually used to single out resonance terms of small degree and remove perturbation up to terms of some finite order, i.e., reduce the equation to the form

$$\dot{x} = Ax + w(x) + o(|x|^N)$$

(where w is the polynomial of resonant monomials) by a true (polynomial, if desired) rather than formal change of variables.

EXAMPLES

1. We consider a vector field in the plane with a node-type singular point with resonance $\lambda_1 = 2\lambda_2$. The Poincaré–Dulac theorem enables us to reduce (formally) the equation to the normal form

$$\begin{cases} \dot{x}_1 = \lambda_1 x_1 + cx_2^2 \\ \dot{x}_2 = \lambda_2 x_2. \end{cases}$$

In this case, the normal form is polynomial, since the number of resonant terms is finite (just one).

2. We consider a vector field in the plane $\mathbb{R}^2$ with a singular point with purely imaginary eigenvalues $\lambda_{1,2} = \pm i\omega$ (center in linear approximation).

We pass to an eigenbasis. The eigenvectors can be chosen complex conjugate. It is customary to denote by z and $\bar{z}$ the coordinates in $\mathbb{C}^2$ with respect to a basis of complex conjugate vectors (these numbers are actually conjugate only in the real plane $\mathbb{R}^2 \subset \mathbb{C}^2$).

Our differential equation on $\mathbb{R}^2$ gives an equation on $\mathbb{C}^2$ which can be written in the form

$$\dot{z} = \lambda z + \cdots, \qquad \dot{\bar{z}} = \bar{\lambda}\bar{z} + \cdots.$$

(The dots denote a series in terms of the powers of z and $\bar{z}$.) Since the second equation is obtained from the first by conjugation, we may omit it.

We have the resonance $\lambda_1 + \lambda_2 = 0$. By the Poincaré–Dulac theorem, our equation can be reduced to the form

$$\dot{\zeta} = \lambda\zeta + c\zeta|\zeta|^2 + O(|\zeta|^5)$$

by a *real* change of variable (cf., Remark 4 in § 22E). Consequently, $r^2 = |\zeta|^2$ is a smooth real-valued function on $\mathbb{R}^2$. For this we have

$$(r^2)^{\cdot} = \dot{\zeta}\bar{\zeta} + \zeta\dot{\bar{\zeta}} = (2\,\mathrm{Re}\,c)r^4 + O(r^6).$$

If the real part of c is negative (positive), then the equilibrium position is stable (unstable).

Consequently, the first few steps of the Poincaré method provide a method for solving the problem of the stability of a singular point neutral in linear approximation. Moreover, it is immaterial whether the construction can be continued and whether the procedure converges. It is only important that the "nonlinear decrement" $\mathrm{Re}\,c$ be different from zero.

Remark. A generalization of Poincaré's theorem is a general theorem in the theory of Lie algebras, the so-called Cartan replicas theorem, which also generalizes the theorem on Jordan normal form.

We consider a finite-dimensional Lie algebra. Let u be an element of this algebra. Commutation with this element defines a linear operator $v \mapsto [u, v]$ in the space of the Lie algebra. The element u is said to be *semisimple* if the operator of commutation with u is diagonalizable (has an eigenbasis). The element u is said to be *nilpotent* if the operator of commutation with u is nilpotent (i.e., all eigenvalues of this operator are equal to zero).

The decomposition theorem asserts that every element of the algebra can be decomposed (in a unique way) into the sum of a semisimple element S and a nilpotent element N commuting with S:

$$u = S + N, \qquad SN = NS.$$

The elements S and N are called the replicas of u.

[In the theory of Jordan normal forms, S is an operator with diagonal matrix and N is the sum of nilpotent Jordan blocks.]

In the Lie algebra of jets of vector fields with a singular point 0, the semisimple fields are the fields which are linear and can be expressed by a diagonal matrix in an appropriate coordinate system. A nilpotent field consists of a nilpotent linear part and terms of higher order. The condition that S and N commute means exactly that in the nonlinear part of the field only resonant terms may be present in the indicated coordinate system.

The Poincaré–Dulac theorem can be derived from the general replicas theorem (which has to be applied to finite-dimensional Lie algebras of jets of vector fields at zero).

§ 24. Poincaré and Siegel Domains

In the study of the convergence of Poincaré's series constructed in the preceding section, two essentially distinct cases arise, depending on the distribution of eigenvalues in the complex plane.

A. Resonant Planes

We consider the n-dimensional complex space $\mathbb{C}^n = \{\lambda = (\lambda_1, \ldots, \lambda_n)\}$ of all possible n-tuples of eigenvalues.

Definition. A hyperplane in $\mathbb{C}^n$ given by an equation

$$\lambda_s = (m, \lambda), \qquad m_k \geqslant 0, \qquad \sum m_k \geqslant 2$$

with integral coefficients is called a *resonant plane*.

Varying the integral vector m and the index s, we obtain countably many resonant planes. We shall study how the set of resonant planes is located in the space $\mathbb{C}^n$ of eigenvalues. It turns out that the resonance planes lie discretely in one part of $\mathbb{C}^n$, but are everywhere dense in another part.

Definition. An n-tuple λ of eigenvalues belongs to the *Poincaré domain* if the convex hull of the n points $(\lambda_1, \ldots, \lambda_n)$ in the complex plane does not contain zero.

An n-tuple λ of eigenvalues belongs to the *Siegel domain* if zero lies inside the convex hull of the n points $(\lambda_1, \ldots, \lambda_n)$.

Remark. For $n > 2$, the Poincaré and Siegel domains are open and separated by a cone. For $n = 2$, the Siegel domain has real codimension 1 in $\mathbb{C}^2$.

B. Resonances in the Poincaré Domain

We assume that the n-tuple λ of eigenvalues belongs to the Poincaré domain.

Theorem 1. *Every point of the Poincaré domain satisfies not more than a finite number of resonance relations $\lambda_s = (m, \lambda)$, $|m| \geqslant 2$, $m_i \geqslant 0$, and has a neighborhood not intersecting the other resonant planes.*

In other words, the resonance planes lie discretely in the Poincaré domain. ◄ By definition, in the plane of complex numbers there exists a real straight line separating the collection of eigenvalues from zero. We consider the orthogonal projections of the eigenvalues onto the line normal to this straight line pointing away from zero. All these projections are not smaller than the distance of the separating line from zero.

On the other hand, the coefficients m_i of the resonance relation are nonnegative. Consequently, for sufficiently large m, the projection of (m, λ) onto the normal will be larger than the largest projection of the eigenvalues onto the normal to the separating line. ►

Theorem 2. *If the eigenvalues λ of the linear part of a field v at O lie in the Poincaré domain, then even in the case of resonance, the field can be reduced to polynomial normal form by a formal change of variables.*

◄ According to Theorem 1, the number of resonant terms is finite, so that Theorem 2 follows from Theorem 1 and the Poincaré–Dulac theorem. ►

Remark. In the Poincaré domain, resonance is possible only if one of the eigenvalues with nonnegative components can be expressed in terms of the remaining ones, *not including the eigenvalue itself*, i.e., if $\lambda_s = (m, \lambda)$, then $m_s = 0$. Indeed, if $m_s > 0$, then $0 = (m, \lambda) - \lambda_s$ has positive projection on the normal to the separating line.

C. Resonances in the Siegel Domain

Now we assume that the collection λ of eigenvalues lies in the Siegel domain.

Theorem 3. *The resonance planes are everywhere dense in the Siegel domain.*

◀ The point 0 lies either inside some triangle with vertices $(\lambda_1, \lambda_2, \lambda_3)$ or on the interval (λ_1, λ_2). In the first case, we consider the angle with vertex 0 formed by the linear combinations of the numbers λ_1 and λ_2 with real nonnegative coefficients.

The negative multiples of the number λ_3 lie in this angle. We divide the angle into parallelograms with vertices at the integral linear combinations of λ_1 and λ_2. Let d be the diameter of such a parallelogram. For any natural number N, the number $-N\lambda_3$ lies within one of our parallelograms. Consequently, it lies not farther than d from one of the vertices, so that

$$|N\lambda_3 + m_1\lambda_1 + m_2\lambda_2| \leqslant d.$$

This inequality implies that the distance of our point from the resonance plane $\lambda_3 = m_1\lambda_1 + m_2\lambda_2 + (N + 1)\lambda_3$ does not exceed d/N. Hence, the theorem is proved if zero lies in the triangle.

If 0 lies on the line segment between λ_1 and λ_2, arbitrarily large integers p_1 and p_2 exist such that $|p_1\lambda_1 + p_2\lambda_2| \leqslant d$.

This yields a resonance plane at a distance from λ smaller than d/p. ▶

Definition. A point $\lambda = (\lambda_1, \ldots, \lambda_n) \in \mathbb{C}^n$ is said to be of type (C, v) if for any s we have

$$|\lambda_s - (m, \lambda)| \geqslant \frac{C}{|m|^v}$$

for all vectors m with nonnegative integral components m_i, $\sum m_i = |m| \geqslant 2$.

Theorem 4. *The measures of the set of points which are not of type (C, v) for any $C > 0$ is equal to zero provided that $v > (n - 2)/2$.*

◀ We fix a ball in $\mathbb{C}^n$ and estimate the measure of the set of points in it which are not of type (C, v). The inequality in the definition determines a neighborhood of width not greater than $C_1 C/|m|^{v+1}$ of the resonance plane. Therefore, the measure of the part of this neighborhood which is contained in the ball does not exceed $C_2 C^2/|m|^{2v+2}$. Summing over m with fixed $|m|$, we obtain not more than $|m|^{n-1} C_3 C^2/|m|^{2v+2}$. Summing over $|m|$, we obtain $C_4(v)C^2 < \infty$ if $v > (n - 2)/2$. Consequently, the set of non-(C, v) points in the ball is covered by sets of arbitrarily small measure. ▶

In the real case, $v > n - 1$ is needed in Theorem 4.

D. Poincaré's and Siegel's Theorems

Now we assume that the vector field is given by a convergent rather than formal series, i.e., we consider a differential equation with a holomorphic right-hand side.

Poincaré's Theorem. *If the eigenvalues of the linear part of a holomorphic vector field at a singular point belong to the Poincaré domain and are non-resonant, then the field is biholomorphically equivalent to its linear part in the neighborhood of the singular point.*

In other words, Poincaré's series constructed in § 23 are convergent provided that the eigenvalues belong to the Poincaré domain.

Siegel's Theorem. *If the eigenvalues of the linear part of a holomorphic vector field at a singular point form a vector of type (C, v), then the field is biholomorphically equivalent to its linear part in the neighborhood of the singular point.*

In other words, Poincaré's series are convergent for almost all (in the sense of measure theory) linear parts of a field at a singular point.

Remark. All nonresonant vectors in the Poincaré domain are vectors of type (C, v) for some $C > 0$. On the other hand, in the Siegel domain, the sets of the following vectors are everywhere dense: vectors of type (C, v), resonant vectors, and nonresonant vectors which are not of type (C, v) for any C and v.

For n-tuples of eigenvalues of the last type, although incommensurable but very close to commensurability, Poincaré's series may be divergent, so that the field may be formally equivalent to its linear part but biholomorphically inequivalent to it.

The proofs of Poincaré's and Siegel's theorems can be obtained, with some simplifications, from the proofs of the analogous theorems for mappings in § 28.

E. The Poincaré–Dulac Theorem

Now we consider the case of resonant eigenvalues.

Theorem. *If the eigenvalues of the linear part of a holomorphic vector field at a singular point belong to the Poincaré domain, then in the neighborhood of the singular point, the field is biholomorphically equivalent to a polynomial field in which all vector-valued monomials with coefficients of degree greater than 1 are resonant.*

In other words, Poincaré's series are convergent if the eigenvalues lie in the Poincaré domain even in the case of resonance.

Remark. In contrast to that, if the eigenvalues lie in the Siegel domain, then the series leading to the formal normal forms are often divergent in the case of resonance. The first example of this kind was constructed by Euler [L. Euler, *De seriebus divergenti bus*, Opera omnia, Ser. 1, **14** (1924), Leipzig–Berlin, 247, 585–617; (cf. p. 601)].

In Euler's example,

$$\begin{cases} \dot{x} = x^2, \\ \dot{y} = y - x, \end{cases}$$

the origin is a saddle-node type singular point. Despite the analyticity of the right-hand side, the separatrix separating both sides of the half-plane $x < 0$ is not analytic, but only infinitely differentiable: $y = \sum (k - 1)!\, x^k$.

Many examples of the divergence of Poincaré's series have been constructed by Brjuno (A. D. Brjuno, *Analytic form of differential equations*, Trudy MMO **25** (1971), 119–262; in this work, the convergence of series is also proved in some cases going beyond the scope of the Siegel theorem).

F. The Real and Nonanalytic Cases

The Poincaré and Poincaré–Dulac theorems can be carried over to the real-analytic case and the case of infinitely differentiable vector fields, or even the case of fields with finite (sufficiently large) smoothness.

In the Siegel case, such a generalization is also possible [cf., for example, S. Sternberg. On the structure of local homeomorphisms of Euclidean n-space, Amer. J. Math. **80**, 3 (1958), 623–631, **81**, 3 (1959), 578–604].

However, one should mention that the situations in which these theorems are applicable are topologically trivial. Indeed, the Poincaré case for a real field can be encountered only if the eigenvalues all lie either in the left half-plane or the right half-plane. In this case (independently of resonances), the system is topologically equivalent to the standard system $\dot{x} = -x$ (or $\dot{x} = +x$) in the neighborhood of a stable point in a real space. All phase curves enter an asymptotically stable equilibrium position as $t \to \infty$ (or emanate from an equilibrium position close to $-\infty$).

In the situation of the Siegel theorem in a real domain, the Grobman–Hartman theorem is applicable (the system is topologically equivalent to a standard saddle). Indeed, if at least one of the nonzero eigenvalues of the linear part lies on the imaginary axis, then the complex conjugate eigenvalue lies on the imaginary axis, too; the pair $\lambda_{1,2} = \pm i\omega$ leads to the resonance $\lambda_1 + \lambda_2 = 0$. A vanishing eigenvalue is always resonant. Consequently, in the real case, Siegel's theorem is applicable only to systems without eigenvalues on the imaginary axis, and such systems are locally topologically equivalent to their linear part (the Grobman–Hartman theorem, § 13).

In contrast with Poincaré's and Siegel's theorems, Poincaré's method is applicable to the study of topologically complicated cases where there are eigenvalues on the imaginary axis, for the method can be applied to the normalization of a finite number of terms in the Taylor series. After this, one shows that the terms of higher order do not change the qualitative picture.

The simplest example of this kind has been analyzed in § 23C. This method is especially useful in bifurcation theory (cf., Chap. 6).

§ 25. Normal Form of a Mapping in the Neighborhood of a Fixed Point

The construction of an appropriate coordinate system for a mapping of a space into itself near a fixed point parallels the theory of normal forms of differential equations in the neighborhood of an equilibrium position. In this section, we discuss what form the basic aspects of the theory of normal forms take in this case.

A. Resonances: Poincaré and Siegel Domains

We consider a formal mapping $F: \mathbb{C}^n \to \mathbb{C}^n$ given by a formal power series $F(x) = Ax + \cdots$. Let $(\lambda_1, \ldots, \lambda_n)$ be the eigenvalues of the linear operator A. A *resonance* is, by definition, a relation

$$\lambda_s = \lambda^m, \qquad \text{where } \lambda^m = \lambda_1^{m_1} \cdots \lambda_n^{m_n}, m_k \geqslant 0, \textstyle\sum m_k \geqslant 2.$$

EXAMPLE

For $n = 1$, resonant eigenvalues are 0 and all roots of unity. All other numbers λ are nonresonant.

Definition. A collection of eigenvalues *belongs to the Poincaré domain* if the moduli of the eigenvalues are all smaller or all greater than 1.

Consequently, a mapping F with eigenvalues of its linear part belonging to the Poincaré domain is a contraction (if $|\lambda| < 1$) in the neighborhood of the origin or its inverse is contractive there (if $|\lambda| > 1$).

Definition. The complement of the Poincaré domain is the *Siegel domain*. For $n = 1$, the Siegel domain reduces to the unit circle $|\lambda| = 1$. In the space $\mathbb{C}^n$ of eigenvalues, the resonance equation $\lambda_s = \lambda^m$ determines a complex hypersurface. It is called *resonant surface*. In the Poincaré domain, resonant surfaces lie discretely. In the Siegel domain, both the resonant and non-resonant points are everywhere dense.

B. Formal Linearization

First of all, we consider the problem of the formal normal form of a mapping at a fixed point.

Theorem. *If the collection of eigenvalues of a mapping F at a fixed point is nonresonant, then the mapping $x \mapsto F(x)$ can be reduced to its linear part $x \mapsto Ax$ by a formal change of variables $x \mapsto \mathscr{H}(y) = y + \cdots:$*

$$F \cdot \mathscr{H} = \mathscr{H} \cdot A.$$

◀ Let $H(y) = y + h(y)$, where h is a homogeneous vector-valued polynomial of degree $r \geqslant 2$. Then

$$H \cdot A \cdot H^{-1}(x) = Ax + [h(Ax) - Ah(x)] + \cdots,$$

where the dots indicate terms of order greater than r. The expression in square brackets is a homogeneous vector-valued polynomial of degree r. This polynomial depends on h linearly. The linear operator

$$M_A : h(x) \mapsto [h(Ax) - Ah(x)]$$

on the space of homogeneous vector-valued polynomials has eigenvalues $\lambda^m - \lambda_s$ and eigenvectors $h(x) = x^m e_s$ (here, as usual, the vectors e_k form an eigenbasis for A, $x^m = x_1^{m_1} \cdots x_n^{m_n}$, and the x_k are coordinates with respect to the basis $\{e_k\}$. The eigenvalues of A are assumed to be distinct for the sake of simplicity.).

Consequently, we obtain the following homological equation in h:

$$M_A h = v.$$

To solve this equation we have to divide the coefficients of the decomposition of v by the numbers $\lambda^m - \lambda_s$. Hence, in our problem, the resonance condition takes the form $\lambda_s = \lambda^m$.

The rest of the proof does not differ from that given for the case of differential equations in § 22. ▶

C. Convergence Problems

Poincaré's and Siegel's theorems can be carried over to the case of discrete time in the following way.

Poincaré's Theorem. *If at a fixed point all eigenvalues of a holomorphic diffeomorphism are smaller (or larger) than 1 in modulus and there are no resonances, then in the neighborhood of the fixed point the mapping can be reduced to its linear part by means of a biholomorphic local diffeomorphism.*

Siegel's Theorem. *For almost all (in the sense of Lebesgue measure) collections of eigenvalues of the linear part of a holomorphic diffeomorphism at a fixed point, the diffeomorphism is biholomorphically equivalent to its linear part at the fixed point.*

Namely, for a diffeomorphism to be equivalent to its linear part, it is sufficient that the eigenvalues satisfy the inequalities

$$|\lambda_s - \lambda^m| \geqslant C|m|^{-\nu}$$

for all $s = 1, \ldots, n$, $|m| = \sum m_k \geqslant 2$, $m_k \geqslant 0$. The eigenvalues satisfying this inequality are called collections of multiplicative type (C, ν). The set of collections λ of eigenvalues which are not of multiplicative type (C, ν) for any C has measure zero if $\nu > (n - 1)/2$.

The proof of Poincaré's and Siegel's theorems can be carried out in almost the same way as for differential equations. Although Siegel's theorem has been known since about 1940, its proof apparently was not published before the late 1970's. We give the proof in § 28.

D. The Case of Resonance

To every resonance $\lambda_s = \lambda^m$ there corresponds the resonant vector-valued monomial $x^m e_s$ (where e_s is a vector in the eigenbasis, $x^m = x_1^{m_1} \cdots x_n^{m_n}$ and x_k are coordinates with respect to the eigenbasis).

Theorem of Poincaré–Dulac. *Any formal mapping $x \mapsto Ax + \cdots$, where the matrix of the operator A is diagonal, can be reduced to the normal form $y \mapsto Ay + w(y)$ by a formal change of variables $x = y + \cdots$, where the series w consists of resonant monomials only. If the eigenvalues of the linear part A are all smaller (or all larger) than 1 in modulus, then the holomorphic mapping $x \mapsto Ax + \cdots$ can be reduced to a polynomial normal form with some of the resonant terms by a biholomorphic substitution.*

In the case of resonance, Poincaré's method is usually used to reduce a finite number of terms of the Taylor series of a mapping at a fixed point to normal form.

EXAMPLE

We consider a mapping of $\mathbb{C}^1$ into itself with fixed point O and eigenvalue λ which is an nth root of 1. Such a mapping can be reduced to

$$x \mapsto \lambda x + cx^{n+1} + O(|x|^{2n+1})$$

by an appropriate choice of coordinates.

For example, if $\lambda = -1$, then the mapping can be reduced to

$$x \mapsto -x + cx^3 + O(|x|^5).$$

This formula enables us to study the stability of the fixed point of a real mapping. Indeed, the square of the mapping assumes the form

$$x \mapsto x - 2cx^3 + O(|x|^5).$$

Consequently, if $c > 0$, then the fixed point O of our mapping is stable.

Hence, the first few steps of the Poincaré method enable us to study the stability of a fixed point in a case when it is neutral in linear approximation.

§ 26. Normal Form of an Equation with Periodic Coefficients

One version of Poincaré's method of normal forms is the reduction of an equation with periodic coefficients to a simpler form.

A. Normal Form of a Linear Equation with Periodic Coefficients

We consider a linear equation

$$\dot{x} = A(t)x$$

in a complex phase space, where the complex linear operator $A(t): \mathbb{C}^n \to \mathbb{C}^n$ depends 2π-periodically on t.

The *monodromy operator* is, by definition, the linear operator $M: \mathbb{C}^n \to \mathbb{C}^n$ converting the initial condition for $t = 0$ into the value of the solution with this initial condition at $t = 2\pi$. (The monodromy mapping is defined not only for linear equations, but also for any equation with periodic coefficients; in this more general case, the monodromy mapping is usually called the Poincaré *return map*, or simply the *Poincaré map*.)

Floquet's Theorem. *If the monodromy operator is diagonal and* $\mu_s = e^{2\pi\lambda_s}$ *are its eigenvalues, then the original linear equation with periodic coefficients can be reduced by means of a linear 2π-periodic substitution $x = B(t)y$ to the equation with constant coefficients:*

$$\dot{y} = \Lambda y,$$

where Λ is a diagonal operator with eigenvalues λ_s.

◀ We consider the operator converting the initial condition of the original equation at $t = 0$ into the value of the solution with this initial condition at time t. Let us denote this operator by $g^t: \mathbb{C}^n \to \mathbb{C}^n$. We denote by $f^t: \mathbb{C}^n \to \mathbb{C}^n$ the analogous operator for the equation $\dot{y} = \Lambda y$. Then $g^0 = f^0 = E$, $g^{2\pi} = f^{2\pi} = M$, the monodromy operator (because of our choice of Λ). We set $B(t) = g^t(f^t)^{-1}$. The operator $B(t)$ defines the desired substitution. ▶

Remark. The proof of Floquet's theorem only uses the representation of the monodromy operator in the form $M = e^{2\pi\Lambda}$. Therefore, a periodic change of variables reduces to an equation with constant coefficients not only a complex equation with a diagonal monodromy operator, but also every equation for which the monodromy operator has a logarithm.

Every nondegenerate complex linear operator has a logarithm (which can easily be seen by writing the matrix of the operator in Jordan form).

Corollary 1. *Every complex linear equation with 2π-periodic coefficients can be reduced to an equation with constant coefficients by a 2π-periodic linear change of variables.*

A real linear operator does not always have a real logarithm, even if its determinant is positive (the determinant of the monodromy operator is always positive). Indeed, consider, for example, a linear operator on the plane with eigenvalues $(-1, -2)$. If this operator is the exponent of another linear operator, then the eigenvalues of the latter operator are complex but not complex conjugate. Therefore, our operator on the real plane does not have a real logarithm.

On the other hand, it is easy to see that the square of a real operator always has a real logarithm. This leads to the following.

Corollary 2. *Every real linear equation with 2π-periodic coefficients can be reduced to an equation with constant coefficients by a 4π-periodic linear change of variables.*

It is usually more convenient to use complex reducibility than real reducibility with doubled period.

B. Deduction of the Homological Equation

We consider a linear equation $\dot{y} = \Lambda y$ with constant coefficients. In this equation we perform a nonlinear change of coordinates

$$x = y + h(y, t)$$

where h is a vector-valued function (or a formal power series in y) with 2π-periodic coefficients.

Lemma. *If $h = O(y^r)$ (or the series h begins with terms of degree not smaller than r) and $r \geqslant 2$, then*

$$\dot{x} = \Lambda x + \left[\frac{\partial h}{\partial x} \Lambda x - \Lambda h + \frac{\partial h}{\partial t} \right] + \cdots,$$

where the dots indicate terms of order greater than r with respect to x.

◀ $\dot{x} = (E + h_y)\Lambda y + h_t = (E + h_y)\Lambda(x - h(x, t)) + h_t + \cdots$
  $= \Lambda x + [h_x\Lambda x - \Lambda h(x, t) + h_t] + \cdots.$ ▶

Definition. The *homological equation* associated with an equation $\dot{y} = \Lambda y$ with 2π-periodic coefficients is the equation

$$L_\Lambda h + h_t = v$$

for a vector field h, 2π-periodic in t, where v is a given 2π-periodic vector field and

$$(L_\Lambda h)(x, t) = \frac{\partial h}{\partial x} \Lambda x - \Lambda h(x, t).$$

We shall also consider the case where h and v are formal series with coefficients 2π-periodic in t.

C. Solution of the Homological Equation

First let v and h be a Taylor–Fourier series:

$$v(x, t) = \sum v_{m, k, s} x^m e^{ikt} e_s, \qquad h = \sum h_{m, k, s} x^m e^{ikt} e_s.$$

The formal solution of the homological equation can be expressed by the formula

$$h_{m,k,s} = \frac{v_{m,k,s}}{ik + (m, \lambda) - \lambda_s},$$

where the λ_j are the eigenvalues of Λ.

The resonance condition is as follows:

$$\lambda_s = (m, \lambda) + ik,$$

$$m_j \geqslant 0, \quad \sum m_j \geqslant 2, \quad -\infty < k < +\infty, \quad 1 \leqslant s \leqslant n.$$

If for given m and s there is no resonance, then the Fourier series $\sum h_{m, k, s} e^{ikt}$ and its t derivative are convergent. Therefore, in the absence of resonances, the homological equation is solvable in the class of homogeneous polynomials with coefficients 2π-periodic in t and, thus, in the class of formal power series with coefficients 2π-periodic in t.

On the other hand, if resonance takes place, then the homological equation is formally solvable if the Taylor–Fourier series of v does not contain resonant terms, i.e., if the coefficients $v_{m, k, s}$ vanish for those terms for which the resonance condition $\lambda_s = ik + (m, \lambda)$ is satisfied.

D. Formal Normal Form

Following the usual method, in the nonresonant case, we can reduce an equation with 2π-periodic formal coefficients to a linear equation $\dot{y} = \Lambda y$ with constant coefficients by a change of variable given by a formal series in y with coefficients 2π-periodic in t.

In the case of resonance, we reduce the equation to the form

$$\dot{y} = \Lambda y + w(y, t),$$

where w is a formal power series in y with coefficients 2π-periodic in t and consisting of only some resonant terms. (We note that the resonant terms of any fixed order in y contain only a finite number of Fourier harmonics, since the resonance condition $\lambda_s = (m, \lambda) + ik$ determines k uniquely.) Usually one normalizes only lower-order terms.

EXAMPLE

We consider an equation with 2π-periodic coefficients. We assume that the dimension n of the phase space equals 2 and that both eigenvalues of the monodromy operator are complex and equal to 1 in modulus.

The linearized complexified equation takes the form

$$\dot{z} = i\omega z$$

in an appropriate coordinate system. (As usual, the equation for $\bar{z}$ is omitted, since it is conjugate to the one above.) The eigenvalues are $\lambda_{1,2} = \pm i\omega$. The resonant terms in the equation for $\dot{z}$ are determined from the condition

$$ik + (m_1 - m_2 - 1)i\omega = 0.$$

If the real number ω is irrational, then $k = 0$ and $m_1 = m_2 + 1$. Consequently, the equation can be reduced to the following time-independent formal normal form:

$$\dot{z} = i\omega z + c_1 z|z|^2 + c_2 z|z|^4 + \cdots.$$

By a true (not formal) change of variable, the equation can be reduced, for example, to the form

$$\dot{z} = i\omega z + c_1 z|z|^2 + \cdots,$$

where the (2π-periodic) dependence on t has been preserved only in terms of order of smallness 5 in z (denoted by dots).

We note that in this case every step of the Poincaré method reduces to averaging with respect to t and $\arg z$ and that the equation thus obtained is invariant under translations of t and rotations of z.

E. The Case of Commensurability

Let us now assume that in the preceding example that ω is rational, $\omega = p/q$. In this case, we obtain from the equation for the resonant terms

$$k = pr, \qquad m_1 = m_2 + 1 - qr.$$

For the study of the normal form, it is convenient to consider a q-sheeted covering on the time axis. We note that the integral curves of the linear part of our equation form a Seifert foliation of type (p, q) (cf., § 21). On the space of the q-sheeted covering, the integral curves form a trivial foliation and we can introduce direct product coordinates. The coordinate on a fiber will be denoted by $t(\text{mod } 2\pi q)$. The coordinate ζ on the base is determined from the condition

$$z = e^{i\omega t}\zeta.$$

In this notation, the linear part of our equation takes the form $\dot{\zeta} = 0$. The normal form is a formal series

$$\dot{\zeta} = \sum w_{k,l}\zeta^k\bar{\zeta}^l$$

independent of t, where $k - l \equiv 1 \bmod q$.

In other words, on the base of the q-sheeted covering, a (formal) equation is obtained which is invariant under rotations by angle $2\pi/q$.

If instead of a complete formal reduction we restrict ourselves to normalization of the first few terms of the series, then we obtain the following equation for ζ with a remainder of order $q + 1$, $2\pi q$-periodic in time:

$$\dot{\zeta} = \zeta a(|\zeta|^2) + b\bar{\zeta}^{q-1} + \cdots.$$

In this case, every step of the Poincaré method reduces to averaging along Seifert's foliation. Therefore, the equation thus obtained is invariant under translations of t and rotations of ζ by angles which are multiples of $2\pi/q$.

A study of the equations thus obtained will be carried out in Chap. 6.

F. Discussion of the Convergence

The Poincaré domain for an equation $\dot{x} = \Lambda x + \cdots$ with periodic coefficients is defined by the following condition: all eigenvalues of the linearized equation lie in the left half-plane $\operatorname{Re}\lambda < 0$ (or all of them lie in the right half-plane).

In this domain: (1) the resonance planes $\{\lambda: \lambda_s = (m, \lambda) + ik\}$ lie discretely; (2) in the case of resonance, the normal form contains only a finite number of terms; (3) Poincaré's series are convergent.

The complement of the Poincaré domain makes up the Siegel domain. In the Siegel domain: (1) the resonance planes form an everywhere dense set; (2) normal forms may contain infinitely many terms; (3) Poincaré's series may be divergent.

Nevertheless, *for almost all* (*in the sense of Lebesgue measure*) *collections of eigenvalues* λ *of the operator* Λ, *in the neighborhood of the null solution, a holomorphic differential equation* $\dot{x} = \Lambda x + \cdots$ 2π-*periodic in* t *can be reduced to the autonomous normal form* $\dot{x} = \Lambda x$ *by a biholomorphic transformation* 2π-*periodic in* t (Siegel's theorem for the case of periodic coefficients).

◀ The proof is as usual, cf., § 28. ▶

G. The Neighborhood of a Closed Phase Curve

We consider an autonomous differential equation $\dot{x} = v(x)$ having a periodic solution and, consequently, a closed phase curve. All of what has been said about the neighborhood of the zero solution of an equation with periodic coefficients immediately carries over to this case.

Indeed, in the neighborhood of a closed phase curve, coordinates may be chosen so that the direction field given by the vector field v will be the direction field of an equation with periodic coefficients, and the dimension of the phase space will be reduced by 1 (the coordinate varying along the phase curve will be called time).

Remark. If the phase space is a manifold, then the neighborhood of a closed phase curve may turn out to be nondiffeomorphic to the direct product of the circle with a transversal disk.

EXAMPLE

The phase space is the Möbius band and the phase curve is the circular axis.

In general, the neighborhood of a circle on a manifold is not a direct product if and only if the manifold is not orientable and the circle is a disorienting path. In this case, one has to resort to a two-sheeted covering of the original circle for the passage to an equation with periodic coefficients.

H. Connection with Poincaré Mappings

The theory of normal forms of equations with periodic coefficients could have been deduced from the theory of normal forms of their Poincaré mappings, i.e., normal forms of diffeomorphisms in the neighborhood of a fixed point. Conversely, the study of a diffeomorphism in the neighborhood of a fixed point can be reduced to the study of an equation with periodic coefficients for which the diffeomorphism is the Poincaré mapping.

In the case of finite or even infinite real smoothness, the construction of a differential equation with periodic coefficients* from a given Poincaré mapping does not present

---

*It is usually not possible to imbed a given mapping in the phase flow of an autonomous equation (an example is a diffeomorphism of the circle with irrational rotation number, which is not equivalent diffeomorphically to a rotation, cf., § 11).

major difficulties. In the analytic or holomorphic case, the situation is more complicated. This problem is equivalent to the problem of analytic (holomorphic) triviality of analytic (holomorphic) foliations over a circular ring under the assumption of topological triviality. Although a positive answer follows essentially from the theory of sheaves and Stein manifolds, as far as this author knows, no proof has been published. (The author is indebted to V. P. Palamodov and Ju. S. Il'jašenko for an explanation of this matter.) We shall not go into this theory, especially because the results necessary for the study of differential equations and diffeomorphisms need not be deduced from each other, but can be obtained independently, using the same method of proof.

I. The Case of Quasiperiodic Coefficients

Poincaré's method admits a generalization to the case of quasiperiodic coefficients. Let us consider the equation

$$\dot{x} = \Lambda x + v(x, \varphi),$$

$$\dot{\varphi} = \omega,$$

where φ is a point of the r-dimensional torus, ω is a constant vector, Λ: $\mathbb{C}^n \to \mathbb{C}^n$ is a linear operator (independent of φ), and v is a vector field whose linear part vanishes at $x = 0$.

We impose the usual conditions of normal incommensurability on the components of the frequency vector ω. In this situation, the resonance conditions take the form

$$\lambda_s = i(k, \omega) + (m, \lambda),$$

where k runs through the lattice of integral points of the r-dimensional space, and m satisfies the usual conditions $m_p \geqslant 0$, $\sum m_p \geqslant 2$.

Let v be analytic (holomorphic) in x and φ of period 2π in $\varphi = (\varphi_1, \ldots, \varphi_r)$. Then one can prove the reducibility of the system to the form

$$\dot{y} = \Lambda y, \qquad \dot{\varphi} = \omega$$

by an analytic (holomorphic) substitution $x = y + h(y, \varphi)$, 2π-periodic in φ [E. G. Belaga, *On the reducibility of a system of differential equations in the neighborhood of a quasiperiodic motion*, Sov. Math. Dokl. **143**, 2 (1962), 255–258.]

This theory is unsatisfactory because of the imperfection of the theory of linear equations with quasiperiodic coefficients. Although for equations with periodic coefficients the constancy of the linear part can be achieved by an appropriate periodic linear transformation of the coordinates, for equations with quasiperiodic coefficients, the assumption on the independence of Λ of φ is an essential restriction.

§ 27. Normal Form of the Neighborhood of an Elliptic Curve

Poincaré's theory of normal forms of differential equations in the neighborhood of a
singular point has its analogue in the theory of normal forms of neighborhoods of
elliptic curves on complex surfaces. In this section, we consider this theory briefly. This
theory is an application of methods of the theory of differential equations to analytic
geometry and itself has applications in the theory of differential equations.

A. Elliptic Curves

An *elliptic curve* is, by definition, a one-dimensional complex manifold homeomorphic
to the torus.

EXAMPLE

We consider the plane $\mathbb{C}$ of a complex variable and two complex numbers (ω_1, ω_2)
whose ratio is not real. We identify every point φ of $\mathbb{C}$ with any point obtained from it
by translation by ω_1 and ω_2 (and consequently, with all points $\varphi + k_1\omega_1 + k_2\omega_2$,
where k_1 and k_2 are integers). After this identification, $\mathbb{C}$ becomes the elliptic curve

$$\Gamma = \frac{\mathbb{C}}{\omega_1\mathbb{Z} + \omega_2\mathbb{Z}}.$$

Consequently, the elliptic curve Γ can be viewed as a parallelogram with sides (ω_1, ω_2),
in which corresponding points are identified on the opposite sides.

It can be proved that all elliptic curves can be obtained by the construction above
(up to biholomorphic equivalence). This fact is by no means an obvious theorem.

For example, consider the strip $0 \leqslant \operatorname{Im} \varphi \leqslant \tau$ and glue all points φ, $\varphi + 2\pi$ and
also points of the boundaries of the strip, identifying a point φ with $\varphi + i\tau + \sigma +
0.5 \sin \varphi$ for real numbers φ. The manifold thus obtained can be mapped biholo-
morphically onto the quotient manifold $\mathbb{C}/\omega_1\mathbb{Z} + \omega_2\mathbb{Z}$; nevertheless, it is not easy
to prove this. Probably, under usual Diophantine conditions, ω_1/ω_2 converges to the
rotation number as $\tau \to 0$.

The numbers ω_1 and ω_2 are called the *periods* of the curve. If we multiply both
periods by the same complex number, we obtain new periods which give an elliptic
curve biholomorphically equivalent to the initial one. Therefore, the periods can always
be chosen so that $\omega_1 = 2\pi$.

In this case, we can denote the second period by ω. We may always assume that
$\operatorname{Im} \omega > 0$. To different ω there correspond generally biholomorphically inequivalent
elliptic curves (more precisely, the curves are inequivalent biholomorphically if the
corresponding lattices $\omega_1\mathbb{Z} + \omega_2\mathbb{Z}$ do not turn into each other upon multiplication
by a complex number).

Problem*. Prove that the phase curves of the one-dimensional Newton equation with

---

*The solution of this and the following problems is based on elementary knowledge of the
topology of Riemann surfaces contained in any course on the theory of functions of a complex
variable.

potential energy of degree 3 or 4 are elliptic curves (if they are considered in a complex domain).

Hint. The role of the coordinate φ in the plane covering the elliptic curve is played by the time t of motion on the phase curve, the time being defined by the relation $dt = dx/y$ (the time is also called an elliptic integral of the *first kind*).

Problem. Let the potential energy be a polynomial of degree 4 with two minima. Prove that the periods of (not necessarily small) oscillations with the same total energy coincide in the two wells.

Hint. The integrals of the first kind on any two meridians of the torus coincide.

Problem. Let the potential energy be a polynomial of degree 3 with a local maximum and minimum. Prove that the period of oscillations in the well is equal to the period of the motion from infinity to infinity on a noncompact phase curve with the same total energy.

Remark. Choosing potential energy of degree 3 or 4, we may obtain any elliptic curve. It follows from results of the preceding problems that an elliptic curve is an algebraic manifold.

B. Simplest Fiber Bundles over an Elliptic Curve

The simplest surface containing an elliptic curve is the direct product of an elliptic curve with the complex line. Besides the direct product of a circle with the line, there is a nontrivial bundle over the circle with fiber equal to the line (the Möbius band); similarly, besides the direct product, there are other fiber bundles over an elliptic curve with fiber $\mathbb{C}$.

We consider a fibering of the plane of two complex variables into complex lines. We shall call the fibers of this bundle *vertical* lines.

We shall denote by (r, φ) the coordinates in $\mathbb{C}^2$, where r is the vertical and φ the horizontal coordinate. Our fibering $\mathbb{C}^2 \to \mathbb{C}$ assigns the point φ of the horizontal complex line to the point (r, φ).

Let Γ be an elliptic curve covered by the horizontal line. The curve Γ is obtained from the horizontal φ-axis by identifying points differing by integral multiples of the periods (ω_1, ω_2).

In the plane $\mathbb{C}^2$ we identify those vertical lines whose projections on the horizontal line differ by integral multiples of the periods. This identification turns $\mathbb{C}^2$ into a fiber bundle over the elliptic curve Γ. The identification of vertical lines itself can be carried out in different ways. (Similarly, upon pasting fibers over a circle, out of a rectangle we may obtain either a cylinder or a Möbius band, depending on how the vertical lines are glued together.)

The simplest method of pasting is to identify (r, φ) with the points $(r, \varphi + \omega_1)$ and $(r, \varphi + \omega_2)$. Then we obtain a direct product. The next simplest method of pasting includes a twisting of the vertical lines being glued together.

EXAMPLE

Let λ be a complex number different from zero and let Γ be the elliptic curve with periods $(2\pi, \omega)$. In the plane $\mathbb{C}^2$ we identify the points with coordinates

$$(r, \varphi), (r, \varphi + 2\pi), (\lambda r, \varphi + \omega).$$

After this identification, $\mathbb{C}^2$ turns into a smooth complex surface Σ and the fibering $\mathbb{C}^2 \to \mathbb{C}$, $(r, \varphi) \mapsto \varphi$ turns into a fiber bundle $\Sigma \to \Gamma$ whose base is the elliptic curve Γ and whose fiber is $\mathbb{C}$. The equation $r = 0$ yields an imbedding of Γ in Σ.

The surface Σ can be imagined in the following way (in the case of a real λ). We consider the real three-dimensional space foliated by vertical lines with horizontal plane $\{\varphi \in \mathbb{C}\}$. Consider the strip $0 \leqslant \operatorname{Im} \varphi \leqslant \operatorname{Im} \omega$. We glue the vertical planes which are the boundaries of this strip, by identifying the point (r, φ) in the vertical plane $\operatorname{Im} \varphi = 0$ (r is the vertical coordinate) with the point $(\lambda r, \varphi + \omega)$ in the plane $\operatorname{Im} \varphi = \operatorname{Im} \omega$. Moreover, we glue the points differring by 2π in the coordinate φ. We obtain a fibering over an elliptic curve whose fiber is a line.

To imagine the complex surface Σ, we only have to replace real vertical lines by complex ones.

Topologically, the fiber bundle thus constructed is a direct product. Nevertheless, from the holomorphic point of view, this bundle is generally nontrivial.

C. Trivial and Nontrivial Fiberings

Theorem. Let $\lambda \neq e^{ik\omega}$, $k \in \mathbb{Z}$. No neighborhood of the elliptic curve Γ on the surface Σ described above can be mapped biholomorphically onto a neighborhood of Γ in the direct product.

◀ In the direct product, Γ can be deformed: for any ε the equation $r = \varepsilon$ defines an elliptic curve in the direct product. Let Γ_1 be an elliptic curve in the total space Σ near Γ, which is the zeroth section of the fiber bundle (the equation of Γ has the form $r = 0$). The curve Γ_1 is given by an equation $r = f(\varphi)$, where $f(\varphi + 2\pi) \equiv f(\varphi), f(\varphi + \omega) \equiv \lambda f(\varphi)$. Expanding f in a Fourier series, $f = \sum f_k e^{ik\varphi}$, we find that $f_k e^{ik\omega} = \lambda f_k$. Consequently, $f_k = 0$ and Γ_1 coincides with Γ. Hence, our elliptic curve is not deformable in a fiber bundle with $\lambda \neq e^{ik\omega}$. ▶

Problem. Prove that for $\lambda = e^{ik\omega}$ the fiber bundle with the projection $\Sigma \to \Gamma$ is a direct product (trivial holomorphically).

Problem. Prove that the fiber bundles $\Sigma_1 \to \Gamma$, $\Sigma_2 \to \Gamma$ given by the complex numbers λ_1, λ_2 are equivalent biholomorphically if and only if $\lambda_1 = \lambda_2 e^{ik\omega}$ for some integer k.

Remark. The classes of biholomorphically equivalent fiber bundles of this form (called one-dimensional bundles of degree 0) over a fixed elliptic curve Γ constitute a group (where multiplication is multiplication of the numbers λ).

It follows from the results of the preceding problems that this group can be identified naturally with the quotient group of the multiplicative group of complex numbers modulo the subgroup of numbers of the form $e^{ik\omega}$. The quotient group $\mathbb{C}^*/\{e^{ik\omega}\}$ itself can be mapped biholomorphically onto the original elliptic curve. This group is also called the *Picard group* or the *Jacobi manifold* of Γ. (These notions are defined for not only elliptic curves, but also for arbitrary algebraic manifolds and, in the general situation, they do not coincide with the original manifold.)

Problem. Consider the fiber bundle given by the identifications

$$(r, \varphi) \sim (\lambda_1 r, \varphi + \omega_1) \sim (\lambda_2 r, \varphi + \omega_2)$$

over an elliptic curve.

Prove that this fiber bundle is biholomorphically equivalent to the fiber bundle with $\omega_1 = 2\pi, \lambda_1 = 1$.

Remark. It can be proved that all topologically trivial one-dimensional vector bundles over an elliptic curve are biholomorphically equivalent to the bundles $\Sigma \to \Gamma$ described above.

D. Topologically Nontrivial Fiber Bundles over an Elliptic Curve

All fibrations described above are topologically trivial (homeomorphic to the direct product). The *index of self-intersection* of the zeroth section is an invariant enabling us to distinguish topologically inequivalent bundles.

Let M_1, M_2 be smooth oriented compact submanifolds of an oriented smooth real manifold M (considering manifolds without boundary). We assume that the dimension of M equals the sum of the dimensions of M_1 and M_2. We also assume that M_1 and M_2 intersect transversally (i.e., at every point of intersection, the sum of the tangent spaces of the two submanifolds equals the tangent space of M).

The *index of intersection* of M_1 and M_2 in M is, by definition, the number of points of intersection counting orientation. (A point of intersection is counted with positive sign if the frame orienting M_1 positively followed by the frame orienting M_2 positively determines a frame orienting M positively.)

Let M_1 be an oriented compact smooth submanifold of dimension half that of M. The index of *self-intersection* of M_1 in M is defined as the index of intersection of M_1 with a manifold M_2 obtained from M_1 by a small deformation and intersecting M_1 transversally. For example, the index of self-intersection of the meridian of a torus is equal to zero, since the neighboring meridians do not intersect.

It can be proved that the index of self-intersection of M_1 in M does not depend on the choice of M_2 if it is obtained from M_1 by a small deformation.

Problem. Find the index of self-intersection of the sphere S^2 in the space of its tangent bundle.

Answer. $+2$. In general, the index of self-intersection of a manifold with its tangent bundle is equal to the Euler characteristic of the manifold.

Now, over an elliptic curve, we consider the one-dimensional vector bundle $\Sigma \to \Gamma$ obtained from the fibering $\mathbb{C}^2 \to \mathbb{C}$ by pasting vertical lines (the vertical coordinate is denoted by r) according to the rule

$$(r, \varphi) \sim (r, \varphi + 2\pi) \sim (\lambda e^{ip\varphi} r, \varphi + \omega),$$

where p is an integer and λ is a complex number different from zero.

Problem. Determine the index of self-intersection of the zeroth section ($r = 0$) in the fiber space thus obtained, assuming that Σ is oriented as a complex manifold. (The orientations of complex manifolds are defined so that the index of intersection of complex planes is always positive: a space with complex coordinates $(z_1, \cdots, z_n)$ is oriented by the coordinates $(\mathrm{Re}\, z_1, \mathrm{Im}\, z_1, \ldots, \ldots, \mathrm{Re}\, z_n, \mathrm{Im}\, z_n)$.

Answer. $-p$ if $\mathrm{Im}\, \omega > 0$.

Remark. All one-dimensional vector bundles over an elliptic curve are exhausted by the bundles above up to biholomorphic equivalence.

E. The Neighborhood of an Elliptic Curve on a Complex Surface

We consider an elliptic curve on a complex surface. The neighborhood of the curve on the surface determines a one-dimensional vector bundle over the curve: the so-called *normal bundle*. The fiber of the normal bundle at a point on the curve is the quotient space of the tangent space of the surface at this point modulo the subspace tangent to the curve.

The space of the normal bundle itself is a complex surface. The initial elliptic curve is imbedded in this surface (as the zeroth section of the bundle).

The problem arises whether a sufficiently small neighborhood of the curve on the original surface can be mapped biholomorphically onto a neighborhood of it in the normal bundle. It turns out that this problem is very close to that of the reducibility of a differential equation (or a smooth mapping) to linear normal form in the neighborhood of a fixed point, and it can be solved by the same methods.

First, we show that a neighborhood of an elliptic curve on a surface may be impossible to fiber holomorphically over the curve.

EXAMPLE

We consider a family of elliptic curves such that neighboring curves of the family are not equivalent to each other biholomorphically. Such a family may be obtained, for example, by identifying the points (φ, ω), $(\varphi + 2\pi, \omega)$ $(\varphi + \omega, \omega)$ in the plane of the two complex variables (φ, ω). Upon identification, the domain $\mathrm{Im}\, \omega > 0$ turns into the union of the elliptic curves $\omega = \mathrm{const}$. No neighborhood of any of these curves can be mapped holomorphically onto this curve so that the curve itself remains fixed.

Indeed, if such a mapping existed, we would obtain a biholomorphic mapping close to the identity mapping of elliptic curves with close distinct ω onto each other, which is impossible.

It turns out that in a certain sense this example is exceptional: the neighborhood of an elliptic curve imbedded in a complex surface with vanishing index of self-intersection is, generically, biholomorphically equivalent to a neighborhood of the curve in the normal bundle (in the same sense as a differential equation is generically equivalent to a linear equation in the neighborhood of a singular point). The exceptional character of the example above is connected with the fact that the normal bundle of each of the elliptic curves in the family constructed above is trivial. (It is a direct product.)

F. Preliminary Normal Form

An elliptic curve can be obtained from an annulus by a holomorphic pasting of the bounding circles. In exactly the same way, the neighborhood of an elliptic curve on a surface can be obtained from a neighborhood of an annulus on the surface by pasting the bounding manifolds holomorphically. These bounding manifolds have real dimension 3; the pasting is continued holomorphically to the neighborhood of the boundary.

It turns out that a sufficiently small neighborhood of a biholomorphic image of a closed annulus on a complex surface can always be mapped biholomorphically onto a neighborhood of an annulus imbedded in the complex line $\mathbb{C}$ in the direct product $\mathbb{C} \times \mathbb{C}$.

Like the results on the holomorphic classification of the one-dimensional vector bundles over an elliptic curve, our result on the neighborhood of an annulus cannot be proved easily: one needs some techniques from the theory of functions of several complex variables (sheaves, elliptic partial differential equations, or something substituting for them).

We shall not prove this; instead, we assume directly that our surface containing the elliptic curve can be obtained by pasting from a neighborhood of an annulus by pasting it in the direct product.

Consequently, we consider a surface whose points are obtained from the points (r, φ) of the plane of two complex variables by the pastings

$$\begin{pmatrix} r \\ \varphi \end{pmatrix} \sim \begin{pmatrix} r \\ \varphi + 2\pi \end{pmatrix} \sim \begin{pmatrix} rA(r, \varphi) \\ \varphi + \omega + rB(r, \varphi) \end{pmatrix},$$

where the functions A and B are 2π-periodic in φ and holomorphic in the neighborhood of the real φ-axis.

Here, one obtains, the annulus from the strip $0 \leqslant \operatorname{Im} \varphi \leqslant \operatorname{Im} \omega$ in the complex φ-axis by the pasting $(0, \varphi) \sim (0, \varphi + 2\pi)$; r and φ are coordinates in the direct product.

The pair (A, B) of functions yielding the pasting determines a neighborhood. The form of the functions A and B can be altered by an appropriate choice of the coordinates (r, φ). We attempt to choose these coordinates so that the functions A and B become the simplest possible.

First, we consider the linear change of coordinates $r_{\text{new}} = C(\varphi)r$, where the function C is holomorphic in the strip $0 \leqslant \operatorname{Im} \varphi \leqslant \operatorname{Im} \omega$ of the φ-axis, has period 2π, and does not vanish in the strip.

Theorem. *The function C defining the linear change of the vertical coordinate can be chosen so that in the identification relation in the new coordinates the function $A(0, \varphi)$ takes the form $\lambda e^{ip\varphi}$ (where p is the integer equal to the negative of the index of self-intersection of the elliptic curve $r = 0$ in the surface under consideration).*

◀ The function $A(0, \varphi)$ defines the normal bundle of the curve $r = 0$ on our surface. This bundle is biholomorphically equivalent to the bundle obtained by the pasting

$$(r, \varphi) \sim (r, \varphi + 2\pi) \sim (\lambda e^{ip\varphi r}, \varphi + \omega)$$

(cf., the Remark in § 27E). The linear change of the variable r reducing the pasting of the normal bundle to this canonical form leads to the form of $A(0, \varphi)$ given above. ▶

Definition. The *preliminary normal form* of the neighborhood of an elliptic curve on a surface where the curve has vanishing index of self-intersection is the pasting

$$\begin{pmatrix} r \\ \varphi \end{pmatrix} \sim \begin{pmatrix} r\lambda(1 + ra(r, \varphi)) \\ \varphi + \omega + rb(r, \varphi) \end{pmatrix} \sim \begin{pmatrix} r \\ \varphi + 2\pi \end{pmatrix},$$

where a and b are functions 2π-periodic in φ and holomorphic in the neighborhood of the real φ-axis. λ is a nonzero complex number.

In the following, we shall not always indicate identification of points differing only by 2π in the coordinate φ, because the functions we encounter are 2π-periodic and φ can be assumed to belong to the cylinder $\mathbb{C}$ mod 2π.

G. Formal Normal Form

Definition. A pair of numbers (λ, ω) is said to be resonant if $\lambda^n = e^{ik\omega}$ for some integers n and k not vanishing simultaneously.

Theorem. *The resonant pairs form an everywhere dense set in the space of all pairs of complex numbers.*

◀ This follows from the fact that the set of numbers of the form $i((k/n)\omega + (m/n)2\pi)$ (where k and m are integers and n is a natural number) is everywhere dense in the complex plane. ▶

Theorem. *A pair (λ, ω) is resonant if and only if the bundle corresponding to the pasting $(r, \varphi) \sim (r, \varphi + 2\pi) \sim (\lambda r, \varphi + \omega)$ is trivial over some cyclic n-sheeted covering of the original elliptic curve.*

◀ If $\lambda^n = e^{ik\varphi}$, then $(r, \varphi) \sim (e^{ik\omega}r, \varphi + n\omega)$ and, consequently, the bundle over $\mathbb{C}/2\pi\mathbb{Z} + n\omega\mathbb{Z}$ is trivial (cf., § 27C). The converse can be proved analogously. ▶

Definition. A *formal pasting* is a "mapping"

$$f\begin{pmatrix} r \\ \varphi \end{pmatrix} = \begin{pmatrix} rA(r, \varphi) \\ \varphi + \omega + rB(r, \varphi) \end{pmatrix},$$

where A and B are formal power series in r with coefficients analytic on the real φ-axis and 2π-periodic in φ and $A(0, \varphi) \neq 0$.
    A *formal change* of variables is a "mapping"

$$g\begin{pmatrix} r \\ \varphi \end{pmatrix} = \begin{pmatrix} rC(r, \varphi) \\ \varphi + rD(r, \varphi) \end{pmatrix},$$

where C and D are formal power series in r with coefficients analytic in the strip $0 \leqslant \operatorname{Im} \varphi \leqslant \operatorname{Im} \omega$ of the complex φ-axis and 2π-periodic in φ and $C(0, \varphi) \neq 0$.

A formal change of variables g acts on a formal pasting f according to the formula $f \mapsto g \circ f \circ g^{-1}$. (The right-hand side is defined by a natural substitution of power series and is itself a formal pasting.)

Theorem. *If a pair (λ, ω) is nonresonant, then every formal pasting*

$$\binom{r}{\varphi} \mapsto \binom{r\lambda(1 + ra(r, \varphi))}{\varphi + \omega + rb(r, \varphi)}$$

can be reduced to the linear normal form

$$\binom{r}{\varphi} \mapsto \binom{\lambda r}{\varphi + \omega}$$

by a formal change of variables.

◀ We shall successively annihilate the terms of degree $1, 2, \ldots$ with respect to r in ra and rb. For this, as usual, we need to solve a linear homological equation. We write the equation for the normalization of terms of degree n.

Lemma. *Consider the equation*

$$\lambda^n u(\varphi + \omega) - u(\varphi) = v(\varphi)$$

in u, where v is a 2π-periodic function analytic in the strip $\alpha \leqslant \operatorname{Im} \varphi \leqslant \beta$. If $\tau = \operatorname{Im} \omega > 0, \lambda \neq 0$, and $\lambda^n \neq e^{ik\omega}$ for any integer k, then the equation has a 2π-periodic solution u analytic in the strip $\alpha \leqslant \operatorname{Im} \varphi \leqslant \beta + \tau$.

◀ Let

$$u(\varphi) = \sum u_k e^{ik\varphi}, \qquad v(\varphi) = \sum v_k e^{ik\varphi}.$$

Then $u_k = v_k / (\lambda^n e^{ik\omega} - 1)$.

As $k \to +\infty$, $|v_k|$ can be estimated from above by a quantity on the order of $e^{k(\alpha - \varepsilon)}$ and $e^{ik\omega}$ converges to zero. Therefore, $|u_k|$ can be estimated from above by a quantity on the order of $e^{k(\alpha - \varepsilon)}$.

As $k \to -\infty$, $|v_k|$ can be estimated from above by a quantity on the order of $e^{-|k|(\beta + \varepsilon)}$ and $|e^{ik\omega}|$ increases as $e^{|k|\tau}$ ($\tau = \operatorname{Im} \omega > 0$). Consequently, $|u_k|$ can be estimated from above by $e^{-|k|(\beta + \tau + \varepsilon)}$. This implies the convergence of the Fourier series of u in the neighborhood of the strip $\alpha \leqslant \operatorname{Im} \varphi \leqslant \beta + \tau$. ▶

Let $ra = r^n a_n(\varphi) + \cdots, rb = r^n b_n(\varphi) + \cdots$, where the dots indicate terms of degree greater than n.

We perform a formal change of variables with

$$C(r, \varphi) = 1 + r^n C_n(\varphi), \qquad rD(r, \varphi) = r^n D_n(\varphi).$$

A direct substitution shows that after the change of variables the coefficients of r^n in ra and rb take the form

$$\tilde{a}_n(\varphi) = a_n(\varphi) + \lambda^n C_n(\varphi + \omega) - C_n(\varphi),$$

$$\tilde{b}_n(\varphi) = b_n(\varphi) + \lambda^n D_n(\varphi + \omega) - D_n(\varphi).$$

We determine C_n and D_n from the equations $\tilde{a}_n = 0$, $\tilde{b}_n = 0$. By the previous lemma, these equations have solutions analytic in the strip $0 \leqslant \operatorname{Im} \varphi \leqslant \tau$. We have constructed a formal change of variables after which the degree of lowest-order terms in ra and rb increases. Repeating this construction for $n = 1, 2, \ldots$, we obtain a formal change of variables after which ra and rb vanish completely. ▶

H. Analytic Normal Form

Definition. A pair (λ, ω) of complex numbers, where $\operatorname{Im} \omega \neq 0$, $\lambda \neq 0$, is said to be *normal* if there exist constants $C > 0$, $v > 0$ such that

$$|\lambda^n e^{ik\omega} - 1| > C(|n| + |k|)^{-v}$$

for all integers k and n ($n \neq 0$).

It is easy to prove the following theorems.

Theorem. *For every fixed ω, the nonnormal pairs form an everywhere dense set of Lebesgue measure 0.*

Theorem. *If (λ, ω) is a normal pair, then every holomorphic pasting*

$$\begin{pmatrix} r \\ \varphi \end{pmatrix} \mapsto \begin{pmatrix} r\lambda(1 + ra(r, \varphi)) \\ \varphi + \omega + rb(r, \varphi) \end{pmatrix}$$

reduces to the linear normal form $(r, \varphi) \mapsto (r\lambda, \varphi + \omega)$ *by a holomorphic change of variables.*

◀ The proof is analogous to that of Siegel's theorem given in § 28. ▶
   We translate this theorem into the language of imbeddings of elliptic curves.

Definition. A holomorphic vector bundle ξ is said to be *rigid* if for every imbedding of its base in a complex manifold such that the normal bundle is ξ, a sufficiently small neighborhood of the base imbedded in the manifold can be mapped biholomorphically onto a neighborhood of the zeroth section of the bundle ξ.

In this terminology, our theorem can be formulated in the following way.

Corollary. *In the sense of Lebesgue measure, almost all one-dimensional vector bundles of degree 0 over an elliptic curve are rigid.*

Remark. For certain nonresonant bundles in which the pairs (λ, ω) are nonnormal, the formal series reducing the pasting to normal form may diverge. These nonnormal pairs (λ, ω) constitute an everywhere dense set of measure zero. This problem will be discussed in more detail in § 36.

I. Negative Neighborhoods

Let us consider the case where the index of self-intersection of the elliptic curve on the surface is different from zero. If the index is negative, then the curve cannot be deformed in the class of holomorphic curves. Indeed, otherwise the deformed curve would intersect the original curve with a positive index of intersection (since both curves are complex).

Consequently, a curve with negative index of self-intersection lies isolated on the surface. Such a curve is called an exceptional curve and its neighborhood a *negative neighborhood*.

Theorem (Grauert). *The normal bundle of an exceptional curve is always rigid*, i.e., *the neighborhood of a curve with negative index of self-intersection on a complex surface is defined by the normal bundle of the curve (up to biholomorphic equivalence).*

We outline a simple proof of this theorem for the case of an elliptic curve.

◀ We begin with the preliminary normal form of the pasting

$$f\binom{r}{\varphi} = \binom{r\lambda e^{ip\varphi}(1 + ra(r, \varphi))}{\varphi + \omega + rb(r, \varphi)}.$$

We assume that the terms of degree in r smaller than n are already annihilated in ra and rb, i.e., that

$$ra = r^n a_n(\varphi) + \cdots, \qquad rb = r^n b_n(\varphi) + \cdots.$$

We perform the following formal change of variables:

$$g(r, \varphi) = (r(1 + r^n C_n(\varphi))), \qquad \varphi + r^n D_n(\varphi).$$

After this change of variables (i.e., for the pasting $g \circ f \circ g^{-1}$), the coefficients of r^n in ra and rb have the form

$$\tilde{a}_n(\varphi) = a_n(\varphi) + \lambda^n e^{ipn\varphi} C_n(\varphi + \omega) - C_n(\varphi) - ipD_n(\varphi),$$

$$\tilde{b}_n(\varphi) = b_n(\varphi) + \lambda^n e^{ipn\varphi} D_n(\varphi + \omega) - D_n(\varphi).$$

We make $\tilde{b}_n$ and $\tilde{a}_n$ vanish. To this end, we first determine D_n from the second and then C_n from the first equation. In both cases we have to solve a homological equation of the form

$$\lambda^n e^{ipn\varphi} u(\varphi + \omega) - u(\varphi) = v(\varphi)$$

with respect to the unknown 2π-periodic function u with the known 2π-periodic function v.

J. Study of the Homological Equation

We consider the Fourier series expansions of the unkown and known functions:

$$u = \sum u_k e^{ik\varphi}, \qquad v = \sum v_k e^{ik\varphi}.$$

For the Fourier coefficients, we obtain the equations

$$\lambda^n e^{i(k-pn)\omega} u_{k-pn} - u_k = v_k.$$

In principle, these equations enable us to calculate successively all unknown coefficients u_k from the first pn coefficients. However, the formal Fourier series which one obtains in this way is not always convergent. It turns out that the negativity of the index of self-intersection of the original elliptic curve on the surface (i.e., the positivity of the number p) guarantees convergence.

Indeed, let us first consider a homogeneous equation, i.e., we assume that all v_k are equal to zero.

Our equation connects the values of u_k with k on an arithmetic progression with step pn. From this progression we calculate all values of u_k with such k in terms of one of them. We have to successively multiply numbers of the form $\lambda^n e^{i(k-pn)\omega}$, where k belongs to our progression. The logarithms of these numbers form an arithmetic progression with difference $ipn\omega$. Consequently, the sums of the logarithms constitute a sequence of the form

$$\alpha s^2 + \beta s + \gamma,$$

where s is the index of a term in the sequence and $2\alpha = ipn\omega$.

If $p > 0$ and $\operatorname{Im}\omega > 0$, then $\operatorname{Re}\alpha < 0$. In this case, the sequence $\left|e^{\alpha s^2 + \beta s + \gamma}\right|$ converges to zero rapidly as $s \to +\infty$ or $s \to -\infty$. It follows that for $p > 0$, the homogeneous homological equation has pn linearly independent solutions rapidly decreasing as $|k| \to \infty$.

Now we shall solve the nonhomogeneous equation. Let us first assume that only one of the Fourier coefficients of the known function is different from zero, say v_m. To the left of m, we set $u_k = 0$. For $k \geqslant m$, we determine u_k from the equation. Consequently, to the right of m, the coefficient u_k will coincide with one of the solutions of the homogeneous equation, and thus, will decrease as $\left|e^{\alpha s^2}\right|$.

In the general case, the solution of the nonhomogeneous equation is constructed as a linear combination of the solutions constructed above with coefficients v_k. Convergence is ensured by the condition $\operatorname{Re}\alpha < 0$, i.e., the negativity of the index of self-intersection of the original elliptic curve on the surface.

Performing the estimates outlined here in detail, we can see the solvability of the homological equation for a negative index of self-intersection (i.e., if p is positive). By the same token, it is proved that a negative normal bundle of an elliptic curve on a surface is formally rigid. A more accurate analysis of our construction proves the analytical rigidity (i.e., Grauert's theorem) as well: the proof of convergence here is simpler than in the case $p = 0$ studied in § 27G and § 27H insofar as Poincaré's theorem is simpler than Siegel's theorem (§ 28). ▶

K. Positive Neighborhoods

Let us assume that the index of self-intersection of the elliptic curve on the surface is positive. In this case, the homological equation studied in the preceding section is, in general, unsolvable, since $\left| e^{\alpha s^2 + \beta s + \gamma} \right|$ increases as $|s| \to \infty$. This means that not only can the neighborhood of an elliptic curve with positive index of self-intersection on a general complex surface not be mapped biholomorphically onto a neighborhood of this curve in the normal bundle, but also such a mapping is impossible on the level of 2-jets (i.e., if we neglect terms of order 3 with respect to the distance from the curve). The neighborhood of an elliptic curve with positive index of self-intersection is said to be *positive*.

According to what has been said above, a positive neighborhood of an elliptic curve has to have moduli and even functional moduli: the "normal form" of the neighborhood has to contain arbitrary functions (and even, apparently, functions of two variables or germs of functions of two variables at several points).

While a curve with negative index of self-intersection lies isolated on the surface, an elliptic curve with positive index of self-intersection can always be deformed.

Theorem (Special Case of the Riemann–Roch Theorem). *If the index of self-intersection of an elliptic curve is equal to p on a surface, then the normal bundle has p linearly independent sections.*

◀ The problem reduces to a homogeneous homological equation of the form

$$u(\varphi + \omega) = \lambda e^{ip\varphi} u(\varphi)$$

which has p linearly independent solutions, as we have seen in § 27J. ▶

Passing to terms of higher order with respect to the distance from the elliptic curve, we see that there exists a p-parameter deformation of the curve in its neighborhood.

From this it follows, among other things, that the neighborhood of an elliptic curve on a surface where the curve has a positive index of self-intersection does not admit, as a rule, fiber structures over the curve. Indeed, upon deformation, the complex structure of an elliptic curve changes generically. Therefore, generically we find, among nearby deformed curves, curves not admitting a biholomorphic mapping onto the original curve.

In the study of the passage of differential equations through resonance (§ 36), we encounter only neighborhoods of elliptic curves on surfaces where they have vanishing index of self-intersection.

L. Elliptic Curve in Space

Much of what has been said about the neighborhood of an elliptic curve on a surface can be carried over to the case of an elliptic curve in a multidimensional space. To that end, the variable r has to be considered multidimensional in the formulas above.

Vector bundles of an arbitrary dimension over an elliptic curve can be described by pastings

$$(r, \varphi) \sim (r, \varphi + 2\pi) \sim (\Lambda(\varphi)r, \varphi + \omega),$$

where $\Lambda(\varphi)$ is a linear operator in Jordan normal form with eigenvalues of the form $\lambda e^{ip\varphi}$.

A bundle is said to be *negative* (*nonpositive, zero*) if all numbers p are positive (nonnegative, zero).

We assume that the normal bundle of our elliptic curve is negative. Then the neighborhood of the curve in a manifold can be mapped biholomorphically onto its neighborhood in the normal bundle (Grauert's theorem), i.e., the normal bundle is rigid. In the class of zero normal bundles, rigidity is violated only with probability zero.[*] The condition of resonance takes the form $\lambda_s = \lambda^n e^{ik\omega}$, where k is an integer, $\lambda^n = \lambda_1^{n_1} \cdots \lambda_m^{n_m}$, $m = \dim \{r\}$, $n_j \geq 0$, $\sum n_j \geq 2$.

A nonresonant bundle is formally rigid. For true holomorphic rigidity, the following inequality (of (C, v)-normality) is sufficient:

$$|\lambda^n e^{ik\omega} - \lambda_s| \geq C(|n| + |k|)^{-v}, \qquad |n| = n_1 + \cdots + n_m$$

for all $n \geq 0$, $\sum n_i \geq 2$ and the integral k. The measure of the vectors λ which are not (C, v)-normal for any $(C, v) > 0$ is equal to zero.

Rigidity (apparently) takes place with probability 1 for nonpositive bundles as well.

Structures of neighborhoods of curves of genus greater than 1 have not been studied very much, except the case where the normal bundle is negative and, consequently, rigid according to Grauert's theorem.

§ 28. Proof of Siegel's Theorem

In this section, we prove the theorem on the holomorphic local equivalence of a mapping and its linear part at a fixed point.

A. Formulation of the Theorem

Definition. An n-tuple $(\lambda_1, \ldots, \lambda_n) \in \mathbb{C}^n$ has *multiplicative type* (C, v) if

$$|\lambda_s - \lambda^k| \geq C|k|^{-v}, |k| = k_1 + \cdots + k_n, \lambda^k = \lambda_1^{k_1} \cdots \lambda_n^{k_n}$$

for any s and any vector k with nonnegative integral components with $|k|$ larger than 1 $(C > 0, v > 0)$.

Theorem. *We assume that in the neighborhood of the fixed point O in C^n, the collection of eigenvalues of the linear part of a holomorphic mapping has multiplicative type (C, v). Then in some neighborhood of O, the mapping is biholomorphically equivalent to its linear part.*

Let A be a biholomorphic mapping leaving O fixed and defined in the neighborhood of $O \in \mathbb{C}^n$. Let the linear operator Λ be the linear part A at O. It is claimed that a dif-

---

[*] For a proof, see Yu. S. Ilyashenko, A. S. Pyartly, *Neighborhoods of zero degree of imbedded complex tori*, Trudy Seminara im. I. G. Petrouskogo **5** (1979), 85–95.

feomorphism H exists biholomorphic in the neighborhood of O leaving O fixed and such that $H \circ A \circ H^{-1} = \Lambda$ in some neighborhood of O.

We shall prove this theorem in the case where all eigenvalues λ_s of Λ are distinct. In this case we may choose the coordinate system so that the matrix of the operator Λ is diagonal. We fix such a coordinate system.

B. Construction of the Change of Coordinates H

We write the given mapping A and substitution H in the form

$$A(z) = \Lambda z + a(z), \qquad H(z) = z + h(z),$$

where the Taylor series of a and h at zero do not contain terms of degree 0 or 1. We write the mapping $H \circ A \circ H^{-1}$ calculating the terms of zeroth and first degree with respect to h and a and obtain

$$(H \circ A \circ H^{-1})(z) = \Lambda z + [a(z) - \Lambda h(z) + h(\Lambda z)] + R([a], [h])(z),$$

where the remainder R has second order of smallness with respect to a and h in a sense which will be defined precisely later. We have enclosed the arguments of R in square brackets in order to emphasize that the operator acts on functions rather than their values at the point z.

We consider the following homological equation with respect to h:

$$\Lambda h(z) - h(\Lambda z) = a(z).$$

Here the Taylor series of the known vector-valued function a and the unknown function h do not have constant or linear terms. In the class of such series, the equation can be solved uniquely, since the collection of eigenvalues is nonresonant.

In § 28C we shall prove that the series thus obtained are convergent if the collection of eigenvalues has multiplicative type (C, v) for some positive C and v. We denote by U the operator converting the right-hand side a of the homological equation into its solution $h = U([a])$.

By induction we define functions a_s, h_s by the formulas

$$h_s = U([a_s]), \qquad a_{s+1} = R([a_s], [h_s])$$

beginning with $a_0 = a$.

We construct the mappings $H_0, H_1, \ldots$ defined by the formulas

$$H_s(z) = z + h_s(z).$$

The desired change of coordinates is given by the formula

$$H = \lim_{s \to \infty} H_s \circ \cdots \circ H_1 \circ H_0$$

which will be proved below.

C. Study of the Homological Equation

We assume that the Taylor series of the right-hand side and the solution of the homological equation

$$\Lambda h(z) - h(\Lambda z) = a(z)$$

do not have constant or linear terms. We set $|z| = \max |z_j|$.

Lemma 1. *Let the collection of eigenvalues of the diagonal linear operator Λ have multiplicative type (C, ν). Assume that the right-hand side a of the homological equation is continuous is the polydisk $|z_j| \leqslant r$ and holomorphic inside this polydisk.*

Then the homological equation has a solution h holomorphic inside the polydisk. For every δ such that $0 < \delta < \frac{1}{2}$, the inequality

$$\max_{|z| \leqslant re^{-\delta}} |h(z)| \leqslant \max_{z \leqslant r} |a(z)|/\delta^{\alpha}$$

is satisfied, where the positive constant $\alpha = \alpha(\Lambda)$ does not depend on a or r.

◀ We expand a and h in Taylor series and denote the coefficients of $z^k e_s$ by a_k^s and h_k^s. Then $h_k^s = a_k^s/(\lambda_s - \lambda^k)$. We estimate the numerators using the Cauchy inequality for the coefficients of a Taylor series and the denominators using the fact that $\{\lambda_s\}$ is of type (C, ν).

Let $\max_{|z| \leqslant r} |a(z)| = M$. According to the Cauchy inequality, we have $|a_k^s| \leqslant M/r^{|k|}$. Consequently, $|h_k^s| \leqslant MC^{-1}|k|^{\nu}/r^{|k|}$. Let us estimate the Taylor sum $\sum h_k^s z^k$. We consider the terms of degree $|k| = p$. Their number does not exceed $c_1(n)p^{n-1}$ and, therefore, $|\sum_{|k|=p} h_k^s z^k| \leqslant Mc_2 p^m |(z/r)^k|$, where $c_2 = c_1/C$ and $m = \nu + n - 1$.

The function $x^m e^{-x}$ has a maximum $(m/e)^m$ and, therefore, $p^m e^{-\delta p/2} \leqslant c_3 \delta^{-m}$, where $c_3 = (2m/e)^m$. Consequently, for $|z| \leqslant re^{-\delta}$, we have

$$|h| \leqslant Mc_3 \delta^{-m} \sum_{p=2}^{\infty} e^{-p\delta/2} = \frac{Mc_3 \delta^{-m}}{e^{\delta} - e^{\delta/2}}, \qquad |h| \leqslant Mc_4 \delta^{-(m+1)},$$

where $c_4 = 4c_2 c_3$ is independent of a, r, and δ. ▶

In what follows, in addition to an estimate of h, we need an estimate of the function $h \circ \Lambda$ defined by the relation

$$(h \circ \Lambda)(z) = h(\Lambda z).$$

Lemma 2. *Under the conditions of Lemma 1, we have*

$$\max_{|z| \leqslant re^{-\delta}} |h(\Lambda z)| \leqslant \max_{|z| \leqslant r} |a(z)|/\delta^{\alpha_0},$$

where the positive constant $\alpha_0 = \alpha_0(\Lambda)$ does not depend on δ, a, or r.

◀ We begin with the following remark.

Let $\{\lambda\}$ have multiplicative type (C, ν). Then a constant C_0 exists, independent of k such that

$$|\lambda_s - \lambda^k| \geqslant C_0 |k|^{-\nu} |\lambda^k|$$

for all s = 1, ..., n and for all vectors k with nonnegative integral components whose sum |k| is not smaller than 2.

◀ We denote $\max |\lambda_s|$ by μ. For $|\lambda^k| \leqslant 2\mu$ we may take $C_0 = C/2\mu$. For $|\lambda^k| > 2\mu$ we may take $C_0 = \frac{1}{2}$. ▶

Using this remark we obtain the following estimate for the Taylor series:

$$|h(\Lambda z)| \leqslant \sum M C_0^{-1} |k|^{\nu} |\lambda^k|^{-1} |\lambda^k z^k / r^k|.$$

The remaining estimates are the same as in the proof of Lemma 1. ▶

D. Order of Operators

For the following estimates, it is convenient to introduce some new notation. Let f be a function continuous in the polydisk $|z| \leqslant r$, holomorphic at the interior points, and vanishing at the center of the polydisk.

For such functions, we introduce the norm

$$\|f\|_r = \sup_{0 \leqslant |z| \leqslant r} \frac{|f(z)|}{|z|}.$$

EXAMPLE

The function $f(z) = \varepsilon z$ has the norm $|\varepsilon|$ independent of the radius of the polydisk.

Remark. The $\| \ \|_r$-norm is convenient because it is invariant with respect to changes of scale: for every dilation coefficient $\varkappa$, we have

$$\|\varkappa f \circ \varkappa^{-1}\|_{\varkappa r} = \|f\|_r.$$

The values of the function f may not only be numbers but elements of a normed space, for example, vectors, matrices, etc.

Let Φ be an operator acting on functions of the class described above*. Let d, α, and β be positive numbers and let $0 < r < 1$.

Definition. The operator Φ *has order* $(d; \alpha | \beta)$ if for every δ in the interval $(0, 1/2)$ and every r in $(0, 1)$,

$$\|\Phi([f])\|_{re^{-\delta}} \leqslant \|f\|_r^d \delta^{-\alpha},$$

provided that $\|f\|_r \leqslant \delta^\beta$.

We shall write this relation in the form $\Phi([f]) \dashv f^d(\alpha | \beta)$ or more briefly, $\Phi([f]) \dashv f^d$.

---

*We shall denote by the same letter operators acting on functions from classes with distinct r under the condition that they coincide as operators on the space of germs of functions. This is the same as in ordinary calculus when the notation of sine is not changed when its domain is changed.

An operator *has order d* if constants α and β exist such that it has order $(d; \alpha|\beta)$. [It is essential that α and β are independent of f, $r \in (0, 1)$, and $\delta \in (0, 1/2)$].

EXAMPLES

1. We consider the operator converting the right-hand side a of the homological equation into its solution h. *This operator has order* 1 *if* λ_s *is of type* (C, v). Indeed, the necessary inequality is furnished by Lemma 1.

   In exactly the same way, it follows from Lemma 2 that *the operator converting the right-hand side a of the homological equation into the function* $h \circ \Lambda$ *has order* $1: h \rightrightarrows a, h \circ \Lambda \rightrightarrows a$.

2. We consider the local diffeomorphism H, $H(z) = z + h(z)$. We write the inverse diffeomorphism in the form $H^{-1}(z) = z - g(z)$. Let us now consider the operator G converting h into g.

   The operator G has order 1, *i.e.*, $g \rightrightarrows h$.

◀ First, we note that the Cauchy estimate implies the following inequality:

$$\left\| \frac{\partial h_i}{\partial z_j} \right\| \leq \frac{\|h\|_r}{1 - e^{-\delta/2}} \tag{1}$$

for $|z| \leq re^{-\delta/2}$.

If $\|h\|_r < \delta^\beta$ and β is sufficiently large, the right-hand side of Eq. (1) is arbitrarily small. Now g is constructed as the limit of the following iterations:

$$g_{s+1} = h(z - g_s(z)), \qquad g_0 = 0.$$

The convergence for $|z| \leq re^{-\delta}$ and the estimate $g \rightrightarrows h$ follow easily from theorems on contraction mappings. ▶

EXAMPLE

3. In the notation of Example 2, we have

$$h - g \rightrightarrows h^2.$$

◀ Indeed, from the definition of the function g, it follow that

$$h(z) - g(z) = h(z) - h(z - g(z)).$$

Using Eq. (1) and the estimate (obtained above)

$$\|g\|_{re^{-\delta}} \leq \|h\|_r \delta^{-\alpha},$$

we see that

$$|h(z) - g(z)| \leq C\|h\|_r (1 - e^{-\delta/2})^{-1} r \|h\|_r \delta^{-\alpha}$$

for $|z| \leq re^{-\delta}$. ▶

We note that in our notation $2f \dashv f, f^2 \dashv f$; if $f_1 \dashv f_2$ and $f_2 \dashv f_3$, then $f_1 \dashv f_3$.

We extend our notation to operators of several functions. Let the operator Ξ convert a pair of functions η, ζ into the function ξ. Let φ be a polynomial. We write

$$\xi \dashv \varphi(\eta, \zeta)$$

if positive constants $(\alpha; \beta_1, \beta_2)$ exist such that for any δ in the interval $(0, \frac{1}{2})$ and any r in $(0, 1)$, we have the inequality

$$\|\Xi([\eta], [\zeta])\|_{re^{-\delta}} \leqslant \varphi(\|\eta\|_r, \|\zeta\|_r)\delta^{-\alpha},$$

provided that $\|\eta\|_r \leqslant \delta^{\beta_1}, \|\zeta\|_r \leqslant \delta^{\beta_2}$. Here the constants α and β do not depend on η, ζ, $r \in (0, 1)$ or $\delta \in (0, \frac{1}{2})$. If $\eta \dashv \psi(\sigma, \tau)$ then $\xi \dashv \varphi(\psi(\sigma, \tau)\zeta)$.

EXAMPLE

4. *We define the operator Ξ by the formula*

$$\xi(z) = \eta(z - \zeta(z)) - \eta(z).$$

We have $\xi \dashv \eta\zeta$.

◀ This can be proved by means of inequality (1) which we have used in Examples 2 and 3. ▶

E. Estimate of the Remainder

We explicitly write out the remainder R defined in section B. We introduce the notation

$$H(z) = z + h(z), \qquad H^{-1}(z) = z - g(z).$$

By definition,

$$R(z) = (H \circ A \circ H^{-1})(z) - \Lambda(z) - [a(z) - \Lambda h(z) + h(\Lambda z)].$$

We write R in the form $R = R_1 + R_2 + R_3$, where

$$R_1(z) = \Lambda(h(z) - g(z)), \qquad R_2(z) = a(z - g(z)) - a(z),$$
$$R_3(z) = h(\Lambda z - \Lambda g(z) + a(z - g(z))) - h(\Lambda z).$$

To simplify our subsequent estimates, we represent R in the form of an operator of *three* arguments a, h, and $u = h \circ \Lambda$.

We introduce the operators

$$G, G([h]) = g, \qquad \Xi, \Xi([\alpha], [g])(z) = a(z - g(z)) - a(z).$$

In this notation, we have

$$R_1([h]) = \Lambda(h - G([h])),$$

$$R_2([a], [h]) = \Xi([a], G([h])),$$

$$R_3([u], [a], [h]) = \Xi([u], [v]),$$

where $v(z) = g(z) - \Lambda^{-1} a(z - g(z))$.

The substitution $u = h \circ \Lambda$ transforms the operator $R_1 + R_2 + R_3$ into the remainder $R([a], [h])$ which is what we are interested in. Let $h \dashv$ id (id is the identical mapping; the condition means that the derivative of h is small).

Estimate 1. The following estimates hold.

$$R_1([h]) \dashv h^2, \qquad R_2([a], [h]) \dashv ah, \qquad R_3([u], [a], [h]) \dashv u(h + a).$$

◀ The estimate $R_1 \dashv h^2$ has been proved in Example 3 of § 28D and the inequality $\Xi([a], [g]) \dashv ag$ in Example 4. According to Example 2, we have $G([R]) \dashv h$, and, therefore, $R_2([a], [h]) \dashv ah$.

From the estimates $g \dashv h$, we obtain $v \dashv h + a$. Consequently, according to the estimate of the operator Ξ from Example 4, $R_3 \dashv u(h + a)$. ▶

Estimate 2. *Let U be the operator solving the homological equation. The operator Φ given by the formula $\Phi([a]) = R([a], U([a]))$ has order 2.*

By Lemmas 1, 2, and 3 of § 28C, we have $h \dashv a$, $h \circ \Lambda \dashv a$, where $h = U([a])$. Comparing this with Estimate 1, we obtain

$$R_1(U([a])) \dashv a^2, \qquad R_2([a], U([a])) \dashv a^2,$$

$$R_3(U([a]) \circ \Lambda, [a], U([a])) \dashv a^2.$$

F. Convergence of the Approximations

The conclusion of the proof of Siegel's theorem is exactly the same as the estimates of § 12.

◀ We choose the following number sequences:

$$\delta_0, \delta_1 = \delta_0^{3/2}, \delta_2 = \delta_1^{3/2}, \dots;$$

$$M_0 = \delta_0^N, \qquad M_1 = M_0^{3/2}, \dots, M_s = M_{s-1}^{3/2} = \delta_s^N, \dots;$$

$$r_0, r_1 = e^{-\delta_0} r_0, \qquad r_2 = e^{-\delta_1} r_1, \dots.$$

These sequences are determined by the choice of δ_0, N, and r_0.

Let δ_0 be so small that all r_s are larger than $r_0/2$. We describe the choice of N. According to Estimate 2, there exist constants α and β such that

$$\|R([a], U([a]))\|_{re^{-\delta}} \leqslant \|a\|_r^2 \delta^{-\alpha}$$

provided that $\|a\|_r \leqslant \delta^\beta$.

Let $a_{s+1} = R([a_s], U([a_s]))$. We assume that $\|a_s\|_{r_s} \leqslant M_s = \delta_s^N$. Then for $N > \beta$, the preceding inequality can be applied with $\delta = \delta_s$ and we obtain

$$\|a_{s+1}\|_{r_{s+1}} \leqslant M_s^2 \delta^{-\alpha} = \delta_s^{2N-\alpha}.$$

If $N > 2\alpha$, then the right-hand side does not exceed $M_{s+1} = \delta_s^{3N/2}$. Therefore, we fix $N > (\beta, 2\alpha)$. If $\|a_0\|_{r_0} \leqslant M_0 = \delta_0^N$, then $\|a_s\|_{r_s} \leqslant M_s = \delta_s^N$ for all s.

Finally, we choose r_0. By assumption, the initial function $a_0 = a$ has a root of order at least 2 at the origin. Consequently,

$$|a(z)| \leqslant K|z|^2$$

in some neighborhood of the point $z = 0$. This implies that

$$\|a_0\|_{r_0} \leqslant Kr_0.$$

Hence the condition $\|a_0\|_{r_0} \leqslant \delta_0^N$ is satisfied for sufficiently small r_0. We fix such an r_0. Now all numbers δ_s, M_s, and r_s are defined. The inequalities $\|a_s\|_{r_s} \leqslant M_s$ are satisfied for all s. They imply estimates for h_s. Therefore, the products $H_s \circ \cdots \circ H_1$ are determined for $|z| \leqslant r_0/2$ and converge to a limit H as $s \to \infty$. It is easy to verify that for the limit diffeomorphism, we have $H \circ A \circ H^{-1} = \Lambda$. ▶

Chapter 6

Local Bifurcation Theory

The word bifurcation means division into two and is used in a wide sense to indicate every qualitative topological metamorphosis of a picture under the variation of parameters on which the object being studied depends. The objects can be diverse: for example, real or complex curves or surfaces, functions or mappings, manifolds or fiber bundles, vector fields or equations, differential or integral.

If an object depends on parameters, we speak of a *family*. If we are interested in a family locally, i.e., in small variations of the parameters in the neighborhood of fixed values, then we speak of *deformations* of the object corresponding to these values of the parameters.

It turns out that in many cases the study of all possible deformations reduces to that of a single one from which all others can be obtained. In some sense, such a deformation is the richest; it has to yield all possible bifurcations of the object. It is called a versal deformation.

In this chapter, we mainly study bifurcations and versal deformations of phase portraits of dynamical systems in the neighborhood of equilibrium positions and closed trajectories.

For more details on nonlocal bifurcations, see "Bifurcation theory" (V. I. Arnold, V. S. Afraimovich, Ju. S. Iljashenko, L. P. Shilnikov, eds.), in: *Modern Problems in Mathematics, Fundamental Directions, Dynamical Systems,* Vol. 5, Moscow, VINITI, 1986 (Springer translation, Encyclopaedia of Mathematical Sciences, 1988). See also the survey "Ordinary differential equations" (Arnold, Iljashenko) in the same series of surveys (Vol. I, Moscow, VINITI, 1985), and the survey "Ergodic theory of smooth dynamical systems" (Ja G. Sinai, Ja. B. Pesin, L. A. Bunimovich, M. V. Jakobson, eds.), Vol. 2, Moscow, VINITI, 1985.

§ 29. Families and Deformations

In this section, we discuss general "heuristic" arguments on which bifurcation theory is based. These arguments are essentially due to Poincaré.

Figure 105.

A. Generic Cases and Singular Cases of Small Codimension

In the study of any kind of analytic objects (for example, differential equations, boundary value problems or optimazation problems), one can usually identify the generic cases. For example, the nodal, focal, and saddle points are generic among the singular points of a vector field in the plane while, say, the centers are destroyed by an arbitrarily small perturbation of the field.

The study of generic cases is always a primary problem in the analysis of the phenomena and processes described by a given mathematical model. Indeed, by an arbitrarily small change in the model, a nongeneric case may turn into a generic case; on the other hand, the parameters of the model are usually determined only approximately.

Nevertheless, there are instances in which it is necessary to study nongeneric cases. Let us assume we do not study an individual object (say, a vector field), but a whole family whose objects depend on a number of parameters.

In order to visualize the situation more clearly, we consider a function space whose points are the objects themselves (say, the space of all vector fields). The nongeneric cases correspond to some hypersurface of codimension 1 in the space. A point can be moved by an arbitrarily small shift off the hypersurface to the domain of generic cases. The hypersurfaces of singular cases form the boundaries of the domains of the generic cases (Fig. 105).

A family with k parameters is represented by a k-dimensional manifold in our function space. For example, a one-parameter family is a curve in the function space (the bold line in Fig. 105).

A curve in our function space may intersect a hypersurface of singular cases. If this intersection occurs at "nonzero angle" (transversally), then it is preserved under a small perturbation of the family: every nearby curve intersects the hypersurface of singular points at some nearby point (the nonbold line in Fig. 105).

Consequently, although every individual member of the family can be made generic by an arbitrarily small perturbation, it is impossible to achieve the condition that all members of the family be generic at the same time: by a deformation of the family, the nongeneric case can be avoided for every fixed value of the parameter, but for some nearby value nongenericity occurs just as well.

The hypersurfaces of singular cases in our function space have, in general, their own singularities (for example, the intersection of two hypersurfaces, which corresponds to the simultaneous occurrence of two degeneracies [cf., Fig. 105]. In the study of generic one-parameter families, such singularities of hypersurfaces of singular cases can be ignored. Indeed, the set of all such singularities has codimension not smaller than 2 in the function space. Therefore, the curve in the function space can be moved by means of an arbitrarily small perturbation off these singularities, so that it will intersect the hypersurfaces of singular cases only at its generic points. Hence, in a generic one-parameter family, the only unremovable degeneracies we may encounter are the simplest ones corresponding to nonsingular points of a hypersurface of singular cases. Such degeneracies are called *degeneracies of codimension 1*. The study of degeneracies of codimension 1 enables us to pass continuously from any generic point of the function space to any other generic point, since the functional space is divided only by sets whose codimension is not greater than 1.

At passage, in general, we have to intersect surfaces of degeneracies of codimension 1. The study of singularities of codimension 1 enables us to describe bifurcations occurring at the intersections of these surfaces.

In the study of generic k-parameter families, the only unremovable degeneracies are those of codimension not exceeding k. All other degenerate objects form a set of codimension greater than k in the function space, and we can get rid of them by an arbitrarily small deformation of the k-parameter family.

The larger the codimension of degeneracy, the more difficult it is to study the degeneracy and the less useful is its study (as a rule). A study of singularities of large codimension k is reasonable only if we are interested in the k-parameter family rather than the individual object. Then the natural object of study is not the individual object (say, a vector field with a complicated singular point), but a family so large that the singularity of the type under consideration does not disappear under a small deformation of the family.

This simple argument of Poincaré shows the futility of such a large number of studies in the theory of differential equations and in other areas of analysis that it is always somewhat dangerous to mention it. In essence, every result concerning a degenerate case has to be accompanied by a calculation of the corresponding codimension and an indication of the bifurcations in the family for which the degeneracy in question is unremovable.

From this point of view, which is based on the study of k-parameter families, we can completely ignore the study of degeneracies of infinite codimension, since we can get rid of them by means of a small perturbation of any k-parameter family for any finite k. Of course, degenerate cases may be useful as easily studied first approximations of perturbation theory.

B. Digression: Cases of Infinite Codimension

Occasionally we have to study degeneracies of infinite codimension, too. For example, Hamiltonian systems or systems with a symmetry group form a submanifold of infinite codimension in the space of all dynamical systems. In these cases, it is often possible to restrict the function space in advance so that the codimensions of the degeneracies under study become finite (for example, we consider only Hamiltonian systems and their Hamiltonian deformations).

However, it is not always easy to restrict the function space. Let us, for example, consider boundary value problems for partial differential equations. We deal with the intersection of two submanifolds in the function space: the space of solutions and the space of functions satisfying the boundary conditions. Both these manifolds have infinite dimension and infinite codimension. An analysis of this situation requires an ability to distinguish between various infinite dimensions and codimensions: the condition that a function of one variable constructed from a given object vanishes singles out a "manifold of (infinite) codimension smaller" than the condition that a function of two variables vanishes.

One of the simplest problems of such a calculation of infinite codimensions corresponding to kernels and cokernels consisting of functions on manifolds of different dimensions is the oblique derivative problem. If we consider this problem on the sphere bounding an n-dimensional ball, a vector field tangent to the n-dimensional ambient space is given. A function harmonic in the ball is to be determined whose derivative in the direction of the field is equal to a given boundary function.

We consider, for example, the case $n = 3$. In this case, a generic field is tangent to the sphere on some smooth curve. There are singular points on this curve where the field is tangent to the curve itself. The structure of the field in the neighborhood of each of these singular points is standard: it can be proved that for any n for a generic field* in the neighborhood of every point of the boundary, the field is given, in an appropriate coordinate system, by a formula of the form

$$x_2\partial_1 + x_3\partial_2 + \cdots + x_k\partial_{k-1} + \partial_k, \qquad k \leqslant n,$$

where $\partial_k = \partial/\partial x_k$ and $x_1 = 0$ on the boundary [cf., S. M. Višik, *On the oblique derivative problem*, Vestnik MGU, Ser. Mathem. **1** (1972), 21–28].

The oblique derivative problem apparently has to be stated according to the following scheme. The manifold of tangency of the field with the boundary, the manifold of tangency of the field with the tangency manifolds, etc. divide the boundary into parts of various dimensions. On some of these parts of the boundary, conditions have to be given; on some other parts, the boundary function itself has to satisfy certain conditions for the existence of the classical solution of the problem.

In spite of the abundance of research concerning the oblique derivative problem (this problem is studied especially thoroughly in the publications of Maz'ya, whose work was preceded by that of Maljutov and Egorov with Kondrat'ev), the program described above has been carried out only in the two-dimensional case (where the boundary is a circle).

---

*That is, for every field in an open everywhere dense set in the function space of smooth fields.

Figure 106.

C. The Space of Jets

The study of bifurcations in generic k-dimensional families is in essence the study of a function space divided into parts corresponding to various degeneracies, neglecting degeneracies of codimension greater than k.

In order to get rid of the infinite dimensionality of the function space itself, a special apparatus of finite-dimensional approximations has been elaborated: manifolds of k-jets (the term jet was introduced by Ehresmann).

In the following, we establish the terminology and notation which we shall use. All assertions of this section and the next one are entirely obvious.

Let $f: M^m \to N^n$ be a smooth mapping of smooth manifolds (it can be assumed that M and N are domains in Euclidean spaces of the appropriate dimensions).

Definition. Two such mappings f_1, f_2 are said to be tangent of order k or *k-tangent at a point x* of M if (Fig. 106)

$$\rho_N(f_1(y), f_2(y)) = o(\rho_M^k(x, y)) \text{ for } y \to x.$$

Here ρ denotes some Riemannian metric; it is easy to see that the property of k-tangency does not depend on the choice of the metrics ρ_M and ρ_N.

Two mappings are 0-tangent at a point x if their values coincide at x. Tangency of order k is an equivalence relation

$$(f_1 \sim f_2 \Rightarrow f_2 \sim f_1, f_1 \sim f_2 \sim f_3 \Rightarrow f_1 \sim f_3, f_1 \sim f_1).$$

Definition. A k-jet of a smooth mapping f at a point x is a class of k-tangent mappings at x.

Notation

$$j_x^k(f) = \{f_1 : f_1 \text{ is } k\text{-tangent to } f \text{ at } x\}.$$

The point x is called the *source* and the point $f(x)$ the *target* of the jet.

We choose coordinates on M and N in the neighborhood of the points x and $f(x)$, respectively. Then the k-jet of any mapping close to f at any point close to x can be given by the collection of Taylor coefficients up to degree k. Consequently, for fixed coordinate systems, a jet of order k can be identified with the collection of Taylor coefficients up to degree k.

EXAMPLE

The 0-jet of a mapping f of the x-axis into the y-axis at point x is given by a pair (x, y) of numbers where $y = f(x)$. A 1-jet is given by a triple (x, y, p), where $p = df/dx$.

Besides tangency of order k, there is another equivalence relation leading to *germs* of mappings instead of jets.

Definition. Two mappings in two neighborhoods of one and the same point *have a common germ* at that point if they coincide in a third neighborhood of this point. (The third neighborhood can be smaller than the intersection of the first two neighborhoods.)

The germ of a mapping at a point is an equivalence class with respect to the equivalence relation introduced above.

As for a mapping, for a germ we can define its 0-jet at a point, 1-jet at a point, etc.

We consider the set of all k-jets of germs* of smooth mappings from M into N at all possible points of M.

Definition. The set of all k-jets of germs of mappings of M into N is called the *space of k-jets* of mappings of M into N and denoted by

$$J^k(M, N) = \text{the space of } k\text{-jets of mappings of } M \text{ into } N.$$

EXAMPLE

$J^1(\mathbb{R}, \mathbb{R})$ is a three-dimensional space with coordinates (x, y, p) (cf., § 3).

The set $J^k(M, N)$ has a natural smooth manifold structure. Indeed, we choose coordinate systems in the neighborhood of a point of M and in the neighborhood of the image of this point in N under some mapping f. Then the k-jet of f and all nearby jets can be given by coordinates of the preimage and the collections of Taylor coefficients up to order k at this point. Thus, we have constructed a chart of the manifold $J^k(M, N)$ of jets in the neighborhood of the point which is the k-jet of f.

It is easy to calculate the dimension of a manifold of jets. For example,

---

* In the real smooth case, it is immaterial whether we consider jets of germs or jets of mappings on the whole of M, since every germ is the germ of a global mapping. On the other hand, in the complex situation, a global mapping with a given jet may not even exist.

$$J^0(M, N) = M \times N, \dim J^0(M, N) = \dim M + \dim N,$$

$$\dim J^1(M, N) = \dim M + \dim N + \dim M \dim N.$$

We have a natural mapping

$$J^{k+1}(M, N) \to J^k(M, N).$$

(A$(k + 1)$-jet determines a k-jet, since $(k + 1)$-tangency implies k-tangency). This smooth mapping is a fibration. We obtain the following chain of fibrations:

$$\cdots \to J^k \to J^{k-1} \to \cdots \to J^1 \to J^0 = M \times N.$$

A fiber of each of these fibrations is diffeomorphic to a linear space but does not have a natural linear structure (for $k > 1$) ("noninvariance of higher differentials").

The manifolds J^k are, in a certain sense, finite-dimensional approximations of the infinite-dimensional function space of smooth mappings from M into N.

D. Groups of Jets of Local Diffeomorphisms and Spaces of Jets of Vector Fields

We consider the space $J^k(M, M)$ of jets. The submanifold of k-jets of diffeomorphisms lies in this manifold. This submanifold is not a group, since jets may be multiplied only if the target of one jet is the source of the other.

We fix a point of M and consider all germs of diffeomorphisms of M leaving this point invariant. Their k-jets form a group.

Definition. The group of k-jets of diffeomorphisms of M leaving a point x fixed is called the *group of k-jets of local diffeomorphisms of the manifold M at the point x* and denoted by $J_x^k(M)$.

EXAMPLE

The group of 1-jets of local diffeomorphisms is isomorphic to a linear group: $J_x^1(M^m) = GL(\mathbb{R}^m)$.

For $k > 1$, we obtain a more complicated Lie group. Since a k-jet determines a $(k - 1)$-jet, we obtain the following chain of mappings:

$$J_x^k(M) \to \cdots \to J_x^1(M) = GL(\mathbb{R}^m).$$

It is easy to see that *these mappings* (the mappings ignoring the terms of degree k in

the Taylor polynomial) *are homomorphisms and their kernels are commutative groups.*
For example, let $m = 1$.
◀ Then: if $f(x) = x + ax^k(\mathrm{mod}(x^{k+1}))$ and $g(x) = x + bx^k \mathrm{mod}(x^{k+1})$, $(f \circ g)(x) = x + ax^k + bx^k(\mathrm{mod}\, x^{k+1})$. ▶

A *vector field* on a manifold M is a section of the tangent bundle $p: TM \to M$, i.e., a smooth mapping $v: M \to TM$ such that the diagram

is commutative.

The definition of germs, jets, and spaces of jets of vector fields mimic the above definitions.

The group of diffeomorphisms of a manifold M acts on the space of all vector fields on M and also on the spaces of k-jets of vector fields on M.

The group of k-jets of local diffeomorphisms of a manifold M at a point acts on the space of $(k - 1)$-jets of vector fields on M at the point; this action is linear.

EXAMPLE

Let $y = a_1 x + a_2 x^2 + \cdots$ be the 2-jet of a local diffeomorphism at zero. The image of the 1-jet $v(x) = v_0 + v_1 x + \cdots$ of a field is given by the formula $w(x) = w_0 + w_1 x + \cdots$, where $w_0 = a_1 v_0$, $w_1 = a_1 v_1 a_1^{-1} + 2a_2 a_1^{-1} v_0$.
◀ This formula can be obtained by writing the equation $\dot{x} = v(x)$ in y coordinates. ▶

E. The Weak Transversality Theorem

Proofs of the possibility of reduction to the generic form can often be replaced by a reference to some standard (and obvious) transversality theorems. Below we formulate and outline the proofs of the most useful transversality theorems. Transversality theorems mainly help economize space: in every concrete case, the corresponding concrete assertion can easily be proved directly.

Definition. Two linear subspaces X and Y of a linear space L are said to be *transversal* if their sum is the whole space:

$$L = X + Y.$$

For example, two planes intersecting at a nonzero angle in the three-dimensional space are transversal and two straight lines are not.

Let A and B be smooth manifolds and let C be a smooth submanifold of B.

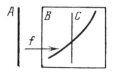

Figure 107.

(Here and in the following the word manifold means a manifold without boundary.)

Definition. A mapping $f: A \to B$ is said to be *transversal to C at a point a* of A if either $f(a)$ does not belong to C or the tangent plane to C at $f(a)$ and the image of the tangent plane to A at a are transversal (Fig. 107):

$$f_* T_a A + T_{f(a)} C = T_{f(a)} B.$$

Definition. A mapping $f: A \to B$ is *transversal* to C if it is transversal to C at any point of the preimage manifold.

For example, an imbedding of a straight line in the three-dimensional space is transversal to another line in the same space if and only if the lines do not intersect.

Remark. A mapping of a straight line in the plane can be nontransversal to a straight line lying in the plane even in the case where the image of the line under the mapping is a line normal to the given line. (The image of the tangent space and the tangent space to the image are not the same.)

We also note that if $f: A \to B$ is transversal to C, then the preimage of C in A is a smooth submanifold and its codimension in A is equal to that of C in B.

One often encounters situations where C is not a smooth submanifold but a submanifold with singularities.

Definition. A *stratified subvariety* of a smooth manifold is a finite union of mutually disjoint smooth manifolds (strata) satisfying the following condition: the closure of every stratum consists of the stratum itself and a finite union of strata of smaller dimensions.

A mapping is said to be *transversal to a stratified subvariety* if it is transversal to every stratum.

EXAMPLE

Let C be the union of two planes intersecting in a straight line in the three-dimensional space. The stratification is the partition into the line of intersection and four half-planes. Transversality to C means transversality to each of the planes and transversality to the line of intersection. For example, a curve

Figure 108.

transversal to the stratified variety C does not intersect the straight line of singularities of C.

Theorem. *Let A be a compact manifold and let C be a compact submanifold in a manifold B. The mappings $f: A \to B$ transversal to C form an open, everywhere dense set in the space of all sufficiently smooth mappings $A \to B$. (The proximity of mappings f is defined as the proximity of the functions determining f and their derivatives up to a sufficiently large order r.)*

This theorem is called the weak transversality theorem. Its assertion means that a mapping not transversal to a fixed submanifold can be turned into a transversal mapping by a small perturbation (Fig. 108). If, on the other hand, transversality is present, then it is preserved under small perturbations.

◀ We consider the special case where B is a linear space, $B = \mathbb{R}^n$ and C is its subspace $\mathbb{R}^{n-k}$.

We represent B in the form of the sum $B = C + D$ of two subspaces of complementary dimensions, $C = \mathbb{R}^{n-k}$ and $D = \mathbb{R}^k$. We project B onto D in the direction of C; we denote this projection by π. Let us consider the mapping $\pi \circ f: A \to D$.

The point 0 is a critical value for this mapping if and only if the mapping $f: A \to B$ is not transversal to the submanifold $C \subset B$. By Sard's lemma (§ 10), almost all points of D are not critical values for $\pi \circ f$. Let ε be a point in D which is not a critical value for $\pi \circ f$. We construct a mapping $f_\varepsilon: A \to B$ by setting $f_\varepsilon(a) = f(a) - \varepsilon$. The mapping f_ε is transversal to C. Since ε can be chosen arbitrarily small, we have proved that the set of transversal mappings is everywhere dense in our special case. That this set is open follows from the implicit function theorem. The general case can easily be reduced to the one just considered. ▶

Remark. If C is not compact, then the term "open" has to be replaced, in general, the term "intersection of countably many open".

EXAMPLE

1. B is the torus, C is a dense winding, and A is a circle.

2. B is the plane, A is a circle imbedded in the plane, and C is a tangent line to A (without the point of tangency). The imbedding is transversal to C but there exist nontransversal mappings to C arbitrarily close.

In order that the mappings of a compact manifold A into B transversal to C form an *open*, everywhere dense set it is sufficient to require, instead of the compactness of C, that every point of B have a neighborhood such that the pair (neighborhood, its intersection with C) be diffeomorphic to the pair $(\mathbb{R}^b, \mathbb{R}^c)$ or to the pair $(\mathbb{R}^b,$ empty set).

If A is not compact, then it is convenient to endow the space of mappings with the "fine topology". In this topology, a neighborhood of a mapping $f: A \to B$ is defined as follows: We fix an open set G in the space $J^k(A, B)$ of jets for some k. The set of C^∞-mappings $f: A \to B$ whose k-jets at every point belong to G is open in the fine topology. These nonempty open sets are taken as a neighborhood basis defining the fine topology in the space of infinitely differentiable mappings.

Consequently, the proximity of two mappings in the fine topology means that the mappings approach each other arbitrarily rapidly "at infinity"; in particular, the graph of a mapping sufficiently close to f lies in a neighborhood of the graph of f which thins out arbitrarily rapidly "at infinity".

This implies that the convergence of a sequence in the fine topology implies the complete coincidence, outside some compact set, of all members of the sequence beginning with some member. Nevertheless, every neighborhood of a given mapping in the fine topology contains mappings which coincide nowhere with the given mapping.

If openness and density everywhere are understood in the sense of the fine topology, then the transversality theorem is true for noncompact A, too. (For openness, C has to be compact or satisfy the condition formulated above.)

The transversality theorem can be extended to the case of a stratified subvariety C in an obvious way. However, in this case, the theorem guarantees that the transversal mappings form only an everywhere dense intersection of a countable number of open sets rather than an open, everywhere dense set.

For the mappings transversal to a stratified variety, to form an open, everywhere dense set it is sufficient that the stratification satisfy the following additional condition: every imbedding transversal to a stratum of smaller dimension is transversal to all adjoining strata of larger dimension at some neighborhood of the stratum of smaller dimension. (It is assumed that the stratification is analytic or at least diffeomorphic to an analytic one.)

EXAMPLE

1. Let C be a finite union of planes in a linear space stratified in a natural way (for example, a pair of intersecting planes in $\mathbb{R}^3$). Transversality to $\mathbb{R}^k$ implies transversality to the ambient $\mathbb{R}^l$. Therefore, our condition is satisfied.

2. Let C be the cone $x^2 = y^2 + z^2$ in $\mathbb{R}^3$ and let the stratification be partition into the point 0 and the two skirts. As is easily seen, our condition is satisfied.

3. Let C be the Whitney–Cayley umbrella given by the equation* $y^2 = zx^2$ in $\mathbb{R}^3$ (Fig. 109). [The portion $z \geqslant 0$ of this stratified manifold is the range of the mapping

---

*The umbrella includes a handle indicated by the bold line in Fig. 109.

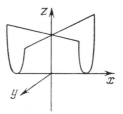

Figure 109.

$\varphi\colon \mathbb{R}^2 \to \mathbb{R}^3$ given by the formulas $x = u$, $z = v^2$, $y = uv$. Whitney has proved that (1) the type of singularity (up to diffeomorphisms of $\mathbb{R}^2$ and $\mathbb{R}^3$) is preserved under a small perturbation of φ; (2) this is the only singularity of mappings of two-dimensional manifolds into three-dimensional manifolds, which is preserved under small perturbations (except for lines of self-intersection); all other singularities split into singularities of this type under a small perturbation.] Transversality to the singular line $x = y = 0$ does not imply transversality to the manifold of regular points of the surface close to this line. (The plane $z = 0$ is transversal to the line but not to the surface.)

If the condition

"transversality to smaller $\Rightarrow$ transversality to larger"

is satisfied on the stratification C, then transversality to the whole stratification can be obtained as follows.

(1) The strata of minimum dimension are smooth; the ordinary theory applies to them. (2) In the neighborhood of strata of minimum dimension, transversality is obtained on all strata and for all mappings close to the given mapping.* (3) From the entire manifold, we remove the closure of the neighborhood of strata of minimum dimension and pass to strata of the next dimension.

EXAMPLE

Let B be the space of linear operators $b\colon \mathbb{R}^n \to \mathbb{R}^n$ and let C be the set of operators of nonmaximum rank. The operators of rank r form a smooth manifold whose codimension in the space B is equal to $(m - r)(n - r)$. The partition into manifolds of operators of distinct ranks defines a stratification on C.

A *mapping* $f\colon A \to B$ is a family of linear operators from $\mathbb{R}^m$ into $\mathbb{R}^n$ smoothly depending on a point of A as a parameter. The manifold A is called the *base* of the family. The transversality theorem immediately implies the following corollary.

_____
* This is the delicate point of the proof where analyticity is involved.

Corollary. *In the space of smooth families of* $m \times n$ *matrices, the families transversal to the stratified variety C of matrices of nonmaximum rank form an everywhere dense set.*

In particular, the values of the parameter to which matrices of rank r correspond form a smooth submanifold of codimension $(m - r)(n - r)$ in the base of the generic family (for families from an everywhere dense intersection of countably many open sets in the space of families).

For example, in a five-parameter generic family of 2×3 matrices, the rank drops to 1 on a three-dimensional smooth submanifold of the parameter space and is not equal to 0 at any point of the parameter space; if this is not so for a given family, then it can be achieved by an arbitrarily small deformation of the family so that it becomes generic.

F. The Thom Transversality Theorem

The Thom transversality theorem is a generalization of the weak transversality theorem, in which the role of the submanifold C is played by a submanifold of a space of jets.

With every smooth mapping $f: M \to N$, we associate its "k-jet extension" $\hat{f}: M \to J^k(M, N)$, $\hat{f}(x) = j_x^k(f)$. (To a point x of M, there corresponds the k-jet of the mapping f at x.)

Theorem. *Let C be a submanifold of the space* $J^k(M, N)$ *of jets. The set of mappings* $f: M \to N$ *whose k-jet extensions are transversal to C is an everywhere dense countable intersection of open sets in the space of all smooth mappings of M into N.*

This theorem means that a smooth mapping can be brought about by a small perturbation to the general position not only with respect to any smooth submanifold in the range space, but also with respect to any condition imposed on derivatives of any finite order.

Remark. The weak transversality theorem can be obtained from the above theorem for $k = 0$. On the other hand, the strong theorem cannot be obtained from the weak theorem. It is possible to apply the weak theorem to a mapping $\hat{f}: M \to J^k$ and obtain a mapping close to $\hat{f}$ and transversal to C. However, this nearby mapping will not generally be a k-jet extension of any smooth mapping of M into N.

The Thom transversality theorem asserts that the transversalizing deformation can be chosen in a smaller class of deformations: it is sufficient to restrict oneself to a deformation of a k-jet extension in the space of k-jet extensions and not in the space of all sections $M \to J^k$. Consequently, the theorem means that the integrability conditions (which are satisfied by the k-jet extensions of mappings of M into N, but not by arbitrary sections $M \to J^k$) do not interfere with transversality.

◀ The essence of the proof consists of the same kind of reduction to Sard's lemma as in the case of the weak transversality theorem. The main difference lies in the fact that the transversalizing deformation is not sought for the class of mappings $f_\varepsilon = f - \varepsilon$, but for the larger class of polynomial deformations $f_\varepsilon = f + \varepsilon_1 e_1 + \cdots + \varepsilon_s e_s$, where e_i are all possible vector monomials of degree not greater than k.

Lemma 1. *We consider a smooth mapping $F: A \times E \to B$ of the direct product of the smooth manifolds A and E into a smooth manifold B.*

We shall consider F as a family of mappings F_ε of A into B depending on the point ε of E as a parameter. If the mapping F is transversal to a submanifold C of B, then almost every member $F_\varepsilon: A \to B$ of the family corresponding to F is transversal to C.

◀ We consider $F^{-1}(C)$. By the implicit function theorem, this is a smooth submanifold in $A \times E$. We consider the projection of this submanifold onto E in the direction of A. By Sard's lemma, almost all values are noncritical. Let ε be a noncritical value. The mapping $F_\varepsilon: A \to B$ is transversal to C (because F is transversal to C and $A \times \varepsilon$ is transversal to $F^{-1}(C)$.) ▶

Lemma 2. *Let f be a smooth mapping of $\mathbb{R}^m$ into $\mathbb{R}^n$. In $\mathbb{R}^m$ and $\mathbb{R}^n$, we fix coordinate systems and consider the smooth mapping of the direct product of $\mathbb{R}^m$ with $\mathbb{R}^s$ into the space $J^k(M, N)$ of k-jets of mappings defined by the formula*

$$(x, \varepsilon) \mapsto (j_x^k f_\varepsilon),$$

where $f_\varepsilon = f + \varepsilon_1 e_1 + \cdots \varepsilon_s e_s$. $e_1, \cdots, e_s$ are all possible products of monomials of degree not greater than k in the coordinates of the point x in $\mathbb{R}^m$ with the basis vectors of $\mathbb{R}^n$.

This mapping does not have critical values (and, consequently, it is transversal to any submanifold of the space of k-jets).

◀ The coordinates in J^k are the coordinates of x in $\mathbb{R}^m$ and the Taylor coefficients of the jet at this point up to order k inclusive. For an appropriate choice of the coefficients $\varepsilon_1, \ldots, \varepsilon_s$, the vector-valued polynomial $\varepsilon_1 e_1 + \cdots + \varepsilon_s e_s$ will have any collection of Taylor coefficients given beforehand up to order k inclusive at any point x given beforehand. The lemma follows immediately. ▶

Let C be a smooth submanifold in $B = J^k(\mathbb{R}^m, \mathbb{R}^n)$. We apply Lemma 1 (in which $A = \mathbb{R}^m$, $E = \mathbb{R}^s$, and $F(x, \varepsilon) = j_x^k f_\varepsilon$) to the mapping of Lemma 2. By Lemma 1, the mapping $F_\varepsilon = F(\cdot, \varepsilon)$ is transversal to C for almost all ε. Choosing ε sufficiently small, we obtain a mapping $f_\varepsilon: \mathbb{R}^m \to \mathbb{R}^n$ arbitrarily close to f (in any finite part of $\mathbb{R}^m$) whose k-jet extension is transversal to C. The passage from this local construction to the global one (replacement of $\mathbb{R}^m$, $\mathbb{R}^n$ by M, N) does not create any difficulty. ▶

G. An Example: Disintegration of Complicated Singular Points of a Vector Field

As an application of the transversality theorems, we consider the problem of characterizing the singular points of a generic vector field.

Definition. A singular point x of a vector field v is said to be *nondegenerate* if the operator of the linear part of the field at the singular point is nonsingular.

From the transversality theorems, we obtain the following corollary.

Corollary. *In the function space of smooth vector fields on a compact manifold, the fields with only nondegenerate (and consequently, isolated) singular points form an open, everywhere dense set.*

◀ The singular points are preimages of a smooth manifold (the zeroth section) in the space of 0-jets of vector fields. Nondegeneracy of a singular point is transversality to this manifold of the 0-jet extension of the field. ▶

Hence, a degenerate singular point disintegrates into nondegenerate points under an arbitrarily small perturbation of the field.

EXAMPLE

We consider the following singular point of "saddle-node" type:

$$\dot{x} = x^2, \qquad \dot{y} = -y.$$

Under the perturbation $\dot{x} = x^2 - \varepsilon$, $\dot{y} = -y$, the saddle-node point disintegrates into two singular points: a saddle and a node.

One then asks into how many singular points a given complicated singular point may disintegrate under small perturbations. As usually happens (say, in the theory of algebraic equations), this problem can be solved in the most natural way in the complex domain.

Definition. The number of nondegenerate (complex) singular points into which a complicated singular point disintegrates under a small perturbation is called the *multiplicity* of the singular point.

Remark. Strictly speaking, multiplicity is defined in the following way: (1) a sufficiently small neighborhood of the singular point in a complex space is fixed; (2) this neighborhood determines the smallness of a perturbation; (3) for the perturbed field, the number of singular points is calculated in the neighborhood of the given point.

Below, a formula will be given for the multiplicity of a singular point in terms of Newton diagrams (Kušnirenko, Bernstein, and Hovanskiï). See "The geometry of formulas" by A. G. Hovanskii in *Singularities of Functions,*

Wave Fronts, Caustics and Multidimensional Integrals (V. I. Arnold, A. N. Varchenko, A. B. Givental, A. G. Hovanskii, eds.), Soviet Scient. Reviews, C: Math. Phys. Reviews **4** (1984), 1–92.

Let $f = \sum f_m x^m$ be a formal scalar-valued series in the variables $x_1, \cdots, x_n$ ($x^m = x_1^{m_1} \cdots x_n^{m_n}$). Consider the octant of the points m of the integral lattice with nonnegative coordinates m_k.

We denote this octant by $\mathbb{Z}_+^n$.

Definition. The *support* of the series f is the set of points m in the octant $\mathbb{Z}_+^n$ for which $f_m \neq 0$. Notation:

$$\operatorname{supp} f = \{m \in \mathbb{Z}_+^n : f_m \neq 0\}.$$

Definition. The *Newton polyhedron* of the series f is the convex hull of the union of octants parallel to $\mathbb{Z}_+^n$ with vertices at the points of the support in the octant $\mathbb{R}_+^n$ of the real linear space. Notation:

Γ_f is the convex hull of the union of $m + \mathbb{Z}_+^n$, $m \in \operatorname{supp} f$.

A Newton polyhedron is said to be *appropriate* if it intersects all coordinate axes.

Theorem. *Let n appropriate Newton polyhedra $\Gamma_1, \ldots, \Gamma_n$ be given.*

Consider a vector field $v_1 \partial/\partial x_1 + \cdots + v_n \partial/\partial x_n$, with $\Gamma_1, \ldots, \Gamma_n$ the Newton polyhedra of the components $v_1, \ldots, v_n$. Then the multiplicity μ of the singular point 0 of our vector field is not smaller than the Newton number $\nu(\Gamma_1, \ldots, \Gamma_n)$ defined below and coincides with it for almost all fields whose components have given Newton polyhedra (for all fields except a hypersurface in the space of fields with the given polyhedra).

Remark. The condition of appropriateness of the polyhedra is not a restriction, since it can be proved that it can be satisfied by adding terms of arbitrarily high degree without changing the multiplicity (provided that it is finite).

For the definition of the Newton number of a system of appropriate polyhedra, we need the notion of mixed volume.

Let Γ be an appropriate Newton polyhedron. By $V(\Gamma)$ we denote the volume of the (nonconvex) domain between zero and the boundary of Γ in the positive octant $\mathbb{R}_+^n$.

Let Γ_1, Γ_2 be two appropriate Newton polyhedra. The sum $\Gamma_1 + \Gamma_2$ is meant as an arithmetic sum, i.e., the set of sums of vectors from Γ_1 and Γ_2. The sum is also an appropriate Newton polyhedron.

Consequently, the appropriate Newton polyhedra form a commutative semigroup. From this semigroup, a group (called the Grothendieck group) can be constructed by the usual method: an element of the group is a formal difference $\Gamma_1 - \Gamma_2$ of two Newton polyhedra and $\Gamma_1 - \Gamma_2 = \Gamma_3 - \Gamma_4$ by definition if and only if $\Gamma_1 + \Gamma_4 = \Gamma_2 + \Gamma_3$.

The group thus constructed determines a linear space over the field of real numbers: if λ is a positive number, then $\lambda\Gamma$ denotes the polyhedron obtained from Γ by a homothety with center at zero and coefficient λ. The volume $V(\Gamma)$ can be extended uniquely to this linear space as a form of degree n. (The proof of this not entirely obvious fact is left to the curious reader as an exercise.)

Every form of degree n can be uniquely represented as the value of a symmetric n-linear form for coinciding arguments. For example,

$$a^2 = ab|_{a=b}, \quad ab = \frac{(a+b)^2 - a^2 - b^2}{2}.$$

Definition. The *mixed* (Minkowski) *volume* of a system $(\Gamma_1, \ldots, \Gamma_n)$ of polyhedra is the value, at the n-tuple $(\Gamma_1, \ldots, \Gamma_n)$, of the unique symmetric n-linear form which coincides with the volume $V(\Gamma)$ for $\Gamma_1 = \cdots = \Gamma_n = \Gamma$.

Notation: $V(\Gamma_1, \ldots, \Gamma_n)$.

EXAMPLE

In the planar case, $n = 2$ and the mixed volume of a pair (Γ_1, Γ_2) is $V(\Gamma_1, \Gamma_2)$ $= [V(\Gamma_1 + \Gamma_2) - V(\Gamma_1) - V(\Gamma_2)]/2$.

Definition. The *Newton number* $v(\Gamma_1, \ldots, \Gamma_n)$ is defined in the following way:

$$v(\Gamma_1, \ldots, \Gamma_n) = n! \, V(\Gamma_1, \ldots, \Gamma_n).$$

EXAMPLE

In the two-dimensional case, let Γ_1, Γ_2 be bounded by straight lines intersecting the coordinate axes at the points (a_1, b_1) for Γ_1 and (a_2, b_2) for Γ_2. Then $v(\Gamma_1, \Gamma_2)$ is equal to $\min(a_1 b_2, a_2 b_1)$. Consequently, the multiplicity of the singular point is almost always equal to

$$\mu = \min(a_1 b_2, a_2 b_1).$$

§ 30. Matrices Depending on Parameters and Singularities of the Decrement Diagram

As a preparation for the study of bifurcations of singular points of vector fields, we consider the problem of the normal form of families of endomorphisms of a linear space.

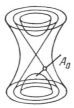

Figure 110.

A. The Problem of Normal Form of Matrices Depending on Parameters

The reduction of a matrix to the Jordan normal form is not a stable operation if there are multiple eigenvalues. Indeed, in the presence of multiple eigenvalues, an arbitrarily small change in the matrix may change the Jordan form completely. Therefore, if the matrix is known only approximately, then its reduction to Jordan normal form is practically impossible in the case of multiple eigenvalues. It is not even necessary, since a generic matrix does not have multiple eigenvalues.

Multiple eigenvalues are unremovable under small perturbations when we are interested in a family of matrices depending on parameters rather than in an individual matrix. In this case, although we can reduce every individual matrix of the family to a Jordan normal form, both this normal form and the transformation leading to it generally depend discontinuously on the parameter.

Consequently, a problem arises: What is the simplest form to which a family of matrices depending smoothly (for the sake of definiteness, holomorphically) on the parameters can be reduced by a change of coordinates depending smoothly (holomorphically) on the parameters?

Let us consider the set of all complex square matrices of order n as a linear space of dimension n^2. The relation of similarity of matrices partitions the whole space $\mathbb{C}^{n^2}$ into manifolds (orbits of a linear group): two matrices lie in the same orbit if their eigenvalues and the dimensions of the Jordan blocks coincide. Because of the eigenvalues, this partition is continuous. As a rough model, one can imagine the division of the three-dimensional space into the strata of the manifolds $x^2 + y^2 - z^2 = C$ (Fig. 110).

A family of matrices is given by a mapping of the space of the parameters of the family into the space $\mathbb{C}^{n^2}$ of matrices. It turns out that from all families of matrices we can select such families that a reduction to them can be effected by a change of basis which now depends on the parameters smoothly (and by a smooth change of parameters). Such families are called versal deformations (the precise definition is given below). Versal deformations with the minimum possible number of parameters are said to be miniversal.

Consequently, miniversal deformations are normal forms with the smallest possible number of parameters in the reduction to which the smooth dependence on the parameters can be preserved.

EXAMPLE

If all eigenvalues of a diagonal matrix are distinct, then we may take the family of all diagonal matrices as its miniversal deformation (the parameters are the eigenvalues).

Below we consider miniversal deformations of arbitrary matrices.

B. Versal Deformations*

Definition. A *family of matrices* is a holomorphic mapping $A : \Lambda \to \mathbb{C}^{n^2}$, where Λ is a neighborhood of the origin of coordinates in some parameter space $\mathbb{C}^l$. The germ of a family A at the point 0 will be called a *deformation* of the matrix $A(0)$.

A deformation A' of the matrix $A(0)$ is said to be *equivalent* to a deformation A if a deformation C of the identity exists such that

$$A'(\lambda) = C(\lambda)A(\lambda)(C(\lambda))^{-1}.$$

Let $\varphi : (M, 0) \to (\lambda, 0)$ be a holomorphic mapping ($M \subset \mathbb{C}^m$ and $\Lambda \subset \mathbb{C}^l$).

Definition. The family *induced from A by the mapping* φ is the family $\varphi^* A$:

$$(\varphi^* A)(\mu) = A(\varphi(\mu)).$$

The induced deformation $\varphi^* A$ of the matrix $A(0)$ is defined by the same formula.

Definition. A deformation A of a matrix A_0 is said to be *versal* if any deformation A' of the matrix A_0 is equivalent to a deformation induced from A. A versal deformation is said to be *universal* if the inducing mapping is determined uniquely by the deformation A'. A versal deformation is said to be *miniversal* if the dimension of the parameter space is the smallest possible for a versal deformation.

EXAMPLE

The family of diagonal matrices with diagonal entries $(\alpha_i + \lambda_i)$, where all α_i are distinct and the λ_i are the parameters of the deformation, is a versal, universal, and miniversal deformation of the matrix (α_i).

The family $\mathbb{C}^{n^2}$ of all matrices determines an n^2-parameter versal deformation of any of its members. However, this deformation is in general neither universal nor miniversal.

The dimension of a miniversal deformation of an arbitrary matrix is given by the following theorem. Denote by α_i the eigenvalues of the matrix A_0 and

---

* Versal deformations were introduced by Poincaré (Lemma IV of his Thesis), he has studied the versal deformations of equilibria and cycles in dynamical systems.

let $n_1(\alpha_i) \geqslant n_2(\alpha_i) \geqslant \cdots$ be the dimensions of the Jordan blocks belonging to α_i, beginning with the largest one.

Theorem 1. *The smallest number of parameters of a versal deformation of the matrix A_0 is equal to*

$$\sum_i [n(\alpha_i) + 3n_2(\alpha_i) + 5n_3(\alpha_3) + \cdots].$$

The miniversal deformations themselves can be chosen in different ways. In particular, the three normal forms described in the following theorem are versal deformations of a matrix reduced to the upper triangular Jordan normal form.

Theorem 2. *Let A be a family of linear operators of $\mathbb{C}^n$ into itself depending holomorphically on a parameter $\lambda \in \mathbb{C}^l$ and let the operator $A(\lambda_0)$ have the eigenvalues α_i and Jordan blocks of orders*

$$n_1(\alpha_i) \geqslant n_2(\alpha_i) \geqslant \cdots .$$

Then a basis in $\mathbb{C}^n$ exists which depends holomorphically on λ varying in some neighborhood of λ_0 and such that the matrix of the operator $A(\lambda)$ has the block-diagonal form

$$A_0 + B(\lambda)$$

in this basis, where A_0 is the Jordan upper triangular matrix of $A(\lambda_0)$ and $B(\lambda)$ a block-diagonal matrix whose blocks correspond to the eigenvalues of A_0.

The block B_i corresponding to the eigenvalue α_i has all zeros except at the places indicated in Fig. 111; these places are taken by holomorphic functions of λ.

In Fig. 111, three normal forms are depicted. In the first two, the number of nonzero entries of B_i is equal to $n_1(\alpha_i) + 3n_2(\alpha_i) + \cdots$; in the third, all entries are equal on every slanted line. We obtain miniversal deformations of A_0 if the indicated entries of the matrices B_i are considered independent variables; in all three cases their number is equal to $\sum [n_1(\alpha_i) + 3n_2(\alpha_i) + \cdots]$. The advantage of the first two normal forms is that the number of nonzero entries is the smallest possible. The advantage of the third form is the ortho-

(a)   (b)   (c)

Figure 111.

gonality of the versal deformation to the corresponding orbit (in the sense of the entry-wise scalar product of matrices).

C. Proof of the Versality

Let $A: \Lambda \to \mathbb{C}^{n^2}$ be a defomation of the matrix $A_0 = A(0)$ with parameter $\lambda \in \Lambda$ such that the mapping A is transversal to the orbit C of the matrix A_0 under the action of the group of linear changes of coordinates. We assume that the number of parameters of deformation is minimal (i.e., is equal to the codimension of the orbit in the space $\mathbb{C}^{n^2}$ of all matrices). Such a deformation is said to be *minitransversal*.

Lemma 1. *A minitransversal deformation A is miniversal.*

For the proof of the lemma we need the following definition.

Definition. The *centralizer* of a matrix u is the set of all matrices commuting with u. Notation:

$$Z_u = \{v: [u, v] = 0\}, \; [u, v] = uv - vu.$$

The centralizer of any matrix of order n is a linear subspace of the space $\mathbb{C}^{n^2}$ of all matrices of order n.

Let Z be the centralizer of the matrix A_0. In the space of nonsingular matrices through the identity matrix e, we pass a smooth surface transversal to $e + Z$ and of dimension equal to the codimension of the centralizer (i.e., the dimension is the smallest possible).

We denote this surface by P and consider the mapping

$$\Phi: P \times \Lambda \to \mathbb{C}^{n^2}, \qquad \Phi(p, \lambda) = pA(\lambda)p^{-1}.$$

Lemma 2. *In the neighborhood of $(e, 0)$, the mapping Φ is a local diffeomorphism on $(\mathbb{C}^{n^2}, A_0)$.*

For the proof of Lemma 2 we consider the mapping ψ of the group of nonsingular matrices into the space $\mathbb{C}^{n^2}$ of all matrices given by the formula $\psi(b) = bA_0b^{-1}$.

1. The derivative of the mapping ψ at the identity is the operator of commutation with A_0: $\psi_*: \mathbb{C}^{n^2} \to \mathbb{C}^{n^2}$, $\qquad \psi_* u = [u, A_0]$.

$$\blacktriangleleft (e + \varepsilon u)A_0(e + \varepsilon u)^{-1} = A_0 + \varepsilon[u, A_0] + \cdots. \blacktriangleright$$

From the above we obtain the following statement.

2. The dimension of the centralizer of A_0 is equal to the codimension of the orbit. The dimension of the transversal to the centralizer is equal to the dimension of the orbit:

$$\dim Z = \dim \Lambda, \qquad \dim P = \dim C.$$

In the space $\mathbb{C}^{n^2}$, we introduce the Hermitian scalar product $\langle A, B \rangle = \mathrm{tr}(AB^*)$, where B^* is the matrix obtained from B by transposition and complex conjugation. The corresponding scalar square is simply the sum of the squares of the absolute values of all the entries of the matrix.

Lemma 3. *A vector B in the tangent space of $\mathbb{C}^{n^2}$ at the point A_0 is perpendicular to the orbit of A_0 if and only if $[B^*, A_0] = 0$.*

◀ The vectors tangent to the orbit are the matrices representable in the form $[X, A_0]$. Orthogonality of B to the orbit means that $\langle [X, A_0], B \rangle = 0$ for any X. In other words,

$$0 = \mathrm{tr}([X, A_0]B^*) = \mathrm{tr}(XA_0B^* - A_0XB^*)$$
$$= \mathrm{tr}([A_0, B^*]X) = \langle [A_0, B^*], X^* \rangle$$

for any X. Since X was chosen arbitrarily, this condition is equivalent to $[A_0, B^*] = 0$.

Hence, the lemma is proved: the orthogonal complement of the orbit of a matrix can be obtained from its centralizer by transposition and conjugation. ▶

It is easy to describe the centralizer of a matrix in Jordan normal form. First, we assume that the matrix has only one eigenvalue and a sequence of upper Jordan blocks of orders $n_1 \geqslant n_2 \geqslant \cdots$.

Lemma 4. *The matrices in Fig. 112 and only they commute with the matrix A_0.*

In Fig. 112, every slanted segment denotes a sequence of equal entries, and we have zeros at the blank spaces. Consequently, the number of slanted segments is equal to the dimension of the centralizer.

◀ Lemma 4 can be proved by a direct calculation of the commutator (cf., for example, F. R. Gantmaher, The Theory of Matrices, Moscow, Nauka, 1967, p. 215–224). ▶

Figure 112.

It follows from Lemma 4 that the dimension of the centralizer of A_0 (equal to the codimension of the orbit and the minimum dimension of a versal deformation) is given by the formula $d = n_1 + 3n_2 + 5n_3 + \cdots$.

If the Jordan matrix A_0 has several eigenvalues, then we divide it into blocks corresponding to the eigenvalues. The matrices commuting with A_0 will be block-diagonal. A block of the form described in Fig. 110 corresponds to every eigenvalue. Therefore, the formula for the dimension of the centralizer (the codimension of the orbit or the dimension of a miniversal deformation) is obtained by summing over all distinct eigenvalues.

Indeed, ψ_* is a linear mapping between spaces of the same dimension, and, therefore, the dimension of the kernel is equal to the codimension of the range.

Proof of Lemma 2. ◀ The derivative of Φ with respect to p at $(e, 0)$ is ψ_*, and the derivative with respect to λ is A_*. According to what has been proved above, these operators isomorphically map the space tangent to P at e and the space tangent to Λ at 0 onto transversal spaces of the same dimensions (the tangent space to the orbit C at A_0 for P and a space transversal to it for Λ). Consequently, the derivative of Φ at $(e, 0)$ is an isomorphism between linear spaces of dimension n^2. By the inverse function theorem, Φ is a local diffeomorphism. ▶

Proof of Lemma 1. We will consider p and λ as coordinates of the point $\Phi(p, \lambda)$. Let $A' : (M, 0) \to (\mathbb{C}^{n^2}, A_0)$ be any deformation of A_0. Let $\mu \in M$ be the parameter of the deformation. We define $\lambda = \varphi(\mu)$ by $\varphi(\mu) = \lambda(A'(\mu))$ and set $B(\mu) = p(A'(\mu))$. Then $A'(\mu) = B(\mu)A(\varphi(\mu))B^{-1}(\mu)$, which proves the versality of the deformation A.

The minimality of the dimension of the basis of this deformation is evident.

As a transversal deformation of A_0, we may take the family of matrices $A_0 + B$, where the matrix B belongs to the orthogonal complement of the orbit of A_0. In this way, we obtain a miniversal deformation of A_0.

If A_0 has only one eigenvalue, the matrix B has the form indicated in Fig. 111(c). Here, every slanted segment denotes the line of equal numbers; the number of parameters is equal to the number of lines and is given by the formula indicated above.

The matrix B has many nonzero entries. It is possible to construct miniversal deformations $A_0 + B$ in which the number of nonzero entries of B is the smallest possible (equal to the number of parameters). To this end, we choose a basis in the centralizer: To every slanted line in Fig. 111(c) we assign a 0-1 matrix in which the 1's stand on the slanted line.

A system of independent equations of the tangent plane of the orbit consists of the following equations: for every slanted line in Fig. 111, the sum of the corresponding entries of the matrix is equal to zero (Lemmas 3 and 4). Consequently, in order to obtain a family $A_0 + B$ transversal to the orbit, it is sufficient to choose, as the family of the matrices B, the matrices in which on every slanted line of Fig. 111(c) an independent parameter stands at one place and zeros at the remaining places. It is possible to choose the nonzero entry at any place on every slanted line. For example, the choice indicated in Theorem 2 of § 30B is suitable.

D. Examples

We shall denote an upper triangular Jordan matrix by the product of the determinants of its blocks. For example, α^2 means a Jordan block of order 2 and $\alpha\alpha$ a 2×2 matrix which is a multiple of the identity matrix.

The first normal form of the theorem of § 30B leads to the following miniversal deformations.

(a) A versal (and universal) two-parameter deformation of the Jordan block α^2 of order 2:

$$\begin{pmatrix} \alpha & 1 \\ 0 & \alpha \end{pmatrix} + \begin{pmatrix} 0 & 0 \\ \lambda_1 & \lambda_2 \end{pmatrix}. \tag{1}$$

(b) A versal (but not universal) four-parameter deformation of the scalar matrix $\alpha\alpha$ of order 2:

$$\begin{pmatrix} \alpha & 0 \\ 0 & \alpha \end{pmatrix} + \begin{pmatrix} \lambda_1 & \lambda_2 \\ \lambda_3 & \lambda_4 \end{pmatrix}.$$

(c) A versal and universal three-dimensional deformation of the Jordan block α^3:

$$\begin{pmatrix} \alpha & 1 & 0 \\ 0 & \alpha & 1 \\ 0 & 0 & \alpha \end{pmatrix} + \begin{pmatrix} 0 & 0 & 0 \\ 0 & 0 & 0 \\ \lambda_1 & \lambda_2 & \lambda_3 \end{pmatrix}.$$

(d) A versal five-parameter deformation of the matrix $\alpha^2\alpha$:

$$\begin{pmatrix} \alpha & 1 & 0 \\ 0 & \alpha & 0 \\ 0 & 0 & \alpha \end{pmatrix} + \begin{pmatrix} 0 & 0 & 0 \\ \lambda_1 & \lambda_2 & \lambda_3 \\ \lambda_4 & 0 & \lambda_5 \end{pmatrix}.$$

For example, every holomorphic family of matrices containing the Jordan block α^2 for the value 0 of the parameter can be reduced to the normal form of (1) for nearby values of the parameter, where λ_1, λ_2 are holomorphic functions of the parameters.

In the study of many problems about operators depending on parameters, the normal forms constructed above enable us to restrict ourselves to special families: miniversal deformations. One of these problems is that of the structure of bifurcation diagrams.

E. Bifurcation Diagrams

A *bifurcation diagram* of a family of matrices is, by definition, a partition of the parameter space Λ according to Jordan types of matrices. A family is a mapping $A : \Lambda \to \mathbb{C}^{n^2}$ of the parameter space into the space of matrices. Therefore, to understand bifurcation diagrams, we have to study the partition of the space of all matrices into matrices with Jordan forms of distinct types. In this partition, we group together the matrices with the same dimensions of Jordan blocks, differring only in the eigenvalues. The partition thus obtained is a finite stratification of the space of matrices.

Every stratum of this stratification is determined by the set of the collections $n_1(i) \geqslant n_2(i) \geqslant \cdots$ of dimensions of the Jordan blocks corresponding to v distinct eigenvalues $(1 \leqslant i \leqslant v)$. The codimension c of such a stratum in the space $\mathbb{C}^{n^2}$ is smaller than the codimension d of the corresponding orbit by the number of distinct eigenvalues, i.e., by v:

$$c = d - v = \sum_{i=1}^{v} [n_1(i) + 3n_2(i) + \cdots - 1].$$

We note that simple eigenvalues contribute nothing to this sum. Applying the weak transversality theorem, we obtain the following.

Theorem. *In the space of families of matrices of order n, the families transversal to the stratification into Jordan types constitute an everywhere dense set.*

This theorem, together with the formulas of versal deformations in § 30D, enables us to describe the bifurcation diagrams of generic families. In particular, for families with a small number of parameters, we obtain the following.

1. One-Parameter Families. From $c = 1$, it follows that the matrix has only one eigenvalue of multiplicity 2 and to it there corresponds a Jordan block of order 2. Such a stratum will be denoted by α^2.

Corollary. *In a generic one-parameter family there are only matrices with simple eigenvalues and, for some isolated values of the parameters, matrices of type α^2 (with one Jordan block of order 2). If in a family there are matrices with a more complicated Jordan structure, then we can remove them by an arbitrarily small perturbation of the family.*

Figure 113.

2. Two-Parameter Families. There exist exactly two Jordan types with $c = 2$: α^3 (one Jordan block of order 3) and $\alpha^2\beta^2$ (two blocks of order 2 with distinct eigenvalues).

Corollary. *The bifurcation diagram of a generic two-parameter family of matrices has the form of a curve whose only singularities are cusps and points of self-intersection (Fig. 113). To the cusps correspond matrices of type α^3 with one Jordan block of order 3; to the points of self-intersection correspond matrices of type $\alpha^2\beta^2$ with two Jordan blocks of order 2 with distinct eigenvalues, and to the points of the curve correspond matrices with one Jordan block of order 2. Matrices with simple eigenvalues correspond to points outside the curve.*

If a family contains matrices of more complicated types or the bifurcation diagram has more complicated singularities, then we can remove them by an arbitrarily small perturbation of the family.

3. Three-Parameter Families. There are four strata with $c = 3$: $\alpha^2\beta^2\gamma^2$ (three 2-blocks), $\alpha\alpha$(two blocks of order 1 with the same eigenvalue), $\alpha^2\beta^3$ (2 blocks of orders 2 and 3), and α^4 (a 4-block).

Consequently, the point singularities of bifurcation diagrams of generic three-parameter families have the form indicated in Fig. 114. The singularity α^4 is called the *swallow tail*: this surface is given by the equation $\Delta(a, b, c) = 0$, where Δ is the discriminant of the polynomial $z^4 + az^2 + bz + c$. Strictly speaking, all of what has been said above concerns the complex case; surfaces in Fig. 114 have to be considered complex.

Versal deformations of real matrices have been constructed by Galin [D. M. Galin, *On real matrices depending on parameters*, Uspekhi Math. Nauka **27**, 1 (1972), 241–242]. The construction is performed in the following

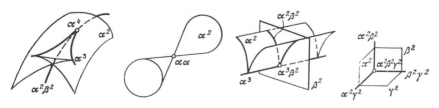

Figure 114.

way. First, let the real operator in $\mathbb{R}^{2n}$ for which the versal deformation is sought have a unique pair of complex-conjugate eigenvalues $x \pm iy$ ($y \neq 0$) with Jordan blocks of dimensions $n_1 \geqslant n_2 \geqslant \cdots$, so that $n_1 + n_2 + \cdots = n$. In some real basis in $\mathbb{R}^{2n}$, the matrix of the operator has the same form as the matrix of the decomplexification of the complex Jordan operator $\hat{A}_0 : \mathbb{C}^n \to \mathbb{C}^n$ with the only eigenvalue $x + iy$ and Jordan blocks of dimensions $n_1 \geqslant n_2 \geqslant \cdots$, i.e., the form

$$A_0 = \begin{pmatrix} X & -yE \\ yE & X \end{pmatrix}, \tag{2}$$

where X is the upper-triangular real Jordan matrix with eigenvalue x and blocks of dimensions $n_1 \geqslant n_2 \geqslant \cdots$, and E is the identity matrix of order n.

It turns out that *we may choose the decomplexification of a minimal complex versal deformation of the complex matrix $\tilde{A}_0$ as a minimal versal deformation of the real matrix A_0.*

Let us, for example, look for a minimal versal deformation of a real matrix of the fourth order with two Jordan blocks of the second order and eigenvalues $x \pm iy$. We may choose the four-parameter deformation which is obtained by decomplexifying the complex versal deformation

$$\begin{pmatrix} z & 1 \\ 0 & z \end{pmatrix} + \begin{pmatrix} 0 & 0 \\ \lambda_1 & \lambda_2 \end{pmatrix}_2,$$

i.e., the following deformation with the parameters $\rho_1, \rho_2, \tau_1, \tau_2$:

$$\begin{pmatrix} x & 1 & -y & 0 \\ 0 & x & 0 & -y \\ y & 0 & x & 1 \\ 0 & y & 0 & x \end{pmatrix} + \begin{pmatrix} 0 & 0 & 0 & 0 \\ \rho_1 & \rho_2 & -\tau_1 & -\tau_2 \\ 0 & 0 & 0 & 0 \\ \tau_1 & \tau_2 & \rho_1 & \rho_2 \end{pmatrix}, \quad \begin{array}{l} z = x + iy \\ \lambda_k = \rho_k + i\tau_k \end{array}.$$

By means of a similarity over the field of real numbers, every real matrix can be reduced to a block-diagonal form where a real Jordan matrix corresponds to every real eigenvalue and a block of the form (2) to every pair of complex-conjugate eigenvalues.

We obtain a real versal deformation of a matrix reduced to this form (with the smallest possible number of parameters) if we replace every block by its minimal versal deformation. The minimum number of parameters of a real versal deformation is thus given by the formula

$$d = \sum_\lambda [n_1(\lambda) + 3n_2(\lambda) + 5n_3(\lambda) + \cdots],$$

where the summation is extended to all ν eigenvalues, both real and complex.

Explicit formulas of versal deformations and tables of bifurcation

diagrams of real matrices are provided in the work of Galin for $d - v \leqslant 3$.
[D. M. Galin, Versal deformations of linear Hamiltonian systems, Trudy
Seminara Im. I. G. Petrovskogo, 1975, 1, p. 63–74, gives tables of versal
deformations of symplectic and Hamiltonian (infinitesimally symplectic)
matrices (we have in mind deformations preserving symplectic character).]
They are important for applications in mechanics.

One of the applications of the bifurcation diagrams obtained above
consists of the following. Let us assume that in the study of a phenomenon
we obtain a bifurcation diagram of structure different from those listed here.
Then it is likely that one of the following two events have occurred: either in
the idealization of the phenomenon something essential was missed which has
changed the structure of the diagram qualitatively or there were some special
reasons for an additional multiplicity of the spectrum or for the nontrans-
versality to the Jordan stratification (for example, symmetry or the Hamiltonian nature of the problem).

F. The Problem of Classification of Singularities of Decrement Diagrams

As an application of versal deformations of matrices, we consider the solution of the
following problem. Let a family of homogeneous linear autonomous differential
equations be given. It is well known that the asymptotics of solutions as $t \to +\infty$ are
determined by that eigenvalue of the operator which has the largest real part. We ask
how this linear part depends on the parameters.

In engineering, the indicated real part (with a minus sign) is called the decrement.
Thus, our problem consists of the study of the behavior of the decrement under the
variation of the parameters of the system.

It is convenient to describe the behavior of the decrement under the variation of the
parameters by level curves (surfaces) of the decrement in the plane (space) of the
parameters. A family of level curves of the decrement in the plane of the parameters will
be called a decrement diagram.

The form of a decrement diagram varies from family to family: in some cases, a
decrement diagram can have very complicated singularities. It turns out, however, that
in generic families only certain simple singularities of decrement diagrams may occur:
all more complicated singularities disintegrate under a small perturbation of the family.

In this section we describe all singularities of decrement diagrams of generic two-
parameter families.

In the study of the dependence of systems on parameters, the classification of singularities of generic decrement diagrams may render the same service as the classification of generic singular points in the study of phase portraits.

The appearance of a nongeneric singularity in a decrement diagram should cause
alarm: it may be explained by a special symmetry of the system or it may show the
inadequacy of the idealization ("ill-posedness"), in which small effects not accounted
for in the equations (for example, "parasitic connections" or "parasitics" in radio-
electronics) are able to change the picture qualitatively.

The classification of singularities of generic two-parameter decrement diagrams
contains, in particular, the study of singularities of the boundary of a domain of
stability in generic three-parameter families of linear equations (surfaces of vanishing
decrements).

These results may also be applied to nonlinear systems having stationary points depending smoothly on the parameters: at these points, the decrement of the linearization of the nonlinear system will have only the simplest singularities as a function of the parameters (in the case of a generic family.)

When we apply these results to nonlinear systems, however, we have to exclude the part of the boundary of the domain of stability corresponding to vanishing roots, since the smooth dependence of the stationary point on the parameters is not preserved. Consequently, the desription of singularities of the boundary of the domain of stability of a generic nonlinear system (and the description of decrement diagrams in the neighborhood of points of this boundary) requires additional analysis. We return to this question in the following subsections.

In the study of iterations of mappings, and also of equations with periodic coefficients or motions in the neighborhood of a periodic trajectory, the role of the decrement is played by the largest absolute value of the eigenvalues. If this absolute value is different from 1, then its singularities (as a function of the parameters in a generic family) are the same as those of the decrement of a generic family. Therefore, in the following, we consider only the decrement.

In the study of the absolute values of eigenvalues in noninear problems of the types just indicated, the results of this section are applicable outside the boundary of stability and at those points of the boundary for which 1 is not an eigenvalue.

G. Decrement Diagrams

In the Euclidean space $\mathbb{R}^n$, we consider a family of operators A depending smoothly on the point λ of a parameter space Λ:

$$A(\lambda): \mathbb{R}^n \to \mathbb{R}^n.$$

Definition. The *increment** of the family is the function f of the parameter whose value at λ is equal to the largest real part of the eigenvalues of $A(\lambda)$:

$$f(\lambda) = \lim_{t \to \infty} \frac{1}{t} \ln \left\| e^{A(\lambda)t} \right\|.$$

The function f is continuous but not necessarily differentiable. Our problem is the study of singularities of f for a generic two-parameter family. Consequently, the parameter space Λ may be assumed to be the plane $\mathbb{R}^2$ or a domain in the plane.

A family of level curves of f in the plane Λ will be called a decrement diagram. A dash across a level curve indicates the direction of the slope, i.e., the direction in which f decreases; in other words, the dash points in the direction of increasing stability.

EXAMPLE

We consider the differential equation

$$\ddot{z} = xz + y\dot{z}$$

---

* In engineering, the quantity $|f|$ is called the *decrement* for $f < 0$ and the *increment* for $f > 0$.

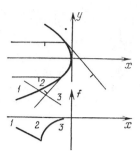

Figure 115.

depending on two parameters (x, y). The matrix of the corresponding system has the form

$$A(x, y) = \begin{pmatrix} 0 & 1 \\ x & y \end{pmatrix}.$$

The decrement diagram is given in Fig. 115. The parabola $4x + y^2 = 0$ divides the (x, y)-plane into two parts. In each of them, the increment is a smooth function. To the left of the parabola the eigenvalues are complex and $f = y/2$. To the right, the eigenvalues are real and $f = (y \pm \sqrt{4x + y^2})/2$. The level curves of the increment are tangent rays to the parabola.

All points of the parabola are singular points of the decrement diagram. To them there correspond matrices A with Jordan blocks of order 2. To the left of the parabola the increment changes linearly, and to the right it changes as the root.

It is clear that the singularity indicated here is unremovable by a small perturbation of the family. There exist other unremovable singularities as well, our purpose is to give their complete list.

H. Strata of Codimension 1 and 2 in the Space of Matrices

If one real eigenvalue or one pair of complex-conjugate eigenvalues* has the maximal real part, then the increment is a smooth function in the neighborhood of the value λ_0 of the parameter.

The smoothness is lost only in the case where the eigenvalue with maximal real part is not unique. The matrices for which several eigenvalues have the maximum real part form a closed semi-algebraic subvariety F in the space $\mathbb{R}^{n^2}$ of all matrices of order n. The codimension of this variety is equal to 1. Its complement consists of two open components:

D_1. Stratum (α). Exactly one real eigenvalue has the maximal real part.

D_2. Stratum $(\alpha \pm i\omega)$. Exactly one complex-conjugate pair has the maximum real part.

---

*Here and in the following we assume that the members of a complex-conjugate pair are not real.

It is easy to stratify the manifold F. The strata of maximal dimension (of codimension 1) are exhausted by the following list:

F_1. *Stratum* (α^2). Exactly two coinciding eigenvalues have the maximal real part: these eigenvalues are real, and a Jordan block of order 2 corresponds to them.

F_2. *Stratum* $(\alpha, \alpha \pm i\omega)$. Exactly three eigenvalues have the maximal real part: one is real and the other two form a complex conjugate pair.

F_3. *Stratum* $(\alpha \pm i\omega_1, \alpha \pm i\omega_2)$. Exactly two distinct complex-conjugate pairs have the maximal real part.

It is clear that the strata F_1, F_2, and F_3 are regular smooth nonclosed disjoint submanifolds of codimension 1 in the space $\mathbb{R}^{n^2}$ of matrices. The remainder $F\backslash(F_1 \cup F_2 \cup F_3)$ of F (the variety of matrices with several eigenvalues with maximal real part) is a closed semi-algebraic* subvariety of codimension 2 in the space $\mathbb{R}^{n^2}$ of all matrices. The strata of maximum dimension of $F\backslash(F_1 \cup F_2 \cup F_3)$ have codimension 2 in $\mathbb{R}^{n^2}$. It is easy to list them:

G_1. *Stratum* (α^3). Exactly three eigenvalues have the maximal real part; they are real and a Jordan block of order 3 corresponds to them.

G_2. *Stratum* $((\alpha \pm i\omega)^2)$. Exactly two coinciding pairs of complex-conjugate numbers have the maximal real part; Jordan blocks of order 2 correspond to them.

G_3. *Stratum* $(\alpha^2, \alpha \pm i\omega)$. Exactly four eigenvalues have the maximal real part; a Jordan block of order 2 corresponds to two real ones and the two complex ones form a complex-conjugate pair.

G_4. *Stratum* $(\alpha, \alpha \pm i\omega_1, \alpha \pm i\omega_2)$. Exactly five eigenvalues have the maximal real part: one real and two distinct complex conjugate pairs.

G_5. *Stratum* $(\alpha \pm i\omega_1, \alpha \pm i\omega_2, \alpha \pm i\omega_3)$. Exactly three distinct complex conjugate pairs have the maximal real part.

The strata G_1–G_5 are regular nonclosed disjoint subvarieties[†] of codimension 2 in the space $\mathbb{R}^{n^2}$ of all matrices. The remainder $F\backslash\cup F_i\backslash\cup G_i$ is a closed semialgebraic subvariety of codimension 3 in $\mathbb{R}^{n^2}$.

The weak transversality theorem (§ 29) implies the following.

Corollary. *In generic two-parameter families of matrices, there are no matrices having collections of eigenvalues with maximal real part other than (D_i, F_i, G_i) listed above; these collections occur only transversally.*

Consequently, in a generic family the codimension 1 eigenvalue collections (F_i) are encountered on smooth curves having singular points only at those points of the parameter plane where the collections (G_i) of codimension 2 occur. The latter phenomenon may occur only at isolated points of the parameter plane.

The segments of F_1 and F_2, if we add their singular points G_i to them, form curves which divide the parameter plane into parts of two types: D_1 and D_2. It is easy to see that all segments F_3 lie in part D_2.

---

*A semi-algebraic subvariety of a linear space is, by definition, a finite union of sets given by finite systems of polynomial equations and inequalities.

[†]All varieties D_i, F_i, G_i are connected for sufficiently large n. Exceptions: for $n = 2$: D_2 and F_1, for $n = 4$: F_3, G_2, and G_3, and for $n = 6$: G_5 have two components.

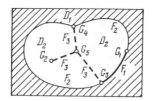

Figure 116.

Moreover, the points

$G_1(\alpha^3)$ lie on the junction of $F_1(\alpha^2)$ and $F_2(\alpha, \alpha \pm i\omega)$.

$G_2((\alpha \pm i\omega)^2)$ adhere to $F_3(\alpha \pm i\omega_1, \alpha \pm i\omega_2)$;

$G_3(\alpha^2, \alpha \pm i\omega)$ are on the junction $F_1(\alpha^2)$, $F_2(\alpha \pm i\omega)$, $F_3(\alpha \pm i\omega_{1,2})$;

$G_4(\alpha, \alpha \pm \omega_{1,2})$ are on the junction of $F_2(\alpha, \alpha \pm i\omega)$ and $F_3(\alpha \pm i\omega_{1,2})$;

$G_5(\alpha \pm i\omega_{1,2,3})$ adhere to $F_3(\alpha \pm i\omega_{1,2})$.

In Fig. 116, we illustrated a (hypothetical) example of a configuration which can be formed by these curves in the parameter plane of a generic family.

I. Structure of Decrement Diagrams near Points of Codimension 0 or 1 Strata

In the complement of the set F of singularities, the increment f is a smooth function of the parameters. Nevertheless, at some points of this complement, the decrement diagrams may have singularities: they are the critical points of the function f.

Outside F, the increment of a generic family has only simple critical points, i.e., points of the following three types (or six types, if we distinguish between the cases (D_1) of real eigenvalues and (D_2) of complex eigenvalues):

D_i^0. *Minimum.* In the neighborhood of the critical point of the parameter plane under consideration, we can choose smooth coordinates (x, y) such that the increment will have the form $f = \text{const} + x^2 + y^2$.

D_i^1. *Saddle Point.* In appropriate coordinates, we have $f = \text{const} + x^2 - y^2$.

D_i^2. *Maximum.* $f = \text{const} - x^2 - y^2$.

Let us now study the behavior of the function f near nonsingular points of F, i.e., near interior points of the curves F_i in the parameter plane. We distinguish between two cases: a point of a curve F_i is either noncritical for the increment viewed as a smooth function on the curve, or it is critical.

The transversality theorem implies that in generic families, critical points of the restrictions of the increment to the curves F_i can only be nondegenerate maxima or minima.

Combining this information with the explicit formulas of versal families of matrices from § 30B, we obtain, without any difficulty, the following normal forms of the increment near points of strata of codimension 1.

Theorem. *In the neighborhood of a noncritical point of the restriction of the increment of a generic family to a curve F_i, smooth coordinates (x, y) can be chosen in the parameter plane so that the increment f assumes one of the following three forms (Fig. 117):*

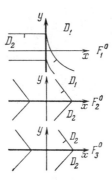

Figure 117.

Case F_1^0 (Jordan Block)

$$f = \text{const} + y + \begin{cases} \sqrt{x} & \text{if } x \geqslant 0, \\ 0 & \text{if } x \leqslant 0. \end{cases}$$

Cases F_2^0 and F_3^0 (Simple Passing)

$$f = \text{const} + x + |y|.$$

The curves F_1^0 and F_2^0 divide the domains D_1 and D_2 of real and complex roots. The level curves of the increment meet F_1 tangentially from the side of the real roots and transversally from the side of the complex roots. The level curves of the decrement meet F_2 and F_3 transversally from both sides at the points F_2^0 and F_3^0. The angle of level curves (smaller than 180°) on the segments F_i contains the direction of the decrease of f along the curve in all cases.

Theorem. *In the neighborhood of a critical point of the restriction of the increment of a generic family, coordinates (x, y) can be chosen so that the increment takes one of the 12 forms below (Fig. 118).*

Cases F_2^k and F_3^k, $k = 1, \ldots, 4$ (Conditional Extremum on Passing)

$$f = \text{const} + \varepsilon x^2 + \varphi(y) + |y|, \qquad \varepsilon = (-1)^k,$$

where $\varphi(y) = ay + \cdots$ is a smooth function and $a > 0, a \neq 1$.

The four values of k are obtained by combining the two signs of ε and the two possibilities for a:

k	1, 2	3, 4
a	(0, 1)	(1, +∞)

The odd k correspond to a conditional maximum and the even k to a minimum. To see the decrement diagram clearly, it is enough to consider the case $\varphi(y) = ay$: in

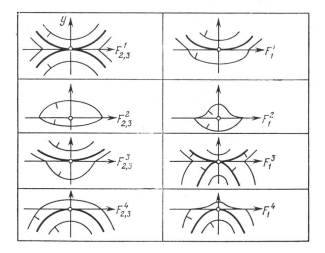

Figure 118.

this case, the level curves of f consist of segments of two parabolas translated along the y-axis.

Cases $F_1^k, k = 1, \ldots, 4$ (Conditional Extremum with Jordan Block α^2)

$$f = \text{const} + \varepsilon x^2 + \varphi(y) + \begin{cases} \sqrt{y} & \text{if } y \geqslant 0, \\ 0 & \text{if } y \leqslant 0. \end{cases}$$

Here $\varepsilon = \pm 1$, $\varphi(y) = ay + \cdots$ *is a smooth function,* $a \neq 0$.

The four values of k are obtained by combining the signs of ε and a:

k	1	2	3	4
sign of ε, sign of a	$--$	$-+$	$+-$	$++$

The odd k correspond to a conditional maximum and the even k to a minimum. To obtain a clear picture of the decrement diagram, it is enough to consider the cases $\varphi(y) = \pm y$.

Our theorem asserts that the increment of a generic two-parameter family has no singularities at the interior points of the F curves other than the 15 types F_i^k listed $(15 = 3 + 12)$: if there are other singularities in a family, then we can remove them by an arbitrarily small perturbation of the family. The singularities F_i^k are obviously unremovable.

J. Structure of Decrement Diagrams near Codimension 2 Strata

In the study of singularities of codimension 2 strata in generic two-parameter families, we may consider only the "most nondegenerate" cases, since every degeneracy increases the codimension and the singularity becomes removable.

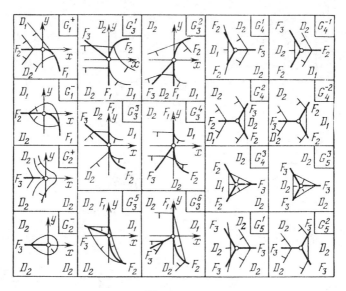

Figure 119.

Combining the transversality theorem and the explicit formulas for versal families of matrices from § 30B, we obtain the following normal forms of the increment near points of codimension 2 strata.

Theorem. *In the neighborhood of a point of every codimension 2 stratum (G_i in the notation of § H) in the plane of parameters of a generic family, smooth coordinates (x, y) can be chosen so that the increment f assumes one of the 18 forms listed below (Fig. 119).*

Cases $G_1^\pm$ (A Jordan Block of Order 3)

$$f = \varphi(x, y) + \lambda(x, y),$$

where λ is the largest among the real parts of the roots of the cubic equation $\lambda^3 = x\lambda + y$ and φ is a smooth function such that $(\partial\varphi/\partial x)(0, 0) = a \neq 0$.

The form of the decrement diagram is determined by the sign of the number a.

The signs "$+$" and "$-$" in $G_1^\pm$ correspond to $a > 0$ and $a < 0$. To clearly see the form of the decrement diagram, it suffices to consider the cases $\varphi = \pm x$. Two singular curves meet tangentially at the point $x = y = 0$: a ray F_2 ($y = 0$, $x < 0$) and one-half of a semicubic parabola F_1 ($4x^3 = 27y^2$, $y < 0$). These two curves separate the (convex) domain D_2 of complex-conjugate roots from the domain D_1 of real roots. In moving along the boundary of the domains D_1 and D_2, the increment f changes monotonically if $a > 0$ and has a minimum at the point G_1^- if $a < 0$. From the side of D_1, the level curves of f are tangent to the semicubic parabola F_1.

Cases $G_2^\pm$ (A Complex Pair of Jordan 2-Blocks)

$$f = \varphi(x, y) + |\text{Re}\sqrt{x + iy}|.$$

Here Re is the real part and φ is a smooth function such that $(\partial\varphi/\partial x)(0, 0) = a \neq 0$.

The form of the decrement diagram is determined by the sign of the number a.

The signs $+$ and $-$ in $G_2^\pm$ correspond to $a > 0$ and $a < 0$. To visualize the form of the decrement diagram clearly, it is enough to consider the cases $\varphi = \pm x$. The ray F_3 ($y = 0, x < 0$) arrives (and ends) at the point $x = y = 0$. In the case $a < 0$, the function f has a minimum at the point $G_2^-(x = y = 0)$. In the case $a > 0$, the point $G_2^+(x = y = 0)$ is topologically nonsingular for f. The level curve of f going through this point has a singularity of semicubic type.

Cases G_3^k ($k = 1, \ldots, 6$; Collision of a Complex Pair with a Jordan Block):

$$f = \text{const} + y + \max \begin{cases} \sqrt{x}, \varphi(x, y) & \text{if } x \geqslant 0, \\ 0, \varphi(x, y) & \text{if } x \leqslant 0. \end{cases}$$

$\varphi(x, y) = ax + by + \cdots$ *is a smooth function with* $a \neq 0, b \neq 0$, *and* $b \neq -1$.

The six values of k are obtained by combining the two possibilities for the sign of a and the three intervals of variation of b:

k	1	2	3	4	5	6
sing of a	$+$	$-$	$-$	$+$	$-$	$+$
interval of b	$(0, +\infty)$	$(0, +\infty)$	$(-1, 0)$	$(-1, 0)$	$(-\infty, -1)$	$(-\infty, -1)$

To visualize the form of the decrement diagram clearly, it is enough to consider a linear function φ. Three smooth rays F_1, F_2, and F_3 enter the origin. F_1 and F_2 come from opposite directions (with tangency of the first order), and F_3 comes transversally from the side D_2 of complex roots. In the case G_3^5 (i.e., where $a < 0, b < -1$), the increment has a minimum at the point $x = y = 0$; in the remaining cases, the point $G_3^k(k \neq 5)$ is a topologically nonsingular point of f.

Cases G_5^k ($k = 1, 2, 3$) (Double Passing)

$$\varphi = \text{const} + x + \max(|y|, \varphi(x, y)),$$

where $\varphi(x, y) = ax + by + \cdots$ *is a smooth function,* $a < 0, b > 0, a + 1 \neq \pm b$.
The three values of k for G_5^k correspond to intervals of variation of a:

k	1	2	3
condition on a	$b - 1 < a$	$-b - 1 < a < b - 1$	$a < -b - 1$

To see the form of the decrement diagram clearly, it suffices to consider a linear function φ.

In each of the three cases ($k = 1, 2, 3$), three smooth branches of the curve F_3 converge transversally at the point G_5^k. In the last case, this point is a minimum of the increment; in the first and second cases, it is a topologically nonsingular point. In approaching the point G_5^k along k of the three rays, the increment decreases. It increases on the remaining ones.

Cases $G_4^k(k = \pm 1, \pm 2, 3)$ (Double Passing with the Participation of a Real Root). The increment is given by the same formula as in cases G_5^k, but we have to distinguish between more cases depending on the question to which of the sectors the real root corresponds.

If k is negative, one obtains cases in which, in approaching the point G_4^k, the increment increases on the curve F_3 (on which the complex pairs collide). The two other rays are branches of F_2.

K. Discussion

Considering the normal forms listed above, we can arrive at a series of conclusions of general character concerning the structure of the decrement diagram both locally and globally. First of all, our theorems imply the following corollary.

Corollary. *The increment* $f: \Lambda \to R$ *of a generic two-parameter family is topologically equivalent to a smooth function having only simple critical points.*

These points are minimum points of the types D_i^0, F_i^2, $G_{1,2}^-$, G_3^5, $G_{4,5}^3$.

The points D_i^1 and F_i^3 are topologically equivalent to a saddle point. In the neighborhood of the maximum points (D_i^2), the increment is a smooth function. The points of the remaining types are topologically nonsingular.

This corollary obviously implies inequalities for the numbers of the singular points of various types. In particular, *if a closed level curve of the increment encloses a simply connected domain, then the total number of points of types* $D_i^{0,2}$, F_i^2, $G_{1,2}^-$, G_3^5, $G_{4,5}^3$ *inside this domain exceeds the number of points* D_i^1, F_i^3 *by 1*. It is not known whether the corollary can be carried over to *l*-parameter families with $l > 2^*$.

The fact that the segments F_1 and F_2 together form closed curves and the description of the singularities at the ends of the segments F_3 imply the following.

Corollary. *If the parameter space* Λ *is a closed two-dimensional manifold, then* (i) *the numbers of points of types* G_1 *and* G_3 *have the same parity and* (ii) *the total number of points of type* G_2, G_3, G_4, *or* G_5 *is even.*

If, for the parameter space Λ, we take a compact domain with boundary transversally intersecting F_i and not going through the points G_i, then the result is changed in the following way: the total number of points of type G_1 or G_3 has the same parity as the total number of points of intersection of the boundary with F_1 or F_2, and the total number of points of type G_2, G_3, G_4, or G_5 has the same parity as the number of points of intersection of the boundary with F_3.

In particular, this study of the increment enables us to study the singularities of the boundary of stability (i.e., curves of increment zero) in the parameter plane of generic two-parameter systems. Our theorems imply the following corollary.

Corollary. *The boundary of stability of a generic two-parameter family of matrices consists of smooth arcs intersecting transversally at their endpoints.*

---

*We note that in the case $l = 2$, the singularities of the increment of a generic family are of the same type as the singularities of the largest real part of the roots of an algebraic equation whose coefficients are generic functions depending on l parameters. For $l \geqslant 3$, this is not so: the increment can have more complicated singularities.

We note that the corner points of the boundary of stability can be of types F_1^0 ("Jordan 2-block") or F_2^0, F_3^0 ("simple passing"), according to the classification of § 30I and § 30J. Therefore, each of the arcs of the boundary of stability can be continued through its endpoints without losing smoothness. Moreover, the total number of corner points of types F_1^0 or F_2^0 on every closed component of the boundary of stability is always even.

We also note that our analysis of the singularities of the increment of two-parameter families is sufficient for the study of the boundary of stability in three-parameter families.

Indeed, according to the transversality theorem, the singular points of strata of codimension 3 and the critical points of the restrictions of the increment to strata of codimension 0, 1, or 2 can be removed from the boundary of stability by a small perturbation of the family. Consequently, the boundary of stability of a general family consists of smooth surfaces, and its singularities lie on curves in which the boundary of stability intersects the surfaces of types F_i and at points of intersection of the boundary of stability with the strata G_i (the latter appear in the form of curves in generic three-parameter families).

Moving along such a curve G_i, we may consider our three-parameter family as a one-parameter family of two-parameter families (two of the parameters are coordinates in a small area transversal to G_i and one is the coordinate t along G_i). Considering the normal forms of § 30J and § 30K, we now have to assume that all constants and the arbitrary functions φ depend smoothly on the parameter t. Moreover, generically, we may use the functions $\varphi(x, y, t)$ as the parameter z. This leads to the following conclusion.

Corollary. *The singularities of the boundary of stability of a generic three-parameter family of matrices are of the same types as the singularities of the graphs of increments of generic two-parameter families*[*].

Up to diffeomorphism[†], *these singularities are described by the following list (Fig. 120):*

A dihedral angle (F_i): $|y| + z = 0$;
A trihedral angle $(G_{3,4,5})$: $z + \max(x, |y|) = 0$.
A dead end on an edge (G_2): $z + |\operatorname{Re}\sqrt{x + iy}| = 0$. (*This surface in* $\mathbb{R}^3$ *is diffeomorphic*[†] *to the surface given by the equality* $XY^2 = Z^2$, *where* $X \geq 0$, $Y \geq 0$).
A broken edge (G_1): $z + \lambda(x, y) = 0$, *where* λ *is the largest of the real parts of the roots of the equation* $\lambda^3 = x\lambda + y$. (*This surface in* $\mathbb{R}^3$ *is diffeomorphic*[‡] *to the surface given by the equation* $X^2Y^2 = Z^2$, *where* $X \geq 0$, $Y \geq 0$.)

The angles of the boundary of stability are always directed outside, driving a wedge into the domain of instability. This is apparently the consequence of a very general principle, according to which everything good is fragile[‡].

What has been stated previously also implies some global properties of the boundary

---

[*] Similarly, the singularities of the boundary of stability of $(n + 1)$-parameter families are of the same types as those of the graphs of increments of n-parameter families.

[†] We have in mind a mapping extendable to a diffeomorphism of the neighborhood of the surface.

[‡] In generic k-parameter families of matrices, only a finite number of singularities of the stability boundary occur for k arbitrarily large (up to a local diffeomorphism of the ambient space (Levantovski)).

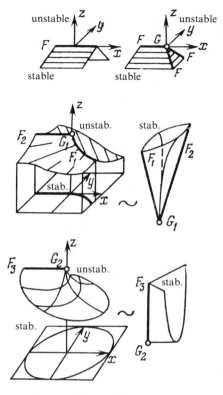

Figure 120.

of stability. For example, if the boundary is closed, then the total number of vertices of types $(G_i, i > 1)$ is even, just as the total number of vertices of type G_1 or G_3 are even.

The proofs of the theorems above may be found in the article V. I. Arnol'd, *Lectures on bifurcations and versal systems*, Uspekhi Mat. Nauk **27**, 5 (1972), 119–184 [Russ. Maths. Surveys, pp. 54–123]. See also, L. V. Levantovski, *On the singularities of the stability boundary*, Vestn. Mosk. Univ. Ser. 1 Mat. Mekh., **1980**, 6, 20–22; and L. V. Levantovski, *On the boundary of the set of stable matrices*, Uspekhi Mat. Nauk, **35**, 2 (1980), 213–214. L. V. Levantovski, *Singularities of the stability boundary*, Funct. Anal. Appl. **16**, 1 (1982), 44–48.

§ 31. Bifurcations of Singular Points of a Vector Field

In this section, we consider one-parameter families of differential equations. We study the bifurcations of singular points in generic families.

A. The Curve of Singular Points

Let us consider a vector field depending smoothly on a parameter. We assume that for some value of the parameter the field has a singular point. We ask: What happens to the singular point if the parameter is changed?

Theorem. *Any singular point of a vector field depending smoothly on a parameter depends smoothly on the parameter itself, provided that all eigenvalues of the linear part of the field at the singular point are different from zero.*

◀ In the neighborhood of the point and the value of the parameter studied, a family of fields in an n-dimensional phase space is given by n functions of $n + 1$ variables (n phase coordinates and the parameter ε). The singular points are given by a system $v(x, \varepsilon) = 0$ of n equations for the $n + 1$ variables. By the implicit function theorem, these equations locally determine a smooth curve $x = \gamma(\varepsilon)$ if at the original point the determinant of $\partial v/\partial x$ is different from zero. On the other hand, this determinant is equal to the product of the eigenvalues of the linearization of the field at the singular point. It is different from zero by assumption. ▶

Remark. The singular points at which all eigenvalues of the linearized field are different from zero are called *nondegenerate.* Consequently, if a field depends smoothly on a parameter, then its singular points depend smoothly on the parameter, provided that they remain nondegenerate.

The above proof remains valid for any dimension of the parameter space.

All singular points of a generic field are nondegenerate. Nevertheless, if we consider a family of vector fields, then degeneracies, which are unremovable by a small perturbation of the family may occur for some values of the parameter.

We study degeneracies in generic one-parameter families for vector fields in an n-dimensional space.

Let us consider the $(n + 1)$-dimensional space which is the direct product of the phase space with the axis of the parameter ε. We shall denote a point of the phase space by x. Our family determines a family

$$\dot{x} = v(x, \varepsilon)$$

of differential equations.

In our $n + 1$-space, we consider the set formed by the singular points of the equations of the family for all values of the parameter (Fig. 121):

$$\Gamma = \{x, \varepsilon : v(x, \varepsilon) = 0\}.$$

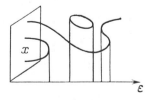

Figure 121.

Theorem. *For a generic family, the set of singular points is a smooth curve.*

Here and in the following, the words "generic families" mean "families from an everywhere dense set in the space of all families"; the everywhere dense set in question is open if the domain of definition of the family is compact or if the families are considered in the fine topology (cf., § 29); in any case, this everywhere dense set is the intersection of a countable number of open sets.

◀ The theorem follows from the transversality theorem (§ 29) or from Sard's lemma (§ 10).

Indeed, by the implicit function theorem, Γ is a locally smooth curve if 0 is not a critical value of the local mapping $\mathbb{R}^{n+1} \to \mathbb{R}^n$, $(x, \varepsilon) \mapsto v(x, \varepsilon)$. On the other hand, for a generic mapping, the value 0 is not critical. ▶

Remark. In this thoerem, the dimension of the parameter space is irrelevant (dim Γ is the dimension of the parameter space).

The theorem just proved immediately excludes certain bifurcations of singular points.

For example, we consider the bifurcations on the left-hand side of Fig. 122. It follows from the theorem that these bifurcations are not preserved under a small perturbation of the family. Indeed, it is easy to see that generically these bifurcations disappear under a small perturbation in one of the ways indicated on the right hand side of Fig. 122. If in a problem there appear bifurcations of the form indicated on the left-hand side at Fig. 122, then this tells us that the family under consideration is not generic. This can be due to some peculiar symmetry of the situation or can indicate the inadequacy of the idealization in which we neglected some small effect which leads to a qualitative change of the behavior of singular points in dependence of the parameter. To see which of the cases occurs in the real system in the idealization of which a nongeneric bifurcation has arisen, it is necessary to calculate some terms of the differential equation which have been omitted in our idealization. The formulas in the following subsections suggest which terms should be calculated.

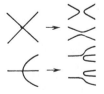

Figure 122.

B. Bifurcation Values of the Parameter

We assume that the set of singular points of the family is a smooth curve $(\operatorname{rank}(\partial v/\partial(x, \varepsilon)) = n)$. We consider the mapping of projection of this smooth curve onto the axis of the values of the parameter. The points at which the curve is projected badly onto the ε-axis are exactly the degenerate singular points. Indeed, by the implicit function theorem, the curve of singular points is the graph of a smooth function of the parameter in the neighborhood of a nondegenerate singular point.

Definition. A value of the parameter to which a degenerate singular point corresponds is called *bifurcation value of the parameter*, and the degenerate singular point itself in the direct product of the phase space with the axis of the values of the parameter is called a *bifurcation point*.

We consider the parameter ε as a function on the curve of singular points. *The bifurcation values of the parameter are the critical values of this function, and the bifurcation points are the critical points of the function* (the points where the differential of the function vanishes).

A critical point of a function is said to be nondegenerate if the second differential of the function at the point is nondegenerate. (In the case under consideration, we speak of functions of one variable. Nondegeneracy of the second differential means that it is different from zero). The corresponding bifurcation point is called a *nondegenerate bifurcation point*.

Definition. A bifurcation value of the parameter is said to be *regular* if exactly one nondegenerate bifurcation point corresponds to it.

Theorem. *For one-parameter generic families, all bifurcation values of the parameter are regular. If the phase space is compact, then the bifurcation values of the parameter are isolated.*

◀ This is a simple consequence of the transversality theorem. The details are left to the reader. ▶

Remark. The theorem states that, as the parameter varies, the singular points of a generic family may only annihilate each other pairwise or be born in pairs, when the parameter passes through bifurcation values (Fig. 121). The bifurcations of this type are stable (are preserved under a small perturbation of the family). All more complicated bifurcations disintegrate into several bifurcations of the type described (Fig. 123) under a small generic perturbation.

C. An Example: Vector Fields on a Line

We consider a one-parameter family of vector fields on a line determining the differential equation

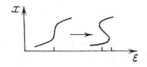

Figure 123.

$$\dot{x} = \pm x^2 + \varepsilon, \qquad x \in \mathbb{R}, \varepsilon \in \mathbb{R}.$$

For $\varepsilon = 0$, this vector field has the simplest nondegenerate singular point $(x = 0)$. If the parameter passes through the regular bifurcation value $\varepsilon = 0$, then, depending on the sign of x^2, either the two singular points (stable and unstable) annihilate each other or a pair of singular points is born whose members run apart immediately (with asymptotics $\sqrt{|\varepsilon|}$).

It is easy to verify that the bifurcation in this example is the only bifurcation that cannot be removed in generic one-parameter families of vector fields on a line.

Definition. Let two parameter-dependent families of vector fields be given. The two families are said to be *topologically equivalent* if there exist a homeomorphism between the parameter spaces and a family of homeomorphisms of the phase space depending continuously on the parameter and mapping a family of oriented phase curves of the first family for every value of the parameter into the family of oriented phase curves of the second family for the corresponding value of the parameter.

We note that the homeomorphisms mentioned in the above definition determine a homemorphism of the direct products of the phase spaces with the spaces of the parameters $(x, \varepsilon) \mapsto (h(x, \varepsilon), \varphi(\varepsilon))$ converting the phase curves of the system $\dot{x} = v(x, \varepsilon)$, $\dot{\varepsilon} = 0$ into phase curves of the same form of the second system.

The equivalence of germs of families at a point is defined in an analogous way. If the pair (x_0, ε_0) consists of a point of the phase space and a point of the parameter space, then the homeomorphisms realizing the equivalence have to determine a homeomorphism $(x, \varepsilon) \mapsto (h(x, \varepsilon), \varphi(\varepsilon))$ of some neighborhood of the point (x_0, ε_0) in the direct product.

Theorem. *In the neighborhood of a nondegenerate bifurcation point, a one-parameter family of vector fields on a line is equivalent to the germ of the family given by the equation $\dot{x} = x^2 + \varepsilon$ at the point $x = 0$, $\varepsilon = 0$.*

◀ The function $v(x, \varepsilon)$ determining the field changes its sign on the curve Γ. We choose the origin of the coordinates (x, ε) at the bifurcation point. In view of the nondegeneracy of this point, the equation of Γ has the form $\varepsilon = Cx^2 + O(|x|^3)$, $C \neq 0$. This immediately implies our state. ▶

The theorem just proved, together with the theorem of § 31B, furnishes a complete topological description of bifurcations of singular points of vector fields on a line in generic one-parameter families.

D. Bifurcations of Periodic Solutions

Bifurcations of fixed points of smooth mappings and bifurcations of periodic solutions of differential equations (bifucations of closed phase or integral curves) can be studied in exactly the same way. The condition of non-degeneracy of a fixed point of a mapping means that all eigenvalues of the linearization are different from 1. In the case of periodic solutions, the eigenvalues of the linearization of the Poincaré mapping (that is, eigenvalues of the monodromy operator determined by the equation in the normal variations along the solution being considered) must not be equal to 1.

In particular, if for $\varepsilon = 0$ the equation $\dot{x} = v(x, \varepsilon)$ has a periodic solution $x = \varphi(t)$ with period T and the above nondegeneracy condition is satisfied, then for small ε a periodic solution $x = \Phi(t, \varepsilon)$ with period $T(\varepsilon)$ and turning into φ for $\varepsilon = 0$ exists and is unique (of course, it is the phase curve which is unique: the initial time may be changed).

Remark. The search for the periodic solution Φ in the form of a series in ε is called the *Poincaré method of a small parameter*. The solution φ is called the *generating solution.* An analogous method can be used in the nonautonomous case where v has period $T(\varepsilon)$ in t and we look for $T(\varepsilon)$-periodic solutions.

Problem. With an error on the order of ε^2, find a 2π-periodic solution of the equation $\ddot{x} = \sin x + \varepsilon \cos t$ with $x \equiv 0$ for $\varepsilon = 0$.

§ 32. Versal Deformations of Phase Portraits

In this section, we determine topologically versal deformations of phase portraits and give their explicit form for the simplest degenerate singular points.

A. Theory of Local Bifurcations and Local Qualitative Theory

As has been mentioned above, unremovable degenerate singular points can occur in the case where we are interested in a family of fields depending on a parameter rather than an individual vector field. Moreover, in generic families, only the simplest degeneracies occur.

The usual methods of the qualitative theory of differential equations (cf., Chap. 3) can be applied to the study of the structure of a vector field near a degenerate singular point. For the simplest degeneracies, these methods

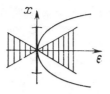

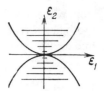

Figure 124.   Figure 125.

enable us to perform a sufficiently complete topological study of the phase portrait. Thus, we are able to study the phase portrait for both general and singular values of the parameter. This is the usual approach to problems concerning families of differential equations.

A consideration of the simplest bifurcations shows that this approach obscures the very essence of phenomena taking place near a critical value of the parameter. This is so because the neighborhood of a nondegenerate singular point in which the phase portrait is given by the local theory shrinks to zero in approaching the singular value of the parameter (Fig. 124) and increases again with a jump at the singular value of the parameter. As a result, the metamorphosis of the phase portrait (say, the approach of a neighboring singular point) remains outside the domain of applicability of the local theory.

Consequently, the local theory excludes bifurcation, the most important phenomenon at a singular value of the parameter.

This leads us to the conclusion that the *study of degenerate singular points represents a real interest only in the case where it is accompanied by the study of families in which this type of degeneracy is unremovable, the study taking place in the neighborhood of the degenerate singular point in the direct product of the phase space with the parameter space.* In other words, the neighborhood of the singular point in the phase space in which the phase portrait has to be studied must not depend on the parameter (must not shrink to zero as the parameter approaches the singular value).

Exactly the same arguments show how dangerous it is to make an error in determining the number of parameters essential for the study of a bifurcation. For example, in the study of an essentially two-parameter phenomenon from the one-parameter point of view, the following phenomenon is typical (Fig. 125). For every value of the parameter forgotten, we succeed in studying bifurcations in a one-parameter family of equations depending on a second parameter. Nevertheless, near the singular value at which we succeed in performing the study, the interval of values of the second parameter will shrink to zero as the parameter not taken into consideration approaches the singular values. If we consider the problem as a two-parameter problem (i.e., in a neighborhood of the singular value of the first parameter not depending on the value of the second parameter), we can study the bifurcations by local methods, which seem global from the one-parameter point of view.

An example of such a two-parameter problem which appears to be a one-parameter one at first glance is the problem of the loss of stability of a closed phase curve. Here the natural parameter is the absolute value of the eigenvalue of the monodromy operator; the second parameter, usually missed, is the argument of the eigenvalue crossing the unit circle. We return to this example in § 34.

B. Topologically Versal Deformations

We consider the family $\dot{x} = v(x, \varepsilon)$ of differential equations.

A *local family* $(v; x_0, \varepsilon_0)$ is, by definition, the germ of the mapping v at the point (x_0, ε_0) of the direct product of the phase space and the parameter space. Thus, every representative of this germ is given in the neighborhood of the point (x_0, ε_0) in the direct product (and not in the neighborhood of the point x_0 in the phase space).

An *equivalence* of the local systems (1) $(v; x_0, \varepsilon_0)$ and (2) $(w; y_0, \varepsilon_0)$ is, by definition, the germ [at the point (x_0, ε_0)] of a continuous mapping h, $y = h(x, \varepsilon)$ such that for representatives of the germ for every ε, $h(\cdot, \varepsilon)$ is a homeomorphism converting phase curves of the system (1) (in the domain of h) into phase curves of the system (2) with $h(x_0, \varepsilon_0) = y_0$ and preseving the direction of motion. We note that for $\varepsilon \neq \varepsilon_0$ the point x_0 does not have to be converted into y_0 by the mapping $h(\cdot, \varepsilon)$.

A local family (3) $(u; x_0, \varepsilon_0)$ is *induced* from the family (1) by means of the germ at the point μ_0 of a continuous mapping φ, $\varepsilon = \varphi(\mu)$, where $\varphi(\mu_0) = \varepsilon_0$ if $u(x, \mu) = v(x, \varphi(\mu))$.

A local family $(v; x_0, \varepsilon_0)$ is called a *topologically orbitally versal* (more briefly, simply, *versal*) *deformation of the germ of the field* $v_0 = v(\cdot, \varepsilon_0)$ *at the point* x_0 if every other local family containing the same germ is equivalent to one induced from the given family.

In the following, we shall occasionally speak of deformations, equivalences, induced and versal deformations of differential equations, having in mind the corresponding concepts for the vector fields determining the equations.

We would like to emphasize that the existence of a topologically versal deformation of a given germ of a vector field is not at all obvious; it is easy to give examples of fields not admitting such a deformation with a finite number of parameters (for example, the zero field). Nevertheless, in the cases where a versal deformation exists, is found, and analyzed, the information thus obtained is quite rich. The determination and study of a versal deformation is a means of concentrated representation of results of a very complete study of bifurcations of phase portraits.

EXAMPLE

The deformation $\dot{x} = \pm x^2 + \varepsilon$ of the differential equation $\dot{x} = \pm x^2$ is versal.

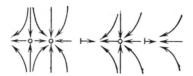

Figure 126.

◀ Cf., the preceding section. ▶

C. Šošitaĭšvili's Reduction Theorem

The bifurcation of the preceding example (birth or annihilation of a pair of singular points) exhausts bifurcations in generic families of vector fields on the line (cf., § 31). In the multi-dimensional case, the birth or annihilation of a pair of singular points is generic, too. Meanwhile, what happens to the phase portraits?

It turns out that if one characteristic number is equal to zero, a topologically versal deformation of a general degenerate singular point in $\mathbb{R}^n$ can be obtained from the equation of the preceding example by the simple suspension

$$\dot{x} = \pm x^2 + \varepsilon, \qquad x \in \mathbb{R}, \quad \varepsilon \in \mathbb{R},$$
$$\dot{y} = -y \qquad\qquad y \in \mathbb{R}^{n-}$$
$$\dot{z} = z, \qquad\qquad z \in \mathbb{R}^{n+}$$

where n_- and n_+ are the numbers of the roots of the characteristic equation in the left and right half-planes, respectively. For example, for $n = 2$, this system describes the coalescence of a nodal and saddle points (Fig. 126). For $\varepsilon = 0$ we obtain a so-called saddle-node.

In § 31 we called bifurcation points those points in the direct product of the phase space with the space of the values of the parameter for which the characteristic equation has a vanishing root.

Theorem. *In the space of one-parameter families of vector fields, an everywhere dense* set is formed by the generic families which are in the neighborhood of every bifurcation point, topologically equivalent to a family* (1) *in the neighborhood of the origin.*

It is convenient to prove this theorem by reducing it to the case $n = 1$, in which case the theorem is obvious (and is proved above). Such a reduction

---

*As usual, the set of generic families is the intersection of a countable number of open sets. It is open if the domain of definition of the family is compact or if we use the fine topology.

enables us to decrease the number of phase coordinates to the necessary minimum and can be performed once and for all in the most general situation.

We consider a local family of vector fields $(v; x_0, \varepsilon_0)$ depending on a finite-dimensional parameter. For the sake of brevity, we shall assume that

$$x_0 = 0 \in \mathbb{R}^n, \qquad \varepsilon_0 = 0 \in \mathbb{R}^k.$$

Suppose $x = 0$ is a singular point of the field $v(\cdot, 0)$ and the corresponding characteristic equation has n_- (correspondingly, n_+, n_0) roots in the left half-plane (correspondingly, in the right half-plane, on the imaginary axis).

Theorem. *Under these assumptions, the family is topologically equivalent to the following suspension over a family with phase space of dimension n_0:*

$$\dot{\xi} = w(\xi, \varepsilon), \qquad \xi \in \mathbb{R}^{n_0}, \quad \varepsilon \in \mathbb{R}^k,$$

$$\dot{y} = -y, \qquad y \in \mathbb{R}^{n_-},$$

$$\dot{z} = z, \qquad z \in \mathbb{R}^{n_+}.$$

The proof of this theorem may be found in A. N. Šošitaĭšvili, *Bifurcations of topological type of a vector field near a singular point*, Trudy Seminara I. G. Petrovskogo 1, (1975, pp. 279–309; cf., also, Funct. Anal. Appl. 6, 2 (1972), 97–98, where the theorem is first formulated).

◀ The proof goes along the same lines as that of Anosov's theorem on Anosov systems: its basic part consists of the construction of five foliations (contracting, expanding, neutral, noncontracting, and nonexpanding) in the direct product of the phase space with the parameter space. (The parameters can be treated as additional phase variables to which the equation $\dot{\varepsilon} = 0$ corresponds; we have to watch out that under equivalences the planes $\varepsilon = $ const turn into planes of the same family.) ▶

The existence of the five foliations was proved by Tikhonova, independently of the needs of bifurcation theory [E. A. Tikhonova, *Analogy and homeomorphism of perturbed and unperturbed systems with a block-triangular matrix*, Differential Equations 6, 7 (1970), 1221–1229], and by Hirsch, Pugh, and Shub [M. W. Hirsch, C. C. Pugh, M. Shub, *Invariant manifolds*, Bull. Am. Math. Soc. 76, 5 (1970), 1015–1019]. The case $n_+ = 0$ was considered earlier by Pliss [V. A. Pliss, *A reduction principle in stability theory of motions*, Izv. AN. USSR, Ser. Mathem. 28, 6 (1964), 1297–1324].

The differential equation $\dot{\xi} = w(\xi, \varepsilon)$ of the reduced system is realized in the original system on some smooth neutral submanifold of dimension n_0, called the *center manifold*, depending smoothly on ε in the phase space. The smoothness of the center manifold is finite (it increases as $\varepsilon \to 0$) and the submanifold is not uniquely determined (as the simplest examples show).

Nevertheless, the behavior of phase curves, including the whole picture of bifurcations, is determined for the entire equation by what takes place on the indicated center manifold (and, in particular, does not depend on the choice of the center manifold).

Šošitaĭšvili also proved that the versality of the original deformation is equivalent to the versality of the reduced deformation (i.e., the versality of the original deformation on the neutral manifold).

Consequently, the topological study of local degeneracies of phase portraits near singular points, including the study of all possible bifurcations, can be restricted to the case where all roots of the characteristic equation lie on the imaginary axis. The passage to the general case can be completed by a simple suspension (direct product with the standard saddle point $\dot{y} = -y, \dot{z} = z$).

EXAMPLE

From what has been said above, it specifically follows that, when a pair of singular points in a generic one-parameter family of vector fields is born, one (and only one) phase curve leads from one of the new singular points to the other (for values of the parameter close to the bifurcation values).

§ 33. Loss of Stability of an Equilibrium Position

Here we study bifurcations of the phase portrait of a differential equation as a pair of roots of the characteristic equation crosses the imaginary axis.

A. Example: Soft and Hard Loss of Stability

We begin with an example of a one-parameter family of vector fields in the plane, going back to Poincaré and Andronov. We write it in the following complex form:

$$\dot{z} = z(i\omega + \varepsilon + cz\bar{z}), \tag{1}$$

where $z = x + iy$ is the complex coordinate on $\mathbb{R}^2$ considered as the plane of the complex variable z.

In the preceding formula, ω and c are real nonzero constants, which can be assumed to be equal to ± 1 if we wish; ε is a real parameter.

For all ε, the point $z = 0$ is a focal type equilibrium position. This focus is stable for $\varepsilon < 0$ and unstable for $\varepsilon > 0$. For $\varepsilon = 0$, linear approximation gives a center; the character of the singular point is determined by the sign of c: $c < 0$ corresponds to stability and $c > 0$ to instability.

In the analysis of singular points performed here locally in z, we observe that the singular point loses stability at the moment $\varepsilon = 0$, but we omit an important phenomenon

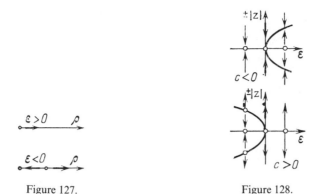

Figure 127.   Figure 128.

connected with the loss of stability: the birth of a limit cycle (cf., Fig. 129). To avoid such a mistake, one must consider a neighborhood of zero in the (z, ε)-space and not in the z-space for fixed ε.

It is convenient to perform the study in the neighborhood of zero in the (z, ε)-space in the following way. We consider the function $\rho(z) = z\bar{z}$. From Eq. (1), we obtain the following equation for ρ:

$$\dot{\rho} = 2\rho(\varepsilon + c\rho), \qquad \rho \geqslant 0.$$

The family of equations thus obtained is easy to study on the ray $\rho \geqslant 0$. In addition to the singular point $\rho = 0$, for every ε there exists another singular point $\rho = -\varepsilon/c$ (if ε and c are of opposite signs). For $c > 0$, the vector field ρ has one of the forms indicated in Fig. 127, depending on the sign of ε.

The point $\rho = 0$ corresponds to the origin on the z-plane, and the point $\rho = -\varepsilon/c$ corresponds to the limit cycle (real only if ε and c have opposite signs).

In order to understand the situation better, we plot ε on one axis and $|z|$ on both sides of the other. Then the behavior of the cycle in the variation of the parameter is illustrated by one of the two diagrams in Fig. 128, depending on the sign of c. The radius of the cycle is proportional to $\sqrt{|\varepsilon|}$.

First, we consider the case $c < 0$. As ε passes through 0, the focus at the origin loses stability. For $\varepsilon = 0$, at the origin the focus is also stable but not robust: the phase curves approach 0 nonexponentially (Fig. 129).

For $\varepsilon > 0$, moving from the focus to a distance proportional to $\sqrt{\varepsilon}$, the phase curves wind onto a stable limit cycle. Consequently, in the case $c < 0$, the loss of stability in the passage of ε through 0 takes place with the birth of a stable limit cycle whose radius increases with $\sqrt{\varepsilon}$.

In other words, the stationary state loses stability and a stable periodic regime arises whose amplitude is proportional to the square root of the deviation of the parameter from the critical value. In this situation, physicists speak of a *soft generation of self-sustained oscillations*.

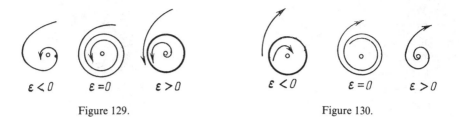

Figure 129.   Figure 130.

In the case $c > 0$ (Fig. 130), a limit cycle exists for $\varepsilon < 0$ and is unstable. As ε converges to 0, the cycle settles onto an equilibrium position, which was a stable focus for $\varepsilon < 0$. For $\varepsilon = 0$, the focus becomes unstable (the instability is weak, nonexponential). For positive ε, the focus is unstable even in linear approximation.

This loss of stability is called *hard* for the following reason.

Imagine that the system is near a stable equilibrium position and that under the variation of the parameter this equilibrium position loses stability. In the case $c > 0$, as ε approaches 0 from the negative side (or even somewhat sooner) the ever present perturbations pull the system out of the neighborhood of the equilibrium position and it jumps into some other regime (for example, a faraway equilibrium position, limit cycle, or a more complicated attracting set). Consequently, under the continuous variation of the parameter, the behavior of the motion changes stiffly, in a jump.

In the case $c < 0$, although the amplitude of the self-sustained oscillations born from the equilibrium does not depend on the parameter smoothly (radical singularity) it does so continuously; in this sense, the behavior of the motion changes softly.

In the study of Eq. (1), we have essentially used the "versal" point of view: if we had considered a neighborhood in the z-space for fixed ε instead of a neighborhood in the (z, ε)-space, we would have missed limit cycles. This agrees with the fact that a degeneracy of codimension k has to be studied in a k-parameter family: our case of codimension 1 is imbedded in a one-parameter family.

In fact, the example just considered exhausts the bifurcations of the phase portrait in generic one-parameter families occurring under the loss of stability of an equilibrium position in the plane and, more generally, under the passage of a pair of roots of the characteristic equation through the imaginary axis.

B. The Poincaré–Andronov Theorem

We consider a one-parameter family of vector fields.

We assume that for the value zero of the parameter, the field has the singular point 0 such that the roots of the characteristic equation are purely imaginary (the dimension of the phase space is equal to 2).

Theorem. *Generic families with the indicated properties are locally topologically equivalent to the family of the preceding example.*

◀ We shall use Poincaré's method for the reduction of the equation to normal form. For the value zero of the parameter there is resonance, which is not present for close nonzero values of the parameter. The corresponding resonant terms for the value zero of the parameter cannot be annihilated, but for nearby values of the parameter they can. If for nonresonant values of the parameter close to zero we annihilate the terms that are resonant for the value zero of the parameter, then our change will discontinuously depend on the parameter. The radius of the neighborhood in which we study the phase portrait will shrink to zero as the parameter approaches the resonant value.

Consequently, we do not annihilate the terms that are resonant for $\varepsilon = 0$ not only for the value zero of the parameter, but also for nearby values as well. As a result, we obtain a change depending smoothly on the parameter, after which only terms which are resonant for the value zero of the parameter and a remainder of arbitrarily high order with respect to the distance from the singular point remain in the system. We intend to study bifurcations in the family thus obtained, omitting the remainder, and then verify that the remainder does not affect the topology of the metamorphosis of the phase portrait (or else take its effect into account).

The program described above is common to many problems of bifurcation theory. Let us see what it comes to in our concrete case of the passage of a pair of roots of the characteristic equation through the imaginary axis. The resonance has the form $\lambda_1 + \lambda_2 = 0$ ($\lambda_{1,2} = \pm i\omega$). Consequently, in eigenvector coordinates in the complexified plane $\mathbb{C}^2$, the normal form can be written as follows (cf., § 23):

$$\dot{z}_1 = \lambda_1(\varepsilon)z_1 + a_1(\varepsilon)z_1^2 z_2 + \cdots,$$
$$\dot{z}_2 = \lambda_2(\varepsilon)z_2 + b_1(\varepsilon)z_1 z_2^2 + \cdots.$$

We note that since the original equation is real, the eigenbasis can be chosen from complex-conjugate vectors, and the normalizing changes can be chosen real. In such a case, the second equation follows from the first by conjugation. Moreover, $z_2 = \bar{z}_1$ in the real plane, and we may write only the first equation, denoting z_1 by z and z_2 by $\bar{z}$. This equation can be considered as another version of the original system in the real plane $\mathbb{R}^2$ written as an (nonholomorphic) equation on the complex line $\mathbb{C}^1$ with coordinate z:

$$\dot{z} = \lambda_1(\varepsilon)z + a_1(\varepsilon)z^2\bar{z} + \cdots.$$

The dots denote a remainder of order 5 with respect to $|z|$.

This leads us to the study of the family

$$\dot{z} = \lambda_1(\varepsilon)z + a_1(\varepsilon)z^2\bar{z}.$$

This can be studied in the same way we studied the special example in § 33A. The correspondence between the two kinds of notation is as follows:

Example of § 33A	$i\omega$	ε	c
Generic family	$\lambda_1(0)$	$\operatorname{Re}\lambda_1(\varepsilon)$	$\operatorname{Re}a_1(0)$

In generic families, we have

$$\lambda_1(0) \neq 0, \qquad \frac{d}{d\varepsilon}\operatorname{Re}\lambda_1(\varepsilon)\big|_{\varepsilon=0} \neq 0, \qquad \operatorname{Re}a_1 \neq 0.$$

The bifurcation consists of the birth or annihilation of a limit cycle (birth in the case where $\operatorname{Re}a_1(0)$ and $\operatorname{Re}d\lambda_1/d\varepsilon\big|_{\varepsilon=0}$ have distinct signs).

Generically, the three quantities above are different from zero and the omitted remainder does not change the picture of bifurcations obtained above. This is easy to prove by considering the derivative of the function $\rho = |z|^2$ along our vector field:

$$\dot\rho = 2\rho(\operatorname{Re}\lambda_1(\varepsilon) + \rho\operatorname{Re}a_1(\varepsilon) + O(\rho^2)).$$

It is easy to see from this formula that $O(\rho^2)$ does not affect the bifurcations of the phase portrait at some neighborhood (independent of ε) of the origin. ▶

The theorem above was essentially known to Poincaré; the explicit formulation and proof were given by Andronov [A. A. Andronov, "Mathematical problems of the theory of self-sustained oscillations." In: *Pervaya vsesojuznaya konferencija po kolebanijam* (I all—Union conference on oscillations), M. L. GTTI, 1933 (reproduced in A. A. Andronov, *Collected Works*, AN SSSR, Moscow, 1956, pp. 85–124). A. A. Andronov, *Application of Poincaré's theorem on "bifurcation points" and "change in stability" to simple autooscillatory systems*, C. R. Acad. Sci. (Paris) **189**, 15 (1929), 559–561; A. A. Andronov, E. A. Leontovič–Andronova, *Some cases of the dependence of periodic motions on a parameter*, Učenye Zapiski Gorki Gosudarstvenny Univ., 1939, Vol. 6, p. 3 (see also A. A. Andronov, *Collected Papers*, pp. 186–216)]. R. Thom to whom I taught this theory in 1965, began to promote it under the name "Hopf bifurcation" (cf., for example, the work of Smale and Hirsch*). Curiously, the 20 pages of bibliography in *The Hopf Bifurcation and Its Applications* by J. E. Marsden and M. McCracken (as well as the 30 pages of bibliography in the second edition of *Foundations of Mechanics* by R. Abraham and J. Marsden) does not contain the basic Andronov–Leontovich–Andronova paper. The theorem is also presented in the well-known Andronov–Vitt–Chaikin book *Theory of oscillations*, Moscow (1937) (English translation by University Press, Princeton (1949)).

C. The Multidimensional Case

Combining the Poincaré–Andronov theorem with the reduction theorem (§ 32), we obtain the following.

---

*M. Hirsch, S. Smale, Differential Equations, Dynamical systems and Linear Algebra, New York, Academic, 1974.

Theorem. *A topologically versal deformation of a singular point of a generic vector field whose characteristic equation has one pair of purely imaginery roots can be obtained by the following simple suspension from the Poincaré–Andronov system:*

$$\dot{z} = z(i + \varepsilon \pm z\bar{z}), \qquad z \in \mathbb{C}^1, \qquad \varepsilon \in \mathbb{R};$$

$$\dot{u} = -u, \qquad\qquad u \in \mathbb{R}^{n_-},$$

$$\dot{v} = v, \qquad\qquad\quad v \in \mathbb{R}^{n_+}, \quad n = n_- + n_+ + 2.$$

The study of this system does not cause any difficulties.

EXAMPLE

Let $n = 3$, $n_+ = 0$, and let the sign in front of $z\bar{z}$ be negative. In this case, the theorem asserts that in the neighborhood of the origin of coordinates not depending on ε, the birth of an invariant cylinder of radius $\sqrt{\varepsilon}$ attracting neighboring trajectories takes place as the pair of eigenvalues passes through the imaginary axis. On the cylinder itself, there is a stable limit cycle onto which all trajectories wind. Consequently, this case corresponds to a soft loss of stability with the appearance of self-sustained oscillations.

This degeneracy has been studied by many authors; in particular, Hopf studied the birth of a cycle in the multidimensional case [E. Hopf, *Abzweigung einer periodischen Lösung von einer stationären Lösung*, Bereich Sächs. Acad. Wiss. Leipzig, Math. Phys. Kl. **94**, 19 (1942), 15–25]. Further results have been obtained by Neĭmark [J. I. Neĭmark, *On some cases of periodic motions depending on parameters*, Dokl. Acad. Nauk SSSR **129** (1959, 736–739] and Brŭslinskaja [N. N. Brŭslinskaja, *Qualitative integration of a system of n differential equations in a region containing a singular point and a limit cycle*, Dokl. Acad. Nauk SSSR **139** (1961), 9–12].

Nevertheless, the general theorem formulated above, covering a complete study of the metamorphoses of the phase portrait (and not only bifurcations of a cycle) has only been proved in Šošitaĭsvilli's work on reduction (cited above) using two-dimensional results of Andronov and Poincaré.

D. Application to the Theory of Hydrodynamic Stability

The phenomena analyzed above are often encountered in various concrete situations: mechanical, physical, chemical, biological, and economical systems often lose stability. Let us, as an example, consider one special problem of this kind: the problem of the loss of stability in the stationary flow of an incompressible viscous fluid.

Let D be a domain filled with fluid and v the field of velocities of the fluid. The motion is described by the Navier–Stokes equations

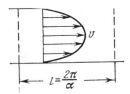

Figure 131.

$$\frac{\partial v}{\partial t} + (v\nabla, v) = v\Delta v - \operatorname{grad} p + f, \qquad \operatorname{div} v = 0,$$

where the coefficient v denotes viscosity; f is the field of nonpotential mass forces; the pressure p is determined from the condition of incompressibility. On the boundary of D, we may have, say, conditions of adhesion ($v|_{\partial D} = 0$).

It is assumed that the initial field of velocities determines the entire motion, so that the equation defines a dynamical system in the infinite-dimensional space of divergence-free vector fields equal to 0 on the boundary of D. [In fact this has only been proven in the two-dimensional case. Extensive literature is devoted to problems of existence, uniqueness, and properties of solutions of the Navier–Stokes equations; nevertheless, the basic problems have remained unsolved.]

Consider, for example, a Poiseuille flow (with parabolic velocity profile; Fig. 131) in a planar canal. For every value of the viscosity v, the Poiseuille flow is a stationary point of our dynamical system in a function space. This equilibrium position is stable for a sufficiently large viscosity, but loses stability for decreasing viscosity. We can study what happens in the meantime by using the theorem of § 33C.

Of course, we have to take special precautions because of the infinite-dimensional character of the problem. There is hope that the infinite-dimensionality is not very dangerous because of the fact that the viscosity extinguishes high harmonics rapidly, so that the system collapses into a finite-dimensional system for any nonzero value of the viscosity coefficient.* Another difficulty lies in the fact that we cannot be sure that our system is actually generic: this has to be verified by calculations. It seems natural that the Navier–Stokes system turns out to be generic in a domain of "general form" and for general mass forces f; nevertheless, a Poiseuille flow is very special: for example, it has a large group of symmetries.

We restrict ourselves to perturbations whose field of velocities is repeated along the flow periodically with wave length l. In order to normalize the velocity of the basic flow, we shall change the external forces proportionally to the viscosity so that the flux Q of the fluid is constant ($f = \operatorname{const} Qv$). In this case, we obtain a two-parameter family with parameters l and v.

---

*According to Yu. S. Ilyashenko, the dimension of any attractor of the Navier–Stokes equation on T^2 does not exceed const R^4. Recently M. I. Vishik and A. Babin proved the following upper estimate of this dimension for 2-manifolds with boundary: dim $\leqslant$ const. $\times R^{2+\varepsilon}$ for every positive ε.

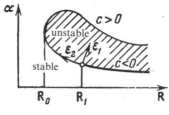

Figure 132.

It is customary to consider as parameters the reciprocals $\alpha = 2\pi/l$ (wave number) and $R = \text{const}\, Q/v$ (Reynolds' number). Consequently, a decrease in viscosity causing instability corresponds to an increase in Reynolds' number.

Calculations (which are practically unfeasible without a computer) show that as the Reynolds' number increases, for some critical value $R_0 = R_0(\alpha)$, a pair of complex roots passes through the imaginary axis from the stable half-plane into the unstable one. Thus we encounter the case of the loss of stability in which a limit cycle is born or dies.

The sign of the coefficient c determining a rough or soft generation of oscillations has also been calculated. For the description of the result, it is convenient to draw the boundary of stability in the (α, R)-plane. It turns out that it has the "tongue" form depicted in Fig. 132; the leftmost point of this tongue is especially important: its R coordinate corresponds to the first loss of stability and its α coordinate determines the wavelength which is the most dangerous for instability.

It turns out that for the entire left and upper parts of the tongue of the boundary of stability, the coefficient c is positive, i.e., a *rough* generation takes place. Consequently, before the Reynolds' number passes through the critical value R_0, somewhere in the phase space away from the stationary point (i.e., from the Poiseuille flow), an oscillating regime* arises into which the system is thrown by small perturbations as the Reynolds' number approaches R_0. This new state can be a stable stationary point (i.e., in hydrodynamical terms, a stationary flow different from the Poiseuille flow) or a limit cycle (in hydromdynamical terms, a periodic flow). It might have a more complicated structure, for example, a quasiperiodic motion on the torus. Moreover, the behavior arising in a rough generation can be an Anosov system or a system of hyperbolic character, i.e., an attracting set with extremely irregular unstable trajectories on it. The spectrum of the corresponding dynamical system can be continuous in spite of the finiteness of the number of degrees of freedom (i.e., the finiteness of the dimension of the attracting set). Experimenters would call such a behavior turbulent.

---

*Another possibility in systems of an arbitrary form is the approach to infinity; in our case, this apparently does not occur, since at infinity the phase velocity is directed backwards to the origin of coordinates because of the damping action of viscosity.

In 1963, Lorenz published a paper [E. N. Lorenz, *Deterministic nonperiodic flow*, J. Atmos. Sci. **20** (1963), 130–141] in which he first observed a nontrivial attracting behavior in a system with three-dimensional phase space modeling the hydrodynamical theory of convection.

Lorenz' system has the form

$$\dot{x} = -\sigma x + \sigma y, \quad \dot{y} = -xz + rx - y, \quad \dot{z} = xy - bz; \quad \sigma = 10, r = 28, b = \tfrac{8}{3}.$$

It seems that all models in which hyperbolic attracting sets have so far been found contain terms of the type of a pump or negative viscosity, which are absent in the Navier–Stokes equation. At any rate when, in 1964, this author attempted to find a hyperbolic attracting set in the six-dimensional phase space of the Galerkin approxima-tion of the Navier–Stokes equation on the two-dimensional torus with sinusoidal external force (using a computer programmed by Vvedenskaya), the attracting set turned out to be apparently the three-dimensional torus (maybe because of the too small Reynolds' number). As far as this author knows, hyperbolic attracting sets for the Navier–Stokes equations or their Galerkin approximations have not yet been found.* On the other hand, the numerical experiment described above has served as a starting point for a series of publications on the application of geodesic flows on groups of diffeomorphisms to hydrodynamics. [V. I. Arnold, *Sur la geometrie differentielle des groupes de Lie de dimension infine et ses applications á l'hydrodynamique des fluides parfaits*, Ann Inst. Fourier, Grenoble **16**, 1 (1966), 319–361; D. G. Ebin, J. Marsden, *Groups of diffeomorphisms and the motion of an incompressible fluid*, Ann. Math. **92** (1970), 102–163; A. M. Lukatskii, *On the curvature of the group of measure-preserving diffeomorphisms of a 2-sphere*, Functional Anal. Appl. **13**, 3 (1979), 23–27; A. M. Lukatskii, *On the curvature of the group of measure preserving diffeomorphisms of an n-torus*, Uspekhi Mat. Nauk **36**, 2 (1981), 187–188. That the curvature is negative indicates exponential instability of the corresponding attractors. Lukatskii found negativeness for flows on S^2 and T^n, for most sections, and negativeness of the Ricci curvature (per unity of dimension). See further applications of these ideas in D. D. Holm, J. E. Marsden, T. Ratiu, A. Weinstein, *Nonlinear stability of fluid and plasma equilibria*, Phys. Rep. **123**, 1&2 (1985), 1–116; A. I. Shnirelman, *On the geometry of the diffeomorphism groups and on ideal fluid dynamics*, Mat. Sbornik **128 (170)**, 1(9), (1985), 82–109; McIntyre and T. G. Sheperd, *An exact conservation theorem for finite-amplitude disturbances to nonparallel shear flows, with remarks on Hamiltonian structure and on Arnold's stability theorem*, J. Fluid Mech. **181** (1987), 527–565.]

A very simple model with unstable trajectories on an attracting set has been proposed by Henon [M. Henon, *A Two-dimensional mapping with a strange attractor*, Comm. Math. Phys. **50** (1976), 69–77]. Henon considers a quadratic "Cremona transformation" on the plane having the form $T = T_1 T_2 T_3$, where

$$T_1(x, y) = (y, x), \qquad T_2(x, y) = (bx, y), \qquad T_3(x_1\, y) = (x, y + 1 - ax^2).$$

Interestingly, the experimentally observed (for $a = 1.4, b = 0.3$) attraction to the set having locally the form of the product of the Cantor set with an interval, could not be

---

* We may think that when the viscosity becomes small (the Reynolds' number becomes large) the minimum of the dimensions of the minimal attractors grows indefinitely. But this is not proved even for the maximum of these dimensions (it has not been proved rigorously that this maximum exceeds one).

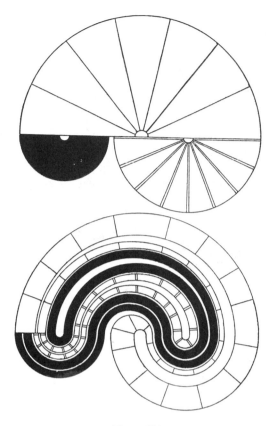

Figure 133.

successfully described in the framework of the existing definitions of hyperbolicity. (It is not even excluded that this set is interspersed with domains of attraction of long cycles.) Therefore, mathematicians do not accept Henon's attracting set as hyperbolic. At the same time, from the experimentalist's point of view the motion of a phase point under the iterations of the transformation T has an obviously stochastic, turbulent character (still another example of the danger of fetishization of axioms).

Examples of true hyperbolic attracting sets in the plane have been constructed by Plykin [R. V. Plykin, *Sources and sinks of axiom A-diffeomorphisms of surfaces*, Mat. Sbornik **94**, 2 (1974), 243–264]. Plykin constructs a diffeomorphism of the closed domain with three holes depicted in the upper part of Fig. 133 onto its striped part (in the lower part of Fig. 133) with the following property: the intersection of the images of the domain under all iterations of the diffeomorphism is an attracting set (the distance of the image of any point under the iterations to this set converges to zero); this intersection is locally the product of the Cantor set with an interval, and the distance between nearby points increases under iterations of the transformation on every interval.

An extensive but incomplete bibliography of works in bifurcation theory and its applications can be found in J. E. Marsden, M. McCracken, *The Hopf Bifurcation and its Applications*, New York, Springer-Verlag, 1976. (over 350 titles). See also B. D.

Hassard, N. D. Kazarinoff, Y.-H. Wan, *Theory and Applications of Hopf Bifurcation*, London Math. Soc. Lect. Notes Series, Vol. 41, Cambridge University Press, Cambridge, 1981.

The determination of the regime occurring after the Poiseuille flow loses stability is, in the opinion of the experts, on the borderline of the capabilities of modern computers.

In this situation one must probably not disregard the qualitative predictions which can be made completely without calculations, relying on the general bifurcation theory expounded above.

In the problem under consideration, there are two parameters, α and R. Consequently, besides singularities of codimension 1, we may also encounter singularities of codimension 2. We direct our attention to the one that is associated with the change of the sign of c. Calculations show that for a sufficiently large Reynolds' number R, the rough generation on the lower side of the tongue of the loss of stability becomes soft. In order to understand what happens at that moment, we have to construct a two-parameter versal family for such a two-fold degeneracy. Such a family can easily be constructed. It has the form

$$\dot{z} = z(i\omega + \varepsilon_1 + \varepsilon_2 z\bar{z} + c_2 z^2 \bar{z}^2), \qquad z \in \mathbb{C}.$$

(The remaining coordinates in the phase space correspond to stable eigenvalues and are not written out.) The meaning of the parameters ε_1 and ε_2 is clear from Fig. 132; the character of the metamorphosis at the point $\varepsilon_1 = \varepsilon_2 = 0$ is determined by the sign of c_2.

As above, setting $\rho = z\bar{z}$, for ρ we obtain the equation

$$\dot{\rho} = 2\rho(\varepsilon_1 + \varepsilon_2 \rho + c_2 \rho^2), \qquad \rho \geqslant 0.$$

Depending on the signs of ε and c, the following cases are possible.

1. $c_2 < 0$, $\varepsilon_2 < 0$. When ε_1 passes from negative to positive values, the systems exits softly to a periodic stable regime of self-sustained oscillations (Fig. 134).

2. $c_2 < 0$, $\varepsilon_2 > 0$. As ε_1 passes from negative to positive values, the system roughly exits to stable periodic self-sustained oscillations; the corresponding cycle is born before the loss of stability of the equilibrium position, together with an unstable cycle landing on the equilibrium position at the moment it loses its stability.

We were able to study the stable limit cycle indicated above near the point where the rough behavior is replaced by soft behavior, because it is then close to an equilibrium position. However, an analytic continuation of this limit cycle may exist (far away from the equilibrium position) for other values of the parameters (α, R), as well; we see that it can be searched

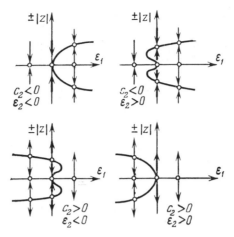

Figure 134.

for by an analytic continuation of an unstable cycle landing on the equilibrium position in the rough loss of stability. The stable cycle indicated above is one of the candidates for the role of the behavior established under or after the loss of stability.

3. $c_2 > 0$, $\varepsilon_2 < 0$. The loss of stability is soft, but the limit cycle dies quickly after its birth, coalescing with the unstable cycle arriving from afar, after which a new behavior is roughly generated in the system.

4. $c_2 > 0$, $\varepsilon_2 > 0$, *Usual Rough Generation*. Consequently, no matter what the sign of c_2, for a suitable sign of ε_2 our analysis enables us to establish a qualitatively new phenomenon compared to the one-parameter analysis: for $c_2 < 0$ we explicitly find the regime established after the rough generation. For $c_2 > 0$, we detect the short life of a regime generated softly. In order to determine which one of the two cases ($c_2 < 0$ or $c_2 > 0$) actually takes place, one has to perform extremely cumbersome calculations.

In the theory of hydrodynamical stability, various singularities of the boundary of stability and the decrement diagrams occur, and the results of § 30 may find an application. For applications of the general theory of bifurcations in the theory of hydrodynamical stability, it would be important to study generic cases in problems with various symmetry groups, since in many hydrodynamical problems, the domain D of the flow exhibits one or another group of symmetries (for example, the group of translations in the problem of a Poiseuille flow; representations of this group play a role in the study in the form of the parameter α).

The behavior of a fluid after the loss of stability of a stationary flow is discussed in many publications [cf., for example, L. D. Landau, E. M. Lifshits, Fluid Mechanics, Reading, Mass., Addison-Wesley (1959), § 27 (Occurrence of Turbulence), based on the work of Landau in 1943 (L. D. Landau, *On the turbulence problem*, Dokl. Acad. Nauk SSSR **44**, 8 (1944), 339–342.)]. In this theory, soft generation of self-sustained oscillations is usually assumed, and the loss of stability of a limit cycle is studied. Landau assumed that in this case quasiperiodic motions arise with a growing number of frequencies; undoubtedly, this can be explained by the fact that other dynamical systems were not known to him.

In 1965, this author lectured on the theory discussed above in the seminar of Thom at IHES at Bures-sur-Yvette. Six years later, Ruelle and Takens constructed [in *On the nature of turbulence*, Comm. Math. Phys. **20** (1971), 167–192; **23** (1971)] examples of the loss of stability of a cycle with the occurrence of a behavior more complicated than a quasiperiodic one; however, their example has an exotic character, since it corresponds to a metrically very skinny (although open) part of the space of the parameters of the deformation.

We would like to mention that for the applicability of the results of the indicated works, it is necessary that the loss of stability take place in a soft regime, whereas the behavior of the loss of stability of the Poiseuille flow turned out to be rough.

For a review of the subsequent experimental work, cf., J. B. McLaughlin, P. C. Martin, *Transition to turbulence of a statically stressed fluid system*, Phys. Rev. Lett. 33 (1974); Phys. Rev. A 12 (1975), 186–203.

E. Degeneracies of Codimension 2

For generic one-parameter families of vector fields, the bifurcations of phase portraits in the neighborhood of a singular point are exhausted by the cases analyzed above (the birth and annihilation of a pair of singular points, the birth or annihilation of a limit cycle from a singular point).

In two-parameter families, these singularites will be encountered along curves of the parameter plane; besides, more complicated degeneracies will be observed at isolated points of the parameter plane. Among these more complicated degeneracies, the following five are unremovable by a small perturbation of the two-parameter family.

1. One Vanishing Root with an Additional Degeneracy. Example:

$$\dot{x} = \pm x^3 + \varepsilon_1 x + \varepsilon_2, \qquad x \in \mathbb{R}$$

(Fig. 135). It is easy to verify that the deformation thus described is (topologically) versal; in the multidimensional case, a versal deformation can be obtained by a suspension of a saddle.

The bifurcation diagram (for the case $+x^3$) is illustrated in the left-hand part of Fig. 135. The semicubic parabola divides the parameter plane into

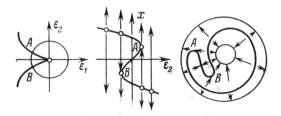

Figure 135.

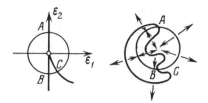

Figure 136.

two parts. In the smaller part, the system has three equilibrium positions near $x = 0$ and in the larger one there is one equilibrium. The metamorphoses of the phase portrait, as the parameter goes around the point $\varepsilon = 0$ on a small circle, are shown on the right-hand side of Fig. 135. The direct product of this circle with the one-dimensional phase space is an annulus, and the equilibrium positions form a closed curve in this ring; the behavior of vectors of the field is clear from Fig. 135.

2. *One Imaginary Pair with an Additional Degeneracy*. Example:

$$\dot{z} = z(i\omega + \varepsilon_1 + \varepsilon_2 z\bar{z} \pm z^2\bar{z}^2), \qquad z \in \mathbb{C}.$$

The bifurcation diagram consists of the line $\varepsilon_1 = 0$ and one-half of a parabola tangent to it at zero; it is depicted in Fig. 136 in the case where we have $+z^2\bar{z}^2$ in the formula.

The metamorphoses of the phase portrait as the parameter goes around 0 on a small circle are shown in the right-hand part of Fig. 136. The ring depicted in Fig. 136 is the direct product of a circle in the parameter plane and a line on which $\pm|z|$ is plotted. The circle in this diagram corresponds to the equilibrium position $z = 0$, and every limit cycle is illustrated by the two points of intersection of the radius with the curve $\varepsilon_1 + \varepsilon_2|z|^2 + |z|^4 = 0$.

The bifurcation diagram and the family over the circle are analogous in the case of $-z^2\bar{z}^2$.

3. *Two Imaginary Pairs*.

4. *An Imaginary Pair and Still Another Vanishing Root*. These cases have not yet been studied sufficiently to describe versal families; moreover, it is not clear whether in the case of two imaginary pairs a two-parameter (or, at least, a finite-parameter) topologically versal family exists (even under the assumption of normal incommensurability of the ratio of frequencies in the case of their simultaneous passage from one half-plane to the other). See, however, E. I. Horozov, *Bifurcations of a vector field near a singular point in the case of two pairs of imaginary eigenvalues*, I and II, Comptes Rendus de l'Académie Bulgare des Sciences, **34** (1981) and **35** (1982), 149–152; and also a paper by H. Żołądek on case 4.

These problems then lead to the problem of bifurcations of phase portraits of planar vector fields when the equations are averaged along the fast rotation (or when finite-order Poincaré normal forms are used). These special vector fields have an invariant line (in the case of one imaginary pair and one zero eigenvalue) or two intersecting invariant lines (in the case of two imaginary pairs) for all values of the parameters, and have zero eigenvalues for zero value of both parameters (at the origin which is a singular point of the vector field). For the case of two pairs the equations are

$$x^{\cdot} = xA(x, y, \varepsilon), \qquad y^{\cdot} = yB(x, y, \varepsilon). \tag{1}$$

Systems of the same type occur in ecology when pedator–prey or competing species interactions are studied (the Lotka–Volterra model); note, however, that it is important to keep the nonlinear terms of A and B, otherwise the bifurcations would not be generic—as happens for the original Lotka–Volterra integrable model, $A = a - py$, $B = -b + qx$.

In terms of the original four-dimensional system, the variables x and y have the same meaning as the squared amplitudes of the eigenoscillations. An equilibrium point in the planar system represents: (a) an equilibrium in the original four-dimensional system if it is the intersection point of both invariant lines; (b) a cycle if it belongs to one of them; and (c) an invariant torus if it lies inside the positive quadrant of the plane (the eigenfrequencies are supposed to be incommensurable or at least low-order resonance free).

The two problems (the relation between the four-dimensional system and the planar system and the bifurcations of the phase portrait of the planar system, which were treated wrongly in many papers by different mathematicians including Guckenheimer, Holmes, and others) can be treated separately. The latter problem is highly nontrivial and was treated incorrectly.

The correct pictures bifurcations of planar fields were produced in 1979 by V. I. Shvetzov in a computer-assisted thesis at Moscow University: the nontrivial point is that the number of limit cycles, born at the origin, does not exceed one (generically) in the planar system.

A rigorous proof (and hence a description of the versal deformations in the class of systems (1)) has recently been found in 1985 by H. Żołądek,* who previously had proved a similar theorem for the case of an imaginary pair and one zero root: see H. Żołądek, *On the versality of one family of symmetric planar vector fields*, Mat. Sbornik **120**, 4 (1983), 473–499 (instead of one or two invariant line conditions one may put the condition of symmetry with respect to one or two lines mathematically; these cases are equivalent).

The number of limit cycles born at the origin in the planar system corresponding to a pair and a zero is still (generically) one.

Finally, there remains the last case of codimension 2.

5. *Two Vanishing Roots.* An example is the family

$$\dot{x}_1 = x_2,$$
$$\dot{x}_2 = \varepsilon_1 + \varepsilon_2 x_1 + x_1^2 \pm x_1 x_2$$

---

*H. Żołądek, *Bifurcations of certain family of planar vector fields tangent to axes*, Journal of Differential Equations **67**, 1 (1987), 1–55.

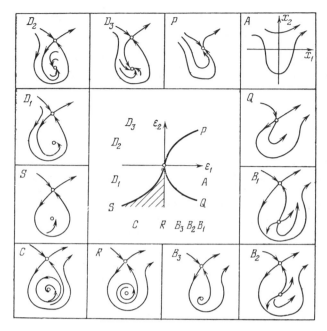

Figure 137.

of equations in the plane with parameters $(\varepsilon_1, \varepsilon_2)$. The bifurcation diagram divides the ε-plane into four parts denoted by A, B, C, and D in Fig. 137, corresponding to the choice $+x_1 x_2$ in the formula.

The phase portraits corresponding to each of the four parts of the ε-plane are shown in Fig. 137. To the branches of the bifurcation diagram there correspond the systems (P, Q, R, S) with degeneracies of codimension 1 depicted in Fig. 137.

We note that the bifurcation on the branch S—the birth of a cycle from a loop of the separatrix—is not included in our classification of singularities of codimension 1, since it is not a local (near a singular point) but a global phenomenon. We see, therefore, that in the local study of bifurcations of singular points with the increase of the number of parameters of the family, global bifurcations of small codimensions begin to play a role. It follows that in a local problem for a sufficiently large number of parameters, we will encounter the same difficulty of structurally stable systems being not everywhere dense, which was discovered by Smale in the global problem of vector fields on a manifold (cf., § 15).

The bifurcations in the case corresponding to the choice " $-$ " of the sign in the formula can be reduced to the preceding ones by changes in the signs of t and x_2.

Theorem. *Vector fields generic among those with two vanishing roots of the characteristic equation at a singular point in the phase plane have a topo-*

logically versal deformation with two parameters equivalent to one of the two deformations considered above.

In other words, *a generic two-parameter family of differential equations in the plane having a singular point with two vanishing roots of the characteristic equation for some value of the parameter can be reduced to the form indicated above by a continuous change of the parameters and a continuous change of the phase coordinates continuously depending on the parameters.*

This theorem, proved by Bogdanov in 1971, was first published in the survey of V. I. Arnold, *Lectures on bifurcations and versal systems*, Uspekhi Math. Nauk **27**, 5 (1972), 119–184 [Russ. Math. Surveys **27**, (1972), 54–123]. Takens announced an analogous result in 1973. The proof of versality is not simple: the main difficulty is the study of uniqueness of the limit cycle. Bogdanov overcomes this by nontrivial arguments on the behavior of elliptic integrals depending on a parameter [cf., R. I. Bogdanov, *Bifurcations of a limit cycle of a family of vector fields in the plane*, Trudy Seminara Imeni I. G. Petrovskogo, 1976, vol. 2, 23–36: *A versal deformation of a singular point of a vector field in the plane in the case of vanishing eigenvalues*, Ibid., pp. 37–65. English translation: Selecta Math. Sovietica, **1**, 4 (1981), 373–388, 389–421].

§ 34. Loss of Stability of Self-Sustained Oscillations

The second most complex problem of bifurcation theory (after the problem of metamorphosis of the phase portraits in the neighborhood of equilibrium positions) is the problem of metamorphoses of a family of phase curves in the neighborhood of a closed phase curve. This problem has not been solved completely, and is apparently unsolvable in some sense. Nevertheless, the general methods of bifurcation theory allow us to obtain significant information on these metamorphoses; in this section, we give a brief survey of the basic results.

A. Monodromy and Multipliers

We consider a closed phase curve of a system of differential equations. We are interested in the metamorphoses of the configuration of phase curves in the neighborhood of the curve under a small change in the equations.

There is a finite number of possibilities for the distribution of phase curves in the neighborhood of a generic closed phase curve (up to a homeomorphism of the neighborhood). To describe them, we choose a point O on the closed phase curve. Through this point we draw a transversal section (of codimension 1 in the phase space) to the closed phase curve. The phase curves starting from points of the transversal section sufficiently close to O intersect the transversal again, making a revolution along the curve. This gives rise to a mapping of the neighborhood of the point O in the section into the section. This mapping is called the *Poincaré map* (Fig. 138).

The point O is a fixed point of the Poincaré map. We consider the linearization of the Poincaré map at the point O. This linear operator is called the *monodromy operator*.

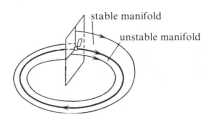

stable manifold

unstable manifold

Figure 138.

The eigenvalues of the monodromy operator are called the *multipliers* of the original closed phase curve. The monodromy operator can be found by solving a linear equation with periodic coefficients (equation in normal variations along our phase curve).

We assume that all multipliers are smaller than 1 in absolute value. Then it can be proved that all neighboring phase curves are attracted to our closed phase curve. If at least one of the multipliers is larger than 1 in absolute value, then phase curves exist which diverge from the closed curve (approach it as $t \to -\infty$).

In the general case, some eigenvalues lie inside the unit circle and some outside. In this case, the phase curves attracted to the given one form, as is easily seen, *stable* manifold* whose intersection with our transversal has dimension equal to the number of multipliers inside the unit circle. Similarly, the phase curves asymptotically approaching the closed curve as $t \to -\infty$ form an *unstable manifold*. The dimension of its intersection with the transversal is equal to the number of unstable multipliers (multipliers outside the unit circle).

In the neighborhood of our closed phase curve, a hyperbolic situation takes place (cf., § 14): all other phase curves diverge from the closed curve as $t \to \infty$ (along the unstable manifold) as well as for $t \to -\infty$ (along the stable manifold). The topological type of a family of phase curves in the neighborhood of a closed phase curve not having multipliers on the unit circle is uniquely determined by the number of stable and unstable multipliers and by the parities of the numbers of negative stable and unstable multipliers.

We will see what changes in this picture under a small change in the system.

B. Simple Degeneracies

A closed phase curve is said to be *nondegenerate* if the unit is not a multiplier. A nondegenerate closed phase curve does not vanish under a small deformation of the system; it only becomes slightly deformed (by the

---

*One should better call it *"contracting manifold"*, because the neighboring phase curves are repelled by this manifold, hence the motion along this manifold is highly *unstable*. The misleading terms "stable", "unstable" and notations "s", "u" were introduced by Smale.

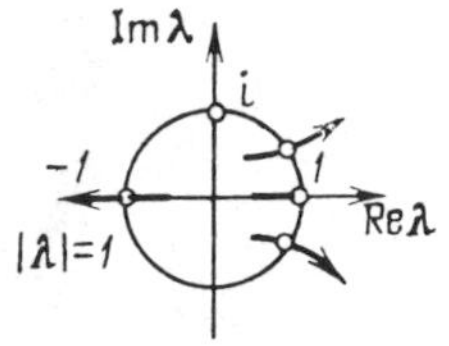

Figure 139.

implicit function theorem applied to the equation $f(x) = x$, where f is the Poincaré map). Under the deformation of a nondegenerate closed phase curve, the multipliers are also deformed only slightly. Consequently, neither the number of stable nor unstable multipliers changes under the deformation if none of the multipliers of the original phase curve lies on the unit circle.

The multipliers of a generic closed phase curve do not lie on the unit circle. Consequently, the distribution of phase curves in the neighborhood of a generic closed phase curve is structurally stable.

However, if we consider a family of systems depending on a parameter rather than an individual system, then for some isolated values of the parameter, the multipliers may fall on the unit circle. Then the problem of bifurcations arises.

As usual, we begin with simple degeneracies, i.e., degeneracies unremovable in one-parameter families. In our case, there are three such degeneracies of codimension 1. Indeed, the characteristic equation of the monodromy operator is real. Therefore, every nonreal multiplier has a complex-conjugate multiplier. Consequently, on the unit circle, we may have either two complex-conjugate multipliers or one real multiplier equal to either 1 or -1. All three cases (a complex pair, $+1$, -1) correspond to manifolds of codimension 1 in the function space.

Let us, for example, consider the boundary of a stability domain of a closed phase curve in the function space. This boundary is a hypersurface in the function space. It consists of three components of codimension 1. The first component corresponds to phase curves with one pair of complex-conjugate multipliers with absolute value 1, the second to ones with the multiplier $+1$, and the third to ones with the multiplier -1; all remaining multipliers lie inside the unit circle (Fig. 139).

These three hypersurfaces of codimension 1 intersect in surfaces of codimension 2 and have further singularities. For example, the self-intersections of the first surface correspond to two pairs of multipliers with absolute value 1 and so on.

The problem of loss of stability of a closed phase curve is therefore a problem of degeneracy of codimension 1. At first glance we have to consider generic one-parameter families in order to examine bifurcations. The matter is, in fact, not this simple: we shall see that in the problem of stability loss, as a pair of multipliers crosses the unit circle, there are *two* essential parameters. However, let us first see to what conclusions the one-parameter point of view leads.

We begin with the case where one of the multipliers is equal to 1. This case is not essentially different from the problem of bifurcations of equilibrium positions in one-parameter families. The generic situation is the birth or death of a pair of closed phase curves. In the meantime, two fixed points are born or die for the Poincaré map.

EXAMPLES

1. We consider the mapping of the x-axis into itself given by the formula $x \mapsto x + x^2$. The point $x = 0$ is fixed and its multiplier is equal to 1. We consider the following one-parameter deformation with parameter ε close to zero:

$$f_\varepsilon = x + x^2 + \varepsilon.$$

This deformation is topologically versal. We consider any mapping of the line into itself having a fixed point with multiplier 1. We call this (degenerate) fixed point *regular* if the second derivative of the mapping at the fixed point is different from zero (in some, and then all, coordinate systems).

If the degenerate fixed point is regular, then a one-parameter topologically versal deformation of the mapping exists. Moreover, both the mapping and its versal deformation are locally topologically equivalent to the deformation f_ε of the special mapping f_0 in the neighborhood of 0.

In order to pass to the multidimensional case, we have to define a suspension over the deformation constructed above.

2. Consider the mapping of a linear space into itself given by the formula

$$(y, z, u, v) \mapsto \left(2y, \ -2z, \frac{u}{2}, \ -\frac{v}{2}\right),$$

where y, z, u, and v are points of four spaces whose direct product is our space. We shall call such a mapping a *standard saddle*. (The dimensions of the spaces to which y and u belong are arbitrary, and the dimensions of the spaces to which z and v belong are equal to 0 or 1.)

We consider any smooth mapping with a fixed point. Let us assume that none of the multipliers lie on the unit circle. Then in the neighborhood of the fixed point, the mapping is topologically equivalent to a standard saddle (which follows easily from the Grobman–Hartman theorem, § 13).

3. Consider the direct product of the deformation of the mapping of the line in Example 1 with a standard saddle. We obtain the following one-

parameter family of mappings with parameter ε and phase coordinates varying in the neighborhood of zero:

$$(x; y, z, u, v) \mapsto \left(x + x^2 + \varepsilon, 2y, - 2z, \frac{u}{2}, - \frac{v}{2}\right).$$

This deformation is called a *suspension* over the deformation of Example 1. It is topologically versal.

Theorem. *Generic one-parameter families of mappings are topologically equivalent to the one described above in the neighborhood of every fixed point with multiplier 1 for values of the parameter close to the one for which the multiplier becomes equal to 1.*

◀ The proof is easy in the one-dimensional case. The multi-dimensional case can be reduced to the one-dimensional case by means of Šošitaišvili's theorem (§ 32), which holds for both differential equations and mappings. ▶

C. The Case of a Multiplier Equal to -1

If the multiplier -1 appears, then the closed phase curve depends smoothly on the parameter and does not bifurcate itself. However, another closed phase curve branches out, winding twice. In order to understand how this occurs, we again appeal to the Poincaré map.

EXAMPLE

1. Consider the following mapping of the line into itself:

$$f_0(x) = -x \pm x^3.$$

The multiplier of the fixed point is equal to -1.
  We imbed f_0 in the family

$$f_\varepsilon(x) = (\varepsilon - 1)x \pm x^3.$$

Theorem. *The deformation f_ε of the mapping f_0 is versal. Any generic one-parameter family is topologically equivalent to the one described above in the neighborhood of a fixed point with multiplier -1 for values of the parameter close to the value for which the multiplier is equal to -1.*

◀ We consider any one-parameter family of mappings of the line in which the multiplier of a fixed point becomes -1 for some value of the parameter.
  The fixed point depends on the parameter smoothly (by the implicit

function theorem). By a change of coordinates smoothly depending on the parameter, we can move the fixed point to zero.

Now we shall make the Poincaré changes (cf., § 25), successively killing the nonresonant terms. These changes will depend on the parameter smoothly if we retain the terms which become resonant for the critical value of the parameter, not only for this value of the parameter (when they cannot be annihilated), but also for neighboring values.

In our case, the resonant terms are all the terms of odd degree. Consequently, the family can be reduced to the form

$$x \mapsto \lambda x + ax^3 + O(|x|^5),$$

where λ, a, and O depend smoothly on the parameter.

In a generic family, the derivative of λ with respect to the parameter for $\lambda = -1$ is different from zero. In this case, we can take $\varepsilon = 1 + \lambda$ as the parameter. Now the deformation assumes the form

$$x \mapsto (\varepsilon - 1)x + a(\varepsilon)x^3 + O(|x|^5).$$

In a generic family, we have $a(0) \neq 0$. By a dilation of the coordinates depending smoothly on the parameter, we can make $a(\varepsilon) = \pm 1$.

We still have to prove that the term O does not affect the topological type of the family. We consider the second iterate of our mapping:

$$x \mapsto (\varepsilon - 1)^2 x + (\varepsilon - 1)ax^3 + a(\varepsilon - 1)^3 x^3 + O(|x|^5).$$

Every point x is shifted by

$$h = -2\varepsilon(1 + \cdots)^2 x + (2a + \cdots)x^3 + O(|x|^5)$$

where $\cdots$ means $O(\varepsilon)$.

It is easy to study the zero level curve of the function h in the (x, ε)-plane (Fig. 140). Figure 140 determines the topological type of the family. ▶

Consequently, in a generic one-parameter family of mappings of the line onto the line, the multiplier of the fixed point becomes equal to -1 at the moment of transversal passage through the unit circle (as opposed to the

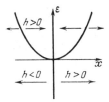

Figure 140.

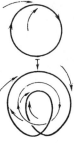

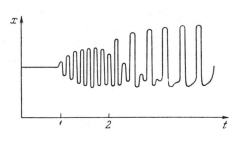

Figure 141.   Figure 142.

case where the multiplier becomes equal to 1, when the multiplier generally does not pass through the circle). At the moment of the passage of the multiplier through -1 from inside to outside, the fixed point loses stability. Then there are two possibilities depending on the sign of the coefficient of x^3: (i) Along with the point which has lost stability (at a distance on the order of the square root of the difference of the parameter from the critical value), a stable cycle of period 2 arises (two fixed points of the square of the mapping). This is a case of the soft loss of stability. (ii) The domain of attraction shrinks to 0 because of the approach of a cycle of order 2 before the loss of stability (rough loss of stability).

The multi-dimensional picture can be obtained by a suspension of a saddle, as described above.

Applying all of what has been said about mappings to the Poincaré map of a closed phase curve, we obtain the picture in Fig. 141 in the case of the soft loss of stability: the original cycle loses stability, but a stable cycle apperars with a period which is approximately twice the original one.

The phenomena described here can well be observed in experiments. The following example is borrowed from a lecture by Barenblat delivered at the seminar of I. G. Petrovskiĭ. We consider a polymer film stretched slowly by a weight. For small expansions, the process is quasistationary. (Time can be considered the parameter; the phase point is in a stable equilibrium position; and all observable quantities are constant for every value of the parameter, i.e., they actually change slowly with the variation of time). However, for some value of the parameter (i.e., for a sufficient expansion of the film), the picture changes and the form of the various physical parameters (say, the tension x of the film) as functions of time becomes such as depicted in Fig. 142. (Every oscillation in this picture can be considered as taking place for a fixed value of the parameter, and in the next oscillation the parameter changes only slightly.)

The interpretation of this evolution of phase variables over time is as follows: Point 1 corresponds to a soft loss of stability of the equilibrium with the onset of oscillations; it is clear that their amplitude grows in proportion the square root of supercriticality. Point 2 corresponds to a soft loss of stability of a cycle with a multiplier passing through -1.

Indeed, assume that the metamorphoses indicated in Fig. 141 take place in the phase space.

Every physically observable quantity is a function on the phase space. As long as the phase point is in an equilibrium position, the quantity is constant. When the phase point moves on a cycle, the quantity x becomes a periodic function of time t (the amplitude of oscillations increases with the size of the cycle dimension). To a doubling of a cycle (see Fig. 141) there corresponds a doubling of the period of dependence of quantity over time. This has been observed in the experiment (Fig. 142).

Here we note that, in general, in the study of self-sustained oscillations, the observed one usually measures time-dependencies of the observed quantities (say, on an electrocardiogram). In many cases, a clearer representation of the character of phenomena can be obtained from the form of a phase curve or its projection onto some plane. This method has long been used for the diagnostics of failure of mechanical oscillating systems such as pumps. The application of this method in electrocardiography has been suggested by physicians.

One of the most striking discoveries of recent years in bifurcation theory is the infinite cascades of doublings phenomenon, discovered in the late 1970s—the so-called Feigenbaum universality.

The phenomenon consists of the repeated doubling of cycles under the variation of a parameter, generating attracting cycles with approximately twice the period, then approximately four times, eight times, and so on: an infinite series of doublings occurs while the parameter changes along a finite interval. The parameter values corresponding to subsequent period doublings are divided by intervals, decreasing asymptotically like a geometrical progression: their ratios converge to a constant value. This constant value is independent of the special properties of the particular (generic) system: it is a universal constant, like π or e. It is called the Feigenbaum constant and its value is (if the preceding interval length is divided by the next one) approximately 4.6692. . . . The universality also means that all the details of the doublings become more and more standard (independent of a particular system) up to an appropriate scale changing, when the variable parameter approaches the limit value, corresponding to an infinity of doublings.

The simplest way to study the cascades of doublings is to consider a family of mappings of a line, or of an interval, into itself. A typical example is the ecological model of a population growth taking the competition into account

$$x \mapsto Axe^{-x}$$

(here the multiple e^{-x} describes the effect of the competition decreasing the Malthusian growth coefficient A). (Fig. 143). For small values of A the fixed point 0 of the mapping is stable (extinction). For larger values of A a positive, stable fixed point appears (Fig. 143), which later, at some value A_1 of the parameter, loses the stability while its eigenvalue goes through -1 (our mapping, of course, is not a diffeomorphism) generating a stable cycle of period 2,* thus starting a sequence of doublings (Fig. 143).

These doublings were found immediately by ecologists when they started computer experiments with the above model and other similar models (see A. P. Shapiro, "Mathematical models of competition," in: *Upraylenie i Informaciya/Control and Information*, Vladivostok: DVNC AN SSSR, **10** (1974), 5–75; R. M. May, *Biological*

---

*This cycle explains, it seems, the well-known observation that the gorbusha (humpbacked salmon) catch oscillates with a two-year period: years with a large gorbusha population are followed by years with a small population, and vice versa.

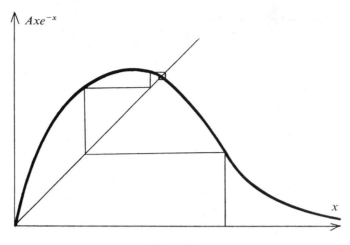

Figure 143.

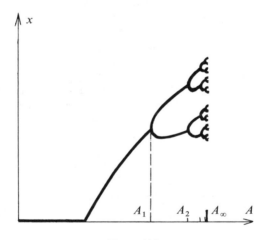

Figure 144.

populations obeying difference equations; *stable points, stable cycles, and chaos*, J. Theor. Biol. **51** (1975), 511–524; *Simple mathematical models with very complicated dynamics*, Nature **261** (1976), 459–466). It was the analysis of these experiments that led M. Feigenbaum to his remarkable discovery of the stability of infinite cascades of doublings and to the even more remarkable universality law.

Let us consider a mapping of a line into inself, $x \mapsto f(x)$, in a neighborhood of the maximum point of function f. The squared (iterated twice) mapping will have a graph with two maxima and one minimum point between them, as is easily seen (Fig. 145). In some neighborhood of the minimum point the graph, after a coordinates dilatation and a sign change, looks very similar to the graph of the original function. A computer experiment shows that this similarity becomes closer and closer under the sequential doublings (squarings) of the mapping.

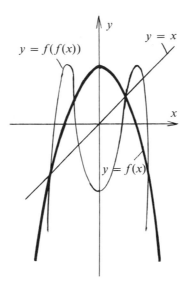

Figure 145.

This leads to an attempt at finding a function, which repeats itself exactly after the squaring of the mappings (followed by a rescaling); that is, to find a fixed point for the doubling operator J, defined on functions as is described below. Such a fixed point happens to exist, namely, the even analytic function

$$\Phi(x) = 1 - 1.52763x^2 + 0.104815x^4 - 0.0267057x^6 + \cdots,$$

satisfying $J\Phi = \Phi$, where

$$(Jf)(x) = -\frac{1}{a}f(f(-ax)).$$

The normalization constant a for Φ is $a = 0.3995\ldots = -\Phi(1)$ (all the numerical coefficients are given approximately).

The operator J, having fixed point Φ, is defined on the even C^1-mappings f: $[-1, 1] \to [-1, 1]$ with one maximum point 0, satisfying the following conditions (see Fig. 146):

$$f'(0) = 0, \quad f(0) = 1, \quad f(1) = -a < 0, \quad b = f(a) > a, \quad f(b) = f(f(a)) < a$$

for some a and b (depending on f). The operator J is a hyperbolic (at point Φ) mapping of the functions space. It has a one-dimensional dilatating invariant manifold (a curve) and a codimension 1 contracting submanifold. The eigenvalue corresponding to the dilatation is the Feigenbaum constant $4.6692\ldots$.

The universality of the period-doubling cascades is explained by this picture as we shall now see. A one-parameter family of mappings (even for the simplicity of the arguments) is represented by a curve in the functions space (γ at Fig. 147). If this curve

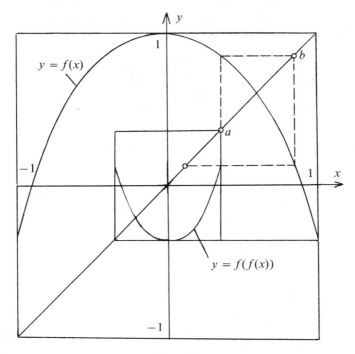

Figure 146.

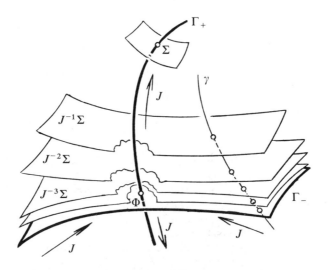

Figure 147.

is not too far from Φ its images become closer and closer to the dilatating invariant curve Γ_+ under consecutive iterations by J.

On the other hand, let us consider the bifurcation surface Σ in the functions space, consisting of the mappings f of a line having a fixed point with eigenvalue -1. It turns out that the dilatating invariant curve Γ_+ intersects this hypersurface Σ transversally. Hence the preimages of the bifurcation surface Σ under J and under its iterations form a sequence of hypersurfaces of codimension 1 in the functions space, converging to the contracting manifold Γ_- of the fixed point Φ when the number of iterations grows (Fig. 147).

The points of intersection of these surfaces, with the curve γ representing our family, define the parameter values corresponding to the period doublings. This explains the origin of the geometrical progression and the universality of its ratio: it depends on the sequence of surfaces, and not on the special choice of the transversal curve. For the (computer-assisted) proof and for a bibliography see P. Collet, J.-P. Eckmann, *Iterated Maps of the Interval as a Dynamical System*. Boston: Birkhauser, 1980, 248 p.

The situation remains more or less the same for the doubling of cycles of differential equations (for instance, the universal constant 4.6... remains unchanged), while some of the details of the cascades of period doublings are different for differential equations and for line mappings. The multiplier (the eigenvalue of the mapping) must change its value from 1 to -1 between two doublings. The reason is that the doubling cycle has (at the moment of doubling) a multiplicator equal to -1. Hence the doubled cycle has (at the moment of its birth) a multiplier $(-1)^2 = 1$. But the path from 1 to -1 along the real line is forbidden for an eigenvalue of a diffeomorphism (and hence for a multiplier of a cycle of a differential equation), since the eigenvalue of a diffeomorphism may never become zero.

Thus the cascades of period-doubling bifurcations in differential equations and diffeomorphisms involve a second multiplier, besides that coming from 1 at some point on the real axis. After a collision both become complex (conjugate) and turn around zero in the upper and lower half-planes, collide once more on the negative part of the real axis, and finally diverge—one going to -1, the other in the opposite direction. The curve, followed by the multiplier, is extremely close to a circle, and displays some universality features for repeated doublings (M. V. Jakobson, 1985).

The universal period-doubling cascades have their close analogs at other resonances. Let us consider, for instance, the 1 : 3 resonance when a multiplier leaves the unit circle through the cubic root of unity. To observe such a phenomenon in a generic family we need at least two parameters. In the parameter plane the moments of tripling are represented by isolated points. These points occur, as for the doubling, in the form of infinite series. The differences of consecutive tripling values of the parameters form asymptotically geometric progressions whose ratio is a universal constant (the same one for all generic families). But in contrast to the case of doublings the universal constant this time is not a real, but a complex number, hence the tripling values of the parameters on the parameters plane are organized in spiraling sequences.

There exist cascades of doublings in Hamiltonian systems too. This time the -1 multiplier is always of even multiplicity, generically of multiplicity 2. In a doubling the doubled cycle is born, having (at the moment of birth) a double multiplier equal to 1. This pair of multipliers moves (until the next doubling) along the unit circle from 1 to -1 in the upper and lower half-plane. The corresponding Feigenbaum constant is much larger than for ordinary doublings (about 8). On the way from 1 to -1 the

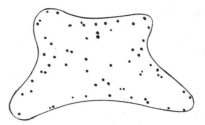

Figure 148.

multipliers visit all the roots of unity. Thus between two doublings a lot of other events occur with the same universality (family independence) as for the doublings: the birth of cycles, whose periods are all kinds of multiples of the original one (at the moment of birth). A newborn cycle of longer period has (at the moment of birth) a pair of multipliers, equal to 1. Not infrequently these multipliers will travel to -1 along the unit circle. Thus, in principle, a very complicated system of bifurcation values occurs. The bifurcation values of the parameter may be enumerated by all the sequences of rational numbers.

The usual cascades of doublings correspond to the series $1, 2, 3, \ldots$; the cascade of doublings of a tripled cycle corresponds to $1/3, 1, 2, 3, \ldots$; we may consider more sophisticated cases, like $1/3, 1/3, \ldots$ (triplings cascade) or $1/3, 1/2, 1/3, 1/2, \ldots$—every such case may lead to new universal constants.

The consequences of these bifurcations were observed in the numerical experiences of the early 1960s. For instance, the hat-shaped invariant curves ("islands"), see Fig. 148) appear as a result of cycle quadrupling and were observed in the first experiences of Hénon, Heiles, and Chirikov* (with no explanation, it seems).

D. Passage of a Pair of Multipliers through the Unit Circle

This case has been understood much less than either of the preceding two. Topologically versal deformations are not known and may not exist. Nevertheless, Poincaré's method allows us to obtain significant information. We begin with the case where the argument of the multiplier falling on the unit circle is incommensurable with 2π. (This case can be considered typical, since the measure of the set of rational numbers is equal to zero.)

We shall assume that the dimension of the space being mapped is equal to 2. In this case, after an appropriate smooth change of coordinates depending smoothly on the parameter, our family of mappings can be reduced to the form

$$z \mapsto \lambda(\varepsilon)z(1 + a(\varepsilon)|z|^2 + O(|z|^4)),$$

---

*M. Hénon, C. Heiles, *The applicability of the third integral of motion: some numerical experiments*, Astron. J. **69** (1964), 73; B. V. Chirikov, *Studies of Nonlinear Resonances and Stochasticity*, Novosibirsk, Izv. Sib. Otd. Akad. Nauk. SSSR, N267, 1969.

where the real number ε is the parameter of the family and $\lambda(0) = e^{i\alpha}$, $\alpha \neq 2\pi p/q$. For a generic family, we have $d|\lambda|/d\varepsilon|_0 \neq 0$, which means we may choose $|\lambda| - 1$ for the parameter.

We assume that the term $O(|z|^4)$ is absent. In this case, it is easy to study the mapping. Indeed, the modulus of the image of a point is determined by the modulus of the preimage, which gives rise to the real mapping:

$$r \mapsto r|\lambda|\,|1 + ar^2|.$$

For $|\lambda| = 1 + \varepsilon$, $|\varepsilon| \ll 1$, $r \ll 1$, we have

$$|\lambda|\,|1 + ar^2| \approx 1 + \varepsilon \operatorname{Re} ar^2 + \cdots.$$

For a generic family, we have $\operatorname{Re} a \neq 0$. In this case, as the parameter ε passes through the value zero, a circle of radius proportional to $\sqrt{|\varepsilon|}$ and invariant with respect to the mapping is being born (or dies at this point) from the fixed point losing stability. In the first case (the birth of a circle), it is stable, and in the second it is unstable. On the circle itself, the mapping reduces to a rotation.

We return to the terms we have omitted and see whether they affect our conclusions.

It can be shown that an invariant closed curve with radius of order $\sqrt{|\varepsilon|}$ actually exists for the complete mapping (cf., R. J. Sacker, *On invariant surfaces and bifurcation of periodic solutions of ordinary differential equations*, New York University, Report IMM–NYU 333, 1964; Comm. Pure Appl. Math. **18**, 4 (1965), 717–732).

The stability of this closed curve is preserved under perturbations. However, for the complete mapping, its structure on the curve itself is different from the structure of the mapping without the remainder. Indeed, on the invariant curve, the complete mapping may have a rational or irrational rotation number. The mapping of the circle thus arising is not bound to be topologically equivalent to a rotation. In the case of a rational rotation number, it will generally have a finite number of periodic points, alternately stable and unstable. For the original mapping of the plane onto itself, these periodic points will be saddle and nodal points, correspondingly. Therefore, in the case of a rational rotation number, our invariant curve consists of a chain of separatrices of saddle points joined at nodes (Fig. 149).

We note that the separatrices of the saddles are smooth. However, in approaching the node from both sides, the two separatrices together form a curve of only finite smoothness generically. Therefore, the invariant curve has only finite smoothness generically. Upon approaching the value of the parameter corresponding to the passage of multipliers through the circle, the smoothness of the invariant curve grows to infinity (as is easily seen).

If our mapping is the Poincaré mapping of a differential equation, then

Figure 149.

in the three-dimensional phase space, the invariant curve of the Poincaré mapping determines an invariant torus consisting entirely of phase curves. Our invariant curve is a transversal section of this torus. The torus has finite smoothness; the nearer the time of birth of the torus to the cycle, the greater the smoothness. As the parameter varies in the family, the rotation number on the torus generically varies and assumes both irrational and rational values.

From the above picture of bifurcations, as a pair of multipliers passes through the unit circle, one also observes, in generic one-parameter families, no branchings of the given periodic solution into periodic solutions of multiplicity other than 2. Indeed, the latter could only take place if a multiplier crossed the unit circle at a point with rational argument, which is an exceptional nongeneric event.

In order to understand how periodic motions with large periods arise, it is necessary to consider families with two parameters.

Indeed, a multiplier can become equal to a root of unity (different from 1 and −1), unremovably only in real two-parameter families. A two-parameter consideration of the loss of stability of a fixed point in the resonant case (i.e., where a multiplier is close to a root of unity) enables us to better understand bifurcations in one-parameter families as multipliers cross the unit circle. Namely, as we shall see, some metamorphoses which seem nonlocal in the one-parameter approach lend themselves to a study by local methods if the problem is considered to be a two-parameter problem. In particular, in this way we can study some cases of the rough loss of stability and see to what regime the system jumps after the rough loss of stability of a cycle.

E. Resonance in Losing the Stability of a Cycle

We consider a mapping of the plane onto itself in the neighborhood of a fixed point with a multiplier equal to a root of unity with index $q > 2$. In accordance with Poincaré's general method (Chap. 5), we can, in an appropriate coordinate system, write the family in the form $z \mapsto z[\lambda + A(|z|^2) + B\bar{z}^{q-1} + O(|z|^{q+1})]$, where λ, A, B, and O depend smoothly on ε.

Instead of studying this mapping, we can do the following. In the case of resonance, every step of the Poincaré method reduces to averaging along the corresponding Seifert foliation (cf., § 21). Therefore, instead of reducing the Poincaré map to normal form, we can, in the neighborhood of a cycle, write the original equation of phase curves as a nonautonomous equation with 2π-periodic coefficients and then reduce it to normal form by changes of coordinates $2\pi q$-periodic over time (§ 26).

As a result of this procedure, we obtain the following equation with $2\pi q$-periodic coefficients in t in the new coordinates (depending smoothly on the parameter):

$$\dot{\zeta} = \varepsilon\zeta + \zeta A(|\zeta|^2) + B\bar{\zeta}^{q-1} + O(|\zeta|^{q+1}).$$

Here ε is a complex parameter, A and B depend holomorphically on ε, $\bar{\varepsilon}$, and the value $\varepsilon = 0$ corresponds to resonance (i.e., the case where a multiplier of the original equation is equal to a qth root of unity).

Remark 1. From the above arguments, it follows in particular that:

(1) Up to terms of degree $q + 1$ (and even up to terms of arbitrarily high degree) the Poincaré map coincides with a transformation of the phase flow of a vector field on the plane.
(2) The indicated vector field is invariant with respect to a cyclic group of diffeomorphisms of the plane (of order q).
(3) Conclusions 1 and 2 hold not only for an individual Poincaré map, but also for a family depending smoothly on the parameters; moreover, both the group and the field invariant under it depend smoothly on the parameters.

Remark 2. In general, the exact Poincaré map is not a transformation of the phase flow of any vector field and does not commute with any finite group of diffeomorphisms.

From the above, it is clear that up to terms of arbitrarily high degree with respect to the distance from the closed phase curve, the bifurcation problem in the case of loss of stability near a resonance of order $q > 2$ reduces to the study of metamorphoses of phase portraits in generic two-parameter families of vector fields in the plane invariant with respect to rotations by an angle $2\pi/q$. The resonance is said to be *strong* if $q \leqslant 4$.

The cases of resonances of order 2 or 1 can also be included in this scheme. Namely, the loss of stability of a cycle as a pair of multipliers crosses the unit circle corresponds to a hypersurface of codimension 1 in the function space. This hypersurface approaches hypersurfaces corresponding to the multipliers 1 and -1 along surfaces of codimension 2. The generic points on these surfaces of codimension 2 correspond to closed phase curves for which the Poincaré map has the double eigenvalue 1 (respectively, -1) with Jordan block of order 2.

Therefore, the study of boundary cases of crossing the unit circle by multipliers reduces, up to terms of arbitrarily high degree, to the study of metamorphoses of phase portraits in generic two-parameter families of vector fields in the plane invariant under rotations by an angle $2\pi q$ ($q = 1, 2$) and having, for some value of the parameter, a singular point whose linear part is a nilpotent Jordan block; the corresponding linear equation can be reduced to the form

$$\dot{x} = y, \qquad \dot{y} = 0.$$

Finally, the problem of metamorphoses in the case of loss of stability near resonances leads to the study of bifurcations of phase portraits in two-parameter families of equivariant vector fields on the plane. We are now going to study this problem.

§ 35. Versal Deformations of Equivariant Vector Fields on the Plane

Metamorphoses of phase portraits of vector fields invariant with respect to some group of symmetries arise naturally in the study of various phenomena in which symmetry appears in the very formulation of the problem.

It is more surprising that metamorphosis problems of symmetric phase portraits arise in an *a priori* nonsymmetric situation, in the study of bifurcations near resonances (cf., § 21 and § 34). In this section, we consider exactly those bifurcations of symmetric phase portraits which are needed for the study of resonances.

A. Equivariant Vector Fields on the Plane

Let F be a vector field on the plane of the complex variable z. We shall consider F as a complex-valued (not necessarily holomorphic) function on $\mathbb{C}$. The Taylor series of this function at zero can be written in the form $\sum F_{k,l} z^k \bar{z}^l$.

Proposition. *Assume that the field F turns into itself under the rotation of the z-plane by an angle $2\pi/q$. Then the coefficients $F_{k,l}$ are different from zero only if $k - l$ is congruent to 1 modulo q.*

◀ The Taylor series is unique, and, therefore, each of its terms has to turn by the angle $2\pi/q$ when z turns by the angle $2\pi/q$. The point $z^k \bar{z}^l$ of the complex plane rotates by an angle $2\pi(k - l)/q$. This rotation coincides with rotation by an angle $2\pi/q$ exactly when the above condition is satisfied. ▶

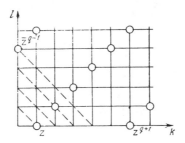

Figure 150.

Corollary. *The differential equations invariant under rotations by an angle* $2\pi/q$ *have the following form:*

$$\dot{z} = zA(|z|^2) + B\bar{z}^{q-1} + O(|z|^{q+1}) \qquad (q > 2).$$

◀ In the (k, l)-plane, we consider the integer points satisfying the congruence $k - l \equiv 1 \pmod{q}$. These points are located on rays parallel to the bisector of the positive quadrant and emanating from points corresponding to the monomials $z, z^{q+1}, z^{2q+1}, \ldots; \bar{z}^{q-1}, \bar{z}^{2q-1}, \ldots$. Among these monomials, we shall seek monomials of smallest degree (Fig. 150).

First, let us successively obtain several monomials on the ray emanating from the point z (i.e., monomials of the form $z|z|^{2k}$) and then the monomial $\bar{z}^{q-1}$; all other monomials have degree not smaller than $q + 1$ (Fig. 150). ▶

Definition. The preceding equation, without the term O, is called the *principal equation* and is invariant with respect to rotation by an angle $2\pi/q$. The right-hand side of a principal equation is called a principal q-equivariant field.

EXAMPLE

The principal equations invariant with respect to the groups of rotations of orders 3 and 4 have the forms

$$\dot{z} = \varepsilon z + Az|z|^2 + B\bar{z}^2; \qquad \dot{z} = \varepsilon z + Az|z|^2 + B\bar{z}^3.$$

To formulate the results of the analysis of metamorphoses of the phase portraits of equivariant vector fields depending on parameters, it is convenient to introduce the following definitions.

B. Equivariant Versal Deformations

We consider a family v_λ of vector fields invariant under the actions of a group G on the phase space and depending on a parameter λ belonging to the neighborhood of 0 in the space $\mathbb{R}^k$ (called the *base* of the family). The dimension of the base is called the *number of parameters* of the family.

The germ of the family at the point $\lambda = 0$ is called an *equivariant deformation of v_0*.

Definition. An equivariant deformation v_λ is called an *equivariantly topologically orbitally versal* (more briefly, *versal*) deformation of v_0 if for any other equivariant deformation w_μ of v_0, there are a continuous mapping φ of the bases of the deformations and a family h_μ of homeomorphisms of the phase space depending continuously on μ and commuting with the action of G such that h_μ turns the phase curves of the field w_μ into phase curves of the field $v_{\varphi(\mu)}$ with preservation of the direction of the motion.

In other words, *an equivariant deformation is versal if every other equivariant deformation is topologically orbitally equivalent to a deformation induced from the versal one.*

Analogous definitions can be given for germs of vector fields and for deformations in a class of fields with special properties (for example, with a fixed linear part at a singular point). Now we pass to the construction of versal deformations.

C. Principal Deformations

We consider a vector field on the plane invariant under rotation by an angle $2\pi/q, q > 2$.

Definition. A field is said to be *singular* if the linear part of the field at zero is equal to zero.

Definition. The *principal deformation* of a q-equivariant principal singular field v_0 $(q > 2)$ is the two-parameter family $v_\varepsilon = \varepsilon z + v_0$, where the parameters are the real and imaginary parts of the complex number ε.

EXAMPLE

In the cases $q = 3, 4$, the principal deformations are given by the equations

$$\dot z = \varepsilon z + Az|z|^2 + B\bar z^2, \qquad \dot z = \varepsilon z + Az|z|^2 + B\bar z^3$$

in which ε is considered a parameter, and the complex coefficients A and B are fixed.

Remark. The objects defined above arise in the study of the loss of stability of a cycle as a pair of complex-conjugate multipliers crosses the unit circle. In the function space of all systems, the systems with such a passage form a hypersurface. This hypersurface is one of the three hypersurfaces bounding the domain of stability. The other two hypersurfaces correspond to the passages of one multiplier through the unit circle at the point 1 and at the point -1.

The boundary of the hypersurface corresponding to the passage of a complex pair of multipliers consists of two parts (of two hypersurfaces of codimension 2 in the function space of all systems). One of these surfaces of codimension 2 corresponds to a pair of multipliers equal to 1 forming a Jordan block of order 2 and the other to such a block with two eigenvalues equal to -1.

The study of the loss of stability in the neighborhood of these codimension 2 surfaces leads to the study of bifurcations in two-parameter families of vector fields on the plane having a nilpotent Jordan block of order 2 as the linear part and invariant under rotation by an angle $2\pi q$, $q = 2$ (for multipliers equal to -1) or $q = 1$ (for multipliers equal to 1). To include these cases in the general scheme, it is convenient to give the following definitions.

D. Cases $q = 1$ and $q = 2$

Definition. A field invariant under rotation of the plane by an angle $2\pi/q$, $q = 1$ or $q = 2$ is said to be *singular* if its linear part at zero is a nilpotent Jordan block of second order.

In other words, for $q = 1$ or $q = 2$, a singular field is a field whose linear part is the field of the phase velocity of the equation $\ddot{x} = 0$ on the phase plane $(x, y = \dot{x})$.

It is easy to prove the following.

Theorem. *A singular field invariant with respect to rotation of the plane by an angle $2\pi/q$, $q = 1$ or $q = 2$, can be reduced to the field of the phase velocity of the equation*

$$\ddot{x} = ax^3 + bx^2y + O(|x|^5 + |y|^5) \qquad (q = 2),$$
$$\ddot{x} = ax^2 + bxy + O(|x|^3 + |y|^3) \qquad (q = 1)$$

on the phase plane $(x, y = \dot{x})$ by a diffeomomphism commuting with the rotation.

◀ The linear part of our field has the form $y\partial/\partial x$. We now construct the homological equation corresponding to this linear field. To do this, we calculate the Poisson bracket of our linear field, $\Lambda = y\partial/\partial x$, with an arbitrary vector field $h = P\partial/\partial x + Q\partial/\partial y$.

We find

$$\left[y\frac{\partial}{\partial x}, P\frac{\partial}{\partial x}\right] = yP_x\frac{\partial}{\partial x}, \quad \left[y\frac{\partial}{\partial x}, Q\frac{\partial}{\partial y}\right] = yQ_x\frac{\partial}{\partial y} - Q\frac{\partial}{\partial x},$$

$$[\Lambda, h] = (yP_x - Q)\frac{\partial}{\partial x} + yQ_x\frac{\partial}{\partial y}.$$

Hence, the homological equation for the unknown functions (P, Q) assumes the form of a system

$$yP_x - Q + u = 0, \qquad yQ_x + v = 0.$$

Here u and v are known functions, namely the components of the vector field $w = u\partial/\partial x + v\partial/\partial y$, which we wish to annihilate by a change of variables.

We study this system. Expressing Q by the first equation and substituting it in the second one, we obtain

$$y^2 P_{xx} = -yu_x - v.$$

In order to achieve divisibility of the right-hand side by y^2, it is sufficient to change the terms of degree 0 and 1 in y in the function v. Adding to v $v_0(x) + yv_1(x)$, we can achieve the solvability of the last equation with respect to P.

Consequently, the homological equation for arbitrary (u, v) is not solvable, but it becomes solvable if v is replaced by its sum with an appropriate nonhomogeneous linear function in y. In other words, the equation

$$[\Lambda, h] + w = (v_0(x) + yv_1(x))\frac{\partial}{\partial y}$$

with appropriate (v_0, v_1) depending on w is solvable for the unknown field h.

Finally, Poincaré's method enables us to annihilate all vector-valued monomials except those of the form $x^k\partial/\partial y$ and $yx^k\partial/\partial y$ in terms of any degree of the field $\Lambda + \cdots$. Thus, in the class of formal power series, our equation can be reduced to the form

$$\ddot{x} = a(x) + yb(x).$$

If the original system were invariant under rotation by an angle π (i.e., if it were odd), then the components of the original vector field would be odd functions. In this case, the transformations in Poincaré's method can also be chosen odd (commuting with the rotation), since in the preceding formulas the degrees of (P, Q) and (u, v) are the same. Then, in the formal normal form, the series of (a, b) will consist solely of terms of odd degree.

Restricting ourselves to a few first approximations in Poincaré's method, we obtain the proposition formulated above. ▶

Definition. The *principal singular equations and fields* for $q = 2$ and $q = 1$ are the equations

$$\ddot{x} = ax^3 + bx^2y \quad (q = 2), \qquad \ddot{x} = ax^2 + bxy \quad (q = 1),$$

and the vector fields defining them in the phase plane $(x, y = \dot{x})$.

Definition. A *principal deformation* of a principal singular field for $q = 2$ and $q = 1$ is a deformation consisting of the addition of the terms $\alpha x + \beta y$ $(q = 2)$, and $\alpha + \beta x$ $(q = 1)$ to the right-hand side of the corresponding second-order equation.

The principal deformations of q-equivariant fields in the cases of a strong resonance, i.e., for $q \leqslant 4$, are listed below:

$$
\begin{aligned}
\dot{z} &= \varepsilon z + Az|z|^2 + B\bar{z}^3, & q &= 4, \\
\dot{z} &= \varepsilon z + Az|z|^2 + B\bar{z}^2, & q &= 3, \\
\ddot{x} &= \alpha x + \beta y + ax^3 + bx^2y, & q &= 2, \\
\ddot{x} &= \alpha + \beta x + ax^2 + bxy, & q &= 1.
\end{aligned}
$$

Here the variables z, ε, A, and B are complex, x, y, α, β, a, and b are real, the parameters of the deformation are denoted by Greek letters, and $y = \dot{x}$.

E. Versality of Principal Deformations

"Theorem". *For every q, all principal singular fields can be divided into degenerate and nondegenerate fields in such a way that*

(1) *the degenerate fields form the union of a finite number of submanifolds in the space of all principal singular fields;*
(2) *the nondegenerate fields form the union of a finite number of open connected domains;*
(3) *the principal deformations of the germs of nondegenerate fields at zero are versal;*
(4) *the principal deformations of the germs of nondegenerate fields are topologically equivalent in every connected component.*

The word "theorem" is in quotation marks because it is not proved for $q = 4$.

Except for the case $q = 4$, the conditions of nondegeneracy can be written explicitly in the following way:

$$a \neq 0, \quad b \neq 0 \quad \text{for} \quad q = 1, 2; \qquad \text{Re } A(0) \neq 0, \quad B \neq 0 \quad \text{for} \quad q \geqslant 3.$$

For $q = 4$, we have to add at least the following conditions to these conditions. (In the section of Additional Problems [(3) and (4)], apparently all other candidates are indicated.) It is likely that the total number of domains of the A-plane for $B = 1$ is 48, see Fig. 156.

$$|A|^2 \neq |B|^2, \qquad |\text{Re } A| \neq |B|,$$
$$|\text{Im } A| \neq (|B|^2 + \text{Re}^2 A)\sqrt{|B^2 - \text{Re}^2 A|}.$$

[See also: A. H. Wan, *Bifurcations into invariant tori at points of resonances*, Arch. Rat. Mech. Anal. **68** (1978) 343–357; A. I. Neĭstadt, *Bifurcations of phase portraits of a system of differential equations, arising in the problem of auto-oscillations stability loss at $1 : 4$ resonance*, Prikl. Math. Mech. **42** (1978), 830–840; F. S. Beresovskaya, A. I. Hibnik, *On separatrices bifurcations in the problem of auto-oscillations stability loss at resonance $1 : 4$*, Prikl. Math. Mech. **44** (1980), 938–943.]

From the theorem, we easily obtain the following corollary.

Corollary. *In the function space of two-parameter families of vector fields invariant under the group of rotations by angles which are multiples of $2\pi/q$, an open*, everywhere dense set is formed by the families for which the fields are singular only at isolated values of the parameters. In the neighborhood of these values, the family is topologically equivalent to a versal deformation of a nondegenerate principal singular field.*

In other words: Let the eigenvalues of the linearization of a vector field in the plane invariant under rotations by angles which are multiples of $2\pi/q$ be equal to zero. We consider a generic field with the indicated properties. We form a generic two-parameter deformation of it in the class of fields having the same symmetry.

It is asserted that there is a neighborhood of the point 0 not depending on the parameters, such that the deformation thus constructed can be reduced to the normal form indicated in § 35C and § 35D by a homeomorphism with the same symmetry depending continuously on the parameters. More precisely, the phase portraits of the corresponding systems are reduced to normal form by homeomorphisms.

Consequently, the above theorem reduces the description of all bifurcations to the description of bifurcations in principal deformations of nondegenerate singular fields.

---

* With the usual qualifications if the base is not compact.

F. Description of Bifurcations

In the case $q = 1$, the above theorem was proved by Bogdanov in 1971 (cf., § 33, where the description of bifurcations is included).

In the case $q = 2$, we can achieve $b < 0$ by a change of time. The bifurcation diagram ("dial") in the (α, β)-plane and the metamorphoses of the phase portrait for the cases $a > 0$ and $a < 0$ are given in Fig. 151.

In the case $q = 3$, by a change of time, we can achieve $\operatorname{Re} A < 0$. The bifurcation diagram and metamorphoses are given in Fig. 152.

For $q \geqslant 5$, we can achieve $\operatorname{Re} A < 0$ by a change of time; the bifurcation diagram and metamorphoses are given in Fig. 153. We note that the region of existence of fixed points approaches the imaginary ε-axis in a narrow

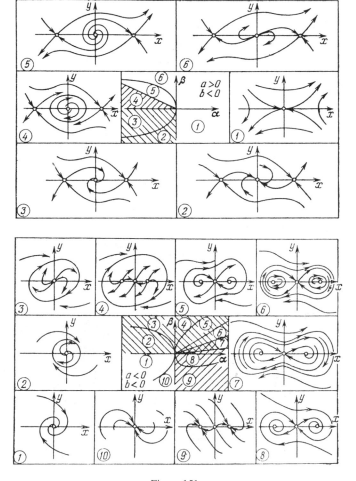

Figure 151.

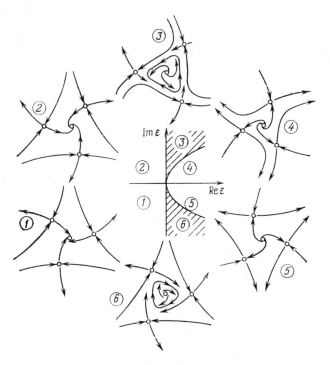

Figure 152.

Figure 153.

tongue, whose sides have the following common tangent:

$$\operatorname{Im} \varepsilon \approx f(\operatorname{Re} \varepsilon) \pm c |\operatorname{Re} \varepsilon|^{(q-2)/2}, \qquad q \geqslant 5.$$

The proofs of the above theorem and statements are simple for $q \geqslant 5$. They are contained in the publications of Bogdanov cited in § 33 in the case $q = 1$, are unknown in the case $q = 4$, and are outlined below in the cases $q = 2, 3$. Some versions of metamorphoses are illustrated below for $q = 4$ in Figs. 157, 158, and 160. [Cf., also, Additional Problems (1)–(4)].

G. The Case of Symmetry of Order 3

1. Let $A = 0$. In this case, the system can be obtained from a certain Hamiltonian system by a rotation of the field. Namely, consider an equilateral triangle formed by singular points (saddle points). The sides of this triangle determine three linear non-homogeneous functions. The product of these three functions is the desired Hamiltonian.

The sign of the derivative of this function along the solution of our system $\dot{z} = \varepsilon z + B\bar{z}^2$ is determined by the sign of the real part of the parameter ε. This enables us to use the indicated function as the Liapunov function. Therefore, for $A = 0$, the principal deformation can be studied without difficulty.

2. In the general case, we reduce the system to the form

$$\frac{dZ}{dT} = EZ + \bar{Z}^2 + |\varepsilon| A Z |Z|^2, \qquad E = \frac{\varepsilon}{|\varepsilon|}$$

by means of the substitutions $t = T/|\varepsilon|$, $z = Z/|\varepsilon|$.

In a domain where $|Z|$ is small compared to $1/\varepsilon$, the third summand can be considered as a small perturbation. In a domain where the argument of E is not close to $\pm \pi/2$, this perturbation does not change the picture obtained in § 35G1. If, on the other hand, the real part of E is small compared to the complex part, then the system can be considered as the small perturbation of the Hamiltonian system:

$$\frac{dZ}{dT} = \pm iZ + \bar{Z}^2.$$

The unperturbed Hamiltonian function H is described in § 35G1.

3. We calculate the rate of change of H along solutions of the system. By integrating along the level line $H = h$, we obtain the condition of the birth of a cycle from this line in the form $\oint \mu r^2 + \lambda r^4 \, d\varphi = 0$, where $\mu = \sigma/\tau$, $\lambda = a\tau$, $\varepsilon = \sigma + i\tau$, $A = -a + ib$, and r and φ are polar coordinates in the plane.

We denote by ρ the radius of inertia of the domain bounded by the elliptic curve $H = h$. Then the condition of the birth of a cycle exactly from the line $H = h$ as $|\varepsilon|$ passes through zero can be written in the form $\sigma = \rho^2 \tau^2 a$.

The largest possible value of ρ corresponds to the triangle formed by the separatrices of the saddles. The function $\rho(h)$ is monotone (cf., § 35I).

4. The remaining proof of the versality of a principal family and of the form of bifurcation diagrams and phase portraits can be carried out as in the case $q = 1$.

H. The Case of Symmetry of Order 2

1. By expansions of x and changes of time and the parameters, the family can be reduced to the form

$$\ddot{x} = \alpha x + 2\beta y + ax^3 + bx^2 y, \qquad a = \pm 1, b = -2.$$

We will study this family with the parameters (α, β).

2. If $|\beta| \leqslant \sqrt{\alpha}$, then we make the substitution $x = \sqrt{|\alpha|/|a|}\, x'$, $t = t'/\sqrt{|\alpha|}$, converting $|\alpha|$ and $|a|$ into 1. We obtain an almost Hamiltonian system. The Hamiltonian has the form

$$H = \frac{y^2}{2} - \operatorname{sgn}\alpha\,\frac{x^2}{2} - \operatorname{sgn}a\,\frac{x^4}{4}.$$

The dissipative terms have the form

$$2\beta' y + b'x^2 y, \qquad \beta' = \frac{\beta}{\sqrt{|\alpha|}}, \qquad b' = \frac{\sqrt{|\alpha|}\,b}{|a|}.$$

3. Integrating the rate of change of H along the level line $H = h$, we obtain the following condition of the birth of a cycle from exactly that line as α and $\beta = u\alpha$ pass through zero: $u = r^2$, where

$$r^2 = \frac{\iint x^2\, dx\, dy}{\iint dx\, dy}$$

is the square of the radius of inertia with respect to the y-axis of the domain bounded by the line $H = h$.

4. If $|\alpha| \ll |\beta|$, we make the substitution $x = \lambda x'$, $t = \varkappa t'$, converting $|a|$ into 1 and $|b|$ into $|\beta|$:

$$\lambda = \sqrt{\left|\frac{\beta}{b}\right|}, \qquad \varkappa = \sqrt{\left|\frac{b}{\beta}\right|}.$$

We obtain $\alpha' = |b/\beta|\alpha$, $b' = \sqrt{|\beta/b|}\,b$, $\beta' = \sqrt{|b/\beta|}\,\beta$, $a' = a$;

$$\ddot{x} = ax^3 + \beta'2y(1 \pm x^2) + \alpha'x.$$

Here both parameters $\beta' \sim \sqrt{|\beta|}$ and $\alpha' \sim \alpha/\beta$ are small. For $\beta' = 0$, we obtain the following Hamiltonian:

$$H = \frac{y^2}{2} - \frac{\alpha'x^2}{2} - \frac{ax^4}{4}.$$

The remainder of the study can be carried out in the usual way.

I. Zeros of Elliptic Intergrals

It has been shown above that the study of the behavior of cycles in our families can be reduced to the solution of special cases of the following "weakened Hilbert's 16th problem":

Let H be a real polynomial of degree n and let P be a real polynomial of degree m in the variables (x, y). How many real zeros can the function

$$I(h) = \iint_{H \leqslant h} P \, dx \, dy$$

have?

For the study of symmetries of order 2, we need the case

$$P = \mu + \lambda x^2, \qquad H = \frac{y^2}{2} - \frac{\alpha x^2}{2} - \frac{ax^4}{4}.$$

For $P = \mu + \lambda x^2$, the problem reduces to the study of the monotonicity of the function $r(h)$, where r is the radius of inertia with respect to the y-axis of a domain bounded by a cycle.

Lemma. *On intervals between critical values of H, the function r behaves in the following way:*

Values of α and a	$-1, +1$	$+1, -1$	$+1, -1$	$-1, -1$
Interval of h	$0, \frac{1}{4}$	$-\frac{1}{4}, 0$	$0, +\infty$	$0, +\infty.$
Behavior of r	$\uparrow$	$\downarrow$	$\downarrow \uparrow$	$\uparrow$

(in the third case, r first decreases and then increases).

An analogous (but weaker) lemma on elliptic integrals has been used in the work of Bogdanov (like earlier). Bogdanov's proof depends on lengthy calculations. Il'jašenko found proofs of both the lemma given here and Bogdanov's lemma based not upon calculations, but upon complex variable topological arguments (monodromy and the Picard–Lefshetz formula). [Cf., Ju. S. Il'jašenko, *On zeros of special Abelian integrals in a real domain*, Funct. Anal. Appl. **11**, 4 (1977), 78–79; *Multiplicity of limit cycles arising under perturbations of Hamiltonian equations $w' = P_2/Q_1$ in real and complex domains*, Trudy Sem. I. G. Petrovskogo 3 (1978), 49–60]. A. N. Varchenko, *Estimation of the number of zeros of an abelian integral, depending on a parameter, and the limit cycles*. Funkt. Anal. i Prilozhen, **18**, 2 (1984), 14–25; A. G. Hovanskii, *Real analytical varieties with a finiteness property and complex abelian integrals*, Funkt. Anal. i Prilozhen **18**, 2 (1984), 40–50; G. S. Petrov, *On the number of zeros of elliptic integrals*, Funkt. Anal. i Prilozhen **18**, 2 (1984), 73–75; *Elliptic integrals and their nonoscillation*, Funkt. Anal. i Prilozhen **20**, 1 (1986), 46–49; V. I. Arnold, *Sturm theorems and symplectic geometry*, Funkt. Anal. i Prilozhen **19**, 4 (1985), 1–10.

The abelian integrals occurring in bifurcation problems always display the minimal number of zeros consistent with the trivial dimension argument (some linear combination of any n function displays $n - 1$ zeros); thus these linear families of abelian integrals are Chebyschev systems (their combinations which are not identically zero have less zeros than the family dimension). The general reason for this nonoscillation property is not clear, but its investigation has led to the above-mentioned papers as well as to the discovery (by B. and M. Shapiro) of the relation of the Chebyschev systems to the natural stratification of the Schubert cells of the flag manifolds and to their natural Bruhat ordering, which we will not discuss here.

J. The Case of Resonance of Order 4

The initial equation is $\dot{z} = \varepsilon z + Az|z|^3 + B\bar{z}^3$.

We assume that $B \neq 0$. Then by expansions and rotations of z and expansion of time, we can reduce the equation to the form

$$\dot{z} = \varepsilon z + Az|z|^3 + \bar{z}^3.$$

By a change of the sign of time and by changing z to $\bar{z}$, we can also achieve Re $A \leqslant 0$, Im $A \leqslant 0$.

First, we study singular points different from zero.

1. Bifurcations of Singular Points. For the study of bifurcations of singular points under the variation of ε, the following auxiliary construction is useful. Let $z = re^{i\varphi}$ be a singular point. Then

$$-\frac{\varepsilon}{r^2} = A + N, \qquad \text{where} \quad N = e^{-4i\varphi}.$$

Let us now consider the circle with radius 1 and center at A (Fig. 154). The value of ε for which the point z is singular lies on the ray opposite the ray connecting zero with the point $A + N$ on our circle. Moreover, the closer the point of the circle to zero, the larger the modulus of ε.

From what has been already stated, it is clear that the cases where $|A|$ is smaller or larger than 1 are very different. If $|A| < 1$, zero lies inside the circle. In this case, for any ε (except zero), the equation has four singular points at the vertices of a square. If ε goes around zero, rotating by 360°, the square of the singular points turns by angle 90° in the opposite direction.

If, on the other hand, $|A| > 1$, then in the plane of the variable ε, there is an angle bounded by the extensions of the tangents to our circle. For ε inside this angle, there are eight singular points and none outisde. When ε turns from one side of the angle to the other, four singular points are born at the vertices of the square. This square is immediately duplicated. Then nearby singular points begin to diverge. When ε

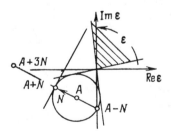

Figure 154.

approaches the other side of the angle, every singular point of the first square dies, colliding with a point of the second square, which initially stood at 90° angle from it (so that one of the squares of singular points turns by 90° with respect to the other).

2. *Types of Singular Points of a Linear Equation.* We begin with a lemma enabling us to easily study types of singular points of the linear vector field in the plane given in complex form:

$$\dot{\xi} = P\xi + Q\bar{\xi}.$$

Lemma. *The type of the singular point 0 does not depend on the argument of Q. This point is a saddle for $|P| < |Q|$, a focus for $|\operatorname{Im} P| > |Q|$, and a node for $|\operatorname{Im} P| < |Q| < |P|$; the focus is stable for $\operatorname{Re} P < 0$ and unstable for $\operatorname{Re} P > 0$ (Fig. 147).*

◀ By multiplying ξ by a complex number λ, the coefficient P does not change, and the coefficient Q is multiplied by $\bar{\lambda}/\lambda$. By changing λ, we can make the argument of Q arbitrary; this proves the first assertion of the lemma. For the proof of the second assertion, we consider the case $Q = 1$. Let $P = \alpha + i\beta$. We write the matrix of the equation in the basis $(1, i)$. This matrix has the form

$$M = \begin{pmatrix} \alpha + 1 & -\beta \\ \beta & \alpha - 1 \end{pmatrix}, \qquad \operatorname{tr} M = 2\alpha, \qquad \det M = \alpha^2 + \beta^2 - 1.$$

The characteristic equation has the roots

$$\lambda_{1,2} = \alpha \pm \sqrt{1 - \beta^2}.$$

they are real if $|\beta| < 1$. The roots have opposite signs if $\alpha^2 + \beta^2 < 1$. The lemma is proved for $Q = 1$. The conditions of a saddle, node, or focus for any $|Q|$ now follow from similarity arguments: for dilations of time, P and Q are multipled by the same real number. ▶

Figure 155 shows clearly the relative location of nodes, foci and saddles in the function space. This is useful to keep in mind in any study of bifurcations of singular point in the plane.

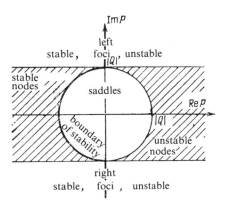

Figure 155.

3. Study of the Saddle Points. We return to the original nonlinear equation. Let $z_0 = re^{i\varphi}$ be a singular point. We linearize the equation at this point. Let $z = z_0 + \xi$.

Retaining the first-order terms in ξ, $\bar{\xi}$ on the right-hand side, we obtain

$$\dot{\xi} = P\xi + Q\bar{\xi}, \qquad P = r^2(A - N), \qquad |Q| = r^2|A + 3N|.$$

Lemma. *If $|A| < 1$, then all singular points are saddles. If $|A| > 1$, then for every ε the singular point with the smaller modulus is a saddle and that with the larger modulus is not.*

By the lemma of § 35J2, the condition for a saddle has the form $|A - N| < |A + 3N|$. We consider the points $A - N$ and $A + 3N$ (Fig. 154). These points are symmetric with respect to $A + N$, and the straight line connecting them goes through A. Which of these points is closer to zero depends on which side of the tangent to our circle at the point $A + N$ lies the point zero. If $|A| < 1$, then zero always lies on the same side of the tangent (where $A - N$ is located). If, on the other hand, $|A| > 1$, then the answer depends on which of the two arcs bounded by the tangents to the circle from the origin lies $A + N$. The arc farther from zero corresponds to saddles and singular points near 0 (cf., § 35J1).

4. Stability of Singular Points. The singular points corresponding to the arc of our circle facing zero can be nodes or foci. The part of the arc adjacent to the tangent from zero corresponds to a node; as one moves along the arc, the node may become a focus, and the focus may change stability. We find out under what condition this change in the stability of the focus takes place.

From the lemma of § 35J2 and the formula in § 35J3, it follows that a change in stability takes place if the point $A - N$ (diametrically opposite the point $A + N$ of our circle) crosses the imaginary axis, while the point $A + N$ lies on the arc facing zero. The boundary separating the points A for which such a phenomenon takes place is determined by the following condition: the diameter drawn through the point of intersection of our circle with the imaginary axis is perpendicular to a tangent from zero. As is easy to calculate, the equation of the boundary has the form

$$|\mathrm{Im}\, A| = \frac{1 + \mathrm{Re}^2\, A}{\sqrt{1 - \mathrm{Re}^2\, A}}.$$

The corresponding line in the plane of the variable A is tangent to the circle $|A| = 1$ at the points $A = \pm i$ and has the straight lines $|\mathrm{Re}\, A| = 1$ as asymptotes (Fig. 156).

5. Behavior at Infinity. For large z, we can "ignore" the term εz. Setting $w = z^2$, we obtain the linear equation

$$\dot{w} = 2Aw + 2\bar{w}.$$

To study this equation, we apply the lemma of § 35J2. Consequently, for $|A| < 1$, the singular point in the w-plane is a saddle point, and for $|A| > 1$, all trajectories from infinity are attracted to a finite domain if $\mathrm{Re}\, A < 0$.

6. Bifurcations of the Phase Portrait. If $|A| < 1$, then the diagram (Fig. 157) is apparently the same as for the third-order resonances (cf., § 35G). If $|\mathrm{Re}\, A| > 1$, then apparently the same occurs as for resonances of order 5 and above (Figs. 158

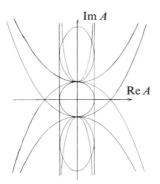

Figure 156.

Figure 157.

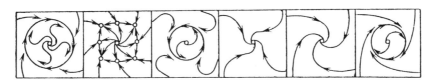

Figure 158.

and 153), although, in this case, singular points may be born not on cycles, as well. [cf., Additional Problems (1), (2), and (3)].

The main difficulties are represented by the case $|\operatorname{Re} A| < 1, |A| > 1$.

7. *New Normalizations and Notation.* For the study of the case $|\operatorname{Re} A| < 1$ it is useful to consider the asymptotics as $|\operatorname{Im} A| \to \infty$. Instead of letting A go to infinity, we may let B converge to zero in the original equation. In order to study this case, we introduce the notation

$$\varepsilon = \sigma + i\tau, \qquad A = -i\alpha - \gamma, \qquad B = \beta$$

and assume that β, $\sigma = u\beta$ and $\gamma = v\beta$ ($\beta \to 0$, $u \sim v \sim 1$) are small parameters of the same order.

We choose the multipliers normalizing the expansion of coordinates and time so that $\alpha = 1$ and $\tau = 1$ and introduce the symplectic polar coordinates $\rho = |z|^2/2$, $\varphi = \arg z$.

This transforms the original equation into the system

$$\dot{\rho} = 2\rho(\sigma - 2\gamma\rho + 2\beta\rho \cos 4\varphi),$$

$$\dot{\varphi} = \tau - 2\alpha\rho - 2\beta\rho \sin 4\varphi.$$

We also introduce the Hamiltonian $H = \tau\rho - \alpha\rho^2 - \beta\rho^2 \sin 4\varphi$ and the potential $\Pi = \sigma\rho^2 - 4\gamma\rho^3/3$. Then

$$\dot{\rho} = -H_\varphi + \Pi_\rho, \qquad \dot{\varphi} = H_\rho.$$

In the case $\tau = \alpha = 1$, $\sigma = u\beta$, $\gamma = v\beta$ which is of interest to us, we have

$$H = H_0 + \beta \dot{H}_1, \qquad \Pi = \beta\Pi_1, \qquad H_0 = \rho - \rho^2.$$

For $\beta = 0$, we obtain an unperturbed motion (rotation with frequency H_ρ). In the domain where $H_\rho \neq 0$ (i.e., where ρ is not close to $\frac{1}{2}$), the basic perturbing effect is given by the dissipative term $\beta\Pi_1$. For $\rho \approx \frac{1}{2}$, we have to take into account βH_1, too.

8. Lemma on the Effect of a Small Dissipation. In the phase plane, we consider the equation $\dot{x} = v + \varepsilon w$, where v is the Hamiltonian field with Hamiltonian function H and w a potential field, $w = \nabla\Pi$ (using the metric given by the symplectic coordinates p, q).

Lemma. *Let* δH *be the increment of* H *under one revolution along the closed phase curve* $H = h$. *Then*

$$\frac{d}{d\varepsilon}\delta H = \iint\limits_{G(h)} \Delta\Pi \, dp \, dq \qquad (\varepsilon \text{ is a small parameter})$$

at $\varepsilon = 0$, *where* $G(h)$ *is the domain bounded by the curve.*
◀ One has

$$\delta H \approx \oint \dot{H} \, dt = \oint H_p(-H_q + \varepsilon\Pi_p) + H_q(H_p + \varepsilon\Pi_q) \, dt$$

$$= \varepsilon \oint \Pi_p \, dp - \Pi_q \, dq = \varepsilon \iint \Delta\Pi \, dp \, dq. \blacktriangleright$$

Applying the lemma to our equation, we find that

$$\delta H_0 = 2\rho(\sigma - 2\gamma\rho) = 2\beta\rho(u - 2v\rho).$$

Consequently, in first approximation, the cycle is given by the formula

$$\rho = \frac{u}{2v} = \frac{\sigma}{2\gamma}.$$

This method enables us to study our system outside the annulus in which ρ is close to $\frac{1}{2}$. Therefore, the case of a σ close to γ has to be considered separately.

9. The Case $\rho \approx \frac{1}{2}$. We make the substitution $\rho = \frac{1}{2} + \sqrt{\beta}\, P$, $t\sqrt{\beta} = T$ and set $\beta = 0$. As the approximate equation, we obtain the Hamiltonian system

$$\frac{dP}{dT} = w + \cos 4\varphi, \qquad \frac{d\varphi}{dT} = -2P$$

with Hamiltonian function

$$H_{00} = P^2 + w\varphi + \frac{\sin 4\varphi}{4}$$

(a pendulum with torque). Here $w = u - v = (\sigma - \gamma)/\beta$.

A potential well exists for $|w| < 1$.

In terms of the notation of § 35J7, we have transferred the field $(\sigma - \gamma)\partial/\partial\rho$ from the potential part to the Hamiltonian part. Therefore, the new Hamiltonian and potential have the form

$$H = \rho - \rho^2 - \beta\rho^2 \sin 4\varphi - \beta w\varphi,$$

$$\Pi = \sigma\rho^2 - \frac{4\gamma\rho^3}{3} - (\sigma - \gamma)\rho.$$

Applying the lemma of § 35J8, we find that

$$\delta H = \oint\!\!\!\oint (2\sigma - 8\gamma\rho)\,d\rho\,d\varphi \text{ inside the well} = 2S(\sigma - 4\gamma\rho_0),$$

where S is the area inside the well in the (ρ, φ)-plane and ρ_0 is the coordinate of the center of gravity of the well.

The condition of the birth of a cycle is $\sigma = 4\gamma\rho_0$. Consequently, we need to calculate ρ_0. We know that $\rho_0 = \frac{1}{2} + \sqrt{\beta}\,\rho_1 + \cdots$. Hence, the condition of the birth of a cycle has the form $\sigma = 2\gamma + 4\gamma\sqrt{\beta}\rho_1 + \cdots, w = (\gamma/\beta) + 4\gamma\rho_1 + \cdots$.

We calculate ρ_1. In the coordinates P, φ, the exact equation of the closed phase curve $H = \text{const}$ has the form

$$P^2 + (\tfrac{1}{2} + \sqrt{\beta}P)^2 \sin 4\varphi + w\varphi = h.$$

Two values of P correspond to every value of φ in the well, with

$$P_1 + P_2 = -\frac{\sqrt{\beta}\sin 4\varphi}{1 + \beta\sin 4\varphi}.$$

From this formula, it follows that correction term ρ_1 is equal to

$$\rho_1 = -\tfrac{1}{2}\overline{\sin 4\varphi}$$

(where the bar means "averaging" along the unperturbed phase curve for $\beta = 0$).

From the location of the well with respect to the maximum and minimum of $\sin 4\varphi$ and its variation as w changes (Fig. 159), we obtain the information on the behavior of ρ_1, on which the diagrams of metamorphoses are based (Fig. 160).

The system of metamorphoses depicted in Fig. 160 is realized if $|\text{Im }A|$ is large compared to $|B|$ and $0 < |\text{Re }A| < B$.

Of course, the above arguments cannot replace proofs and are only first steps in the study of bifurcations in a principal family of 4-symmetric equations.

Results for the cases of symmetry of order $q \neq 4$ have apparently been known to specialists for a long time; see, e.g., V. K. Melnikov, *Qualitative description of resonance phenomena in nonlinear systems*, Dubna, OIJaF, P-1013 (1962), 1–17. Takens announced them in a preprint in 1973 but his proofs have not yet appeared. Our exposition is based

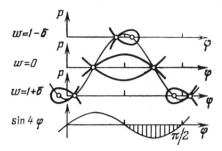

Figure 159.

Figure 160.

on this author's article in Funct. Anal. Appl. **11** No. 2 (1977) 1–10; detailed proofs have been given by E. I. Horozov, *Versal deformations of equivariant vector fields for the cases of symmetries of order 2 and 3*, Trudy Sem. I. G. Petrovskogo **5** (1979), 163–192.

K. The Poincaré Map

The applications of our constructions to the study of the loss of stability of a cycle are based on the following lemma.

Lemma 1. *Consider a mapping* $f: (\mathbb{R}^2, 0) \to (\mathbb{R}^2, 0)$ *having the fixed point* 0 *with eigenvalues* $e^{\pm 2\pi i p/q}$ *(and with a Jordan block of order 2 if* $q = 1$ *or* $= 2$*). For every N in a sufficiently small neighborhood of* 0, *the iterate* f^q *can be represented in the form of the sum* $f^q = g + h$, *where* $h(z) = O(|z|^N)$ *and* g *is a transformation of the phase flow of a vector field invariant with respect to a finite cyclic group of diffeomorphisms* γ *of order* q.

We note that, in particular, g commutes with rotation by an angle $2\pi/q$. The mapping f^q itself does not, in general, commute with the action of any finite group and cannot be interpolated by a flow. Lemma 1 shows, however, that, at the level of formal series, f^q can be interpolated by a flow and commutes with a finite group.

The proof of the lemma can be carried out along the usual scheme of the construction of Poincaré–Dulac–Birkhoff normal forms (cf., Chap. 5).

Lemma 2. *Consider a deformation* f_λ *of a mapping* $f_0 = f$ *satisfying the hypotheses of Lemma 1. For every N and in a sufficiently small neighborhood of* O, *the iterate* f_λ^q *can be represented as a sum* $f_\lambda^q = g_\lambda + h_\lambda$, *where* $h_\lambda(z) = O(|z|^N)$ *and* g_λ *is a transformation of the phase flow of a vector field* v_λ *invariant under a finite cyclic group* γ_λ *of diffeomorphisms. Here* f_λ, g_λ, h_λ, v_λ, *and* γ_λ *depend smoothly on the parameter* λ *varying in the neighborhood of zero.*

◄ The proof relies on the fact that the reduction to normal form of terms of degree not greater than N with retention of the resonant terms is implemented by diffeomorphisms depending smoothly on the parameters. ►

Combining Lemma 2 with the description of bifurcations of phase flows from the preceding subsections, we obtain information on the loss of stability of the fixed point 0 of the mapping f (or of the periodic motion for which f is the Poincaré map).

Remark. We can directly reduce, to normal form, a family of differential equations in the neighborhood of a (p, q)-resonant periodic solution in the space of the q-sheeted covering. In this situation, the application of the standard Poincaré–Dulac–Birkhoff method reduces a family of vector fields 2π-periodic over time to the sum of a q-symmetric field independent of time and a remainder $O(|z|^N)$ of period $2\pi q$.

L. Discussion

1. For the translation of the results obtained above into the language of bifurcations of periodic solutions, the fixed points in the plane have to be replaced by closed trajectories in space, the separatrices of points by invariant stable and unstable manifolds of these closed trajectories, and the limit cycles in the plane by invariant tori. The situation will be significantly different only for the metamorphoses of separatrices: while, in the plane,

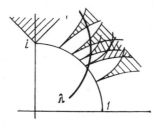

Figure 161.

the separatrices go through each other instantly in bifurcations, in space this process is prolonged with the formation of a homoclinic (or heteroclinic*) picture (Fig. 103).

The invariant tori in space disintegrate before the cycle reaches the loop of separatrices; however, all these purely three-dimensional effects are weak (they are caused by terms of arbitrarily high degree in the normal forms) compared to the two-dimensional ones considered above.

2. The consideration of the loss of stability as a two-parameter rather than a one-parameter phenomenon enables us to easily understand some otherwise surprising phenomena.

Let us consider a two-parameter family in which we take the multiplier as the parameter. In the plane of the parameter, we draw the domains of existence of periodic solutions closing for q rotations along the base solution and making p rotations across it. This domain touches the unit circle at the point $e^{2\pi i p/q}$ in a narrow (for $q > 4$) tongue (its width at a distance σ from the circle on the order of $\sigma^{(q-2)/2}$, cf., § 35F). Therefore, a generic curve in the plane of the multiplier intersects an infinite number of tongues near the unit circle (Fig. 161).

Consequently, in a generic one-parameter family in which a cycle loses stability without strong resonance, an infinite number of cycles with long periods are born and die near the time of the loss of stability.

A proof of this fact not depending on the existence of a weak resonance at the time of the loss of stability has been given by Kozjakin [V. S. Kozjakin, *Subfurcation of periodic oscillations*, Dokl. Akad. Nauk USSR **232**, 1 (1977), 25–27].

3. Considering generic curves on the bifurcation diagrams of strong resonances (§ 35F), we can desribe sequences of metamorphoses which are universal but seem nonlocal from the one-parameter point of view.

For example, in the case $q = 2$, one of the possibilities is the following sequence of events: a stable cycle loses stability softly with the formation

---

*A homoclinic (heteroclinic) picture is, by definition, the net formed in a section plane by the intersecting traces of the stable and unstable invariant manifolds of one (two) closed trajectory (trajectories).

of a torus, which soon becomes pinched along a parallel, so that the form of the meridian of the torus approaches the shape of a figure eight, in approaching the center (which represents an unstable cycle), the attracting set, remaining close to the torus with the meridian almost contracted into the eight, is destroyed near a homoclinic separatrix (Neĭmark).

In this case, the phase trajectory loops around one and then the other side of the collapsed torus, jumping over from one side to the other seemingly at random.

This description is similar to the phenomena observed in a numerical experiment of Hercenšteĭn and Šmidt [S. Ja. Hercenšteĭn, V. M. Šmidt, *Nonlinear evolution and interaction of perturbations of finite amplitude under the convective instability of a rotating planar layer*, Dokl. Akad. Nauk USSR, **225**, 1 (1975), 59–62].

§ 36. Metamorphoses* of the Topology at Resonances

Resonances among eigenvalues of the linear part of a vector field at a stationary point interfere with the choice of coordinates making the field linear. Even if there are no resonances, but the eigenvalues are close to resonant ones, the Poincaré series may diverge and the system cannot be converted into a linear system by an analytic change of the coordinates.

At the same time, the topological type of the phase portrait in the real neigoborhood of a stationary point at resonances does not change generically. For example, if the real parts of all eigenvalues are negative, then the stationary point is attracting and, independently of resonances, the system is topologically equivalent to a standard linear system.

It turns out that a metamorphosis of the topology does take place at resonance, but, in general, in the complex domain.

A generic system is not resonant. We may encounter resonances in an unremovable manner in one-parameter families. Therefore, in the study of the effect of resonances on the metamorphoses of the topology, we have to consider one-parameter families of vector fields. In accordance with what has been said above, we let all phase variables and the time and the parameter be complex.

A. Resonances in a Poincaré Domain

We consider complex phase curves in the neighborhood of the singular point O. These curves constitute a foliation with leaves of two real dimensions, with a singularity at zero. In order to analyze the structure of this singularity, we intersect the foliation by a sphere of small radius and center at the origin.

We assume that, in the coordinates $(z_1, \ldots, z_n)$, the linear part of our system is diagonal: $\dot{z}_j = \lambda_j z_j + \cdots, j = 1, \ldots, n$.

* In Russian: "perestroika".

Theorem. *If a collection* $\{\lambda_j\}$ *of eigenvalues belongs to the Poincaré domain, then every sphere* $|z_1|^2 + \cdots + |z_n|^2 = r^2$ *of sufficiently small radius intersects the foliation transversally.*

◀ First, we consider a linear system. We have

$$dr^2 = \sum z_j \, d\bar{z}_j + \bar{z}_j \, dz_j = A \, dt + \bar{A} \, d\bar{t}, \qquad A = \sum |z_j|^2 \lambda_j.$$

One obtains transversality with the sphere if the 1-form dr^2 does not vanish on the tangent plane of a leaf. On the other hand, the form $A \, dt + \bar{A} \, d\bar{t}$ is a zero form only for $A = 0$. The relation $A = 0$ is not satisfied in the Poincare case (and only in the Poincaré case) for any $z \neq 0$. Hence, in the linear case, the theorem is proved: the leaves intersect the sphere at a nonzero angle $\alpha(z)$.

We consider the minimum α_0 of the angle $\alpha(z)$ on the sphere $|z| = r$. The qunatity α_0 does not depend on r (since $\alpha(cz) = \alpha(z)$). Hence, $\alpha(z) \geqslant \alpha_0 > 0$ for all $z \neq 0$.

Now we turn to the nonlinear system. The angle between the direction fields of the nonlinear system and its linear part is small with $|z|$. Therefore, in a sufficiently small neighborhood of zero, it is smaller than α_0, and the phase curves of the nonlinear system intersect the sphere transversally. ▶

Corollary. *The intersections of complex phase curves with a sphere of sufficiently small radius form a one-dimensional foliation without singular points on this sphere. The foliations obtained on all spheres of sufficiently small radii are diffeomorphic. The differentiable type of the foliation on a sphere does not change under deformations of the sphere, provided that the sphere remains transversal to the complex phase curves.*

Therefore, in the neighborhood of the singular point, the two-dimensional foliation under study is homeomorphic to a cone over a one-dimensional foliation on a sphere. This foliation on a sphere is the partition of the sphere into phase curves of a vector field (since the sphere and the complex foliation are orientable).

Remark. According to Poincaré's theorem, system is linear in the non-resonant case in a sufficiently small neighborhood of the singular point in an appropriate coordinate system. This implies that in the nonresonant case the differentiable type of the foliation on a sphere is the same as that of the linear system.

We conclude that the differentiable type of the foliation on a sphere remains the same as that of the linear system not only in the neighborhood of the origin of coordinates where the Poincaré series are convergent, but also far beyond its limits.

Indeed, in approaching resonance, the domain of convergence of the Poincaré series shrinks to zero, whereas the radius of the domain of transversality remains bounded from below. Consequently, we can study the passage through a resonance of a complex system, investigating the variation of a foliation on a sphere of a fixed (independent of the parameter) small radius.

B. The Resonance $\lambda_1 = 2\lambda_2$

As an example, we consider the topology of the foliation on S^3 in the passage of the resonance $\lambda_1 = 2\lambda_2$ in the system

$$\dot{z}_1 = \lambda_1 z_1 + \cdots, \qquad \dot{z}_2 = \lambda_2 z_2 + \cdots.$$

We find ourselves in the Poincaré domain if the ratio $\lambda = \lambda_1/\lambda_2$ is not a negative real number. First, we consider the foliation on S^3 corresponding to the linear part of the system.

The separatrices $z_1 = 0$, $z_2 = 0$ intersect the sphere in large circles which are cycles of the system on S^3. Their linkage coefficient is equal to 1.

If λ is not a real number (the case of a "focus"), then all remaining curves of the foliation on the sphere wind away from one cycle and wind onto the other. We study the Poincaré map of the cycles.

We note that these maps can be assumed to be holomorphic. Indeed, they are equivalent, with respect to real diffeomorphisms, to complex Poincaré maps mapping a holomorphic transversal to the separatrix into itself. Consequently, they become holomorphic by an appropriate choice of the complex structure on the real two-dimensional transversal to a cycle in S^3. It also follows that the multipliers of our cycles are equal to $e^{\pm 2\pi i \lambda}$ and $e^{\pm 2\pi i \lambda^{-1}}$

The foliations on S^3 corresponding to all foci are homeomorphic but not diffeomorphic to each other: $\lambda^2 + \lambda^{-2}$ is an invariant under diffeomorphisms.

If λ is real and positive (the case of a node), we also find ourselves in the Poincaré domain. In this case, the part of S^3 between two linked cycles is foliated into two-dimensional tori filled out by windings with the rotation number λ the same on all tori.

Now we consider a nonlinear system. In the case of a focus, resonance is impossible. Therefore, in the nonlinear case, the foliation on a sphere is diffeomorphic to the foliation described above, constructed for a linear system. The same is true for a nonresonant node, i.e., for all $\lambda > 0$, excluding the cases where λ or $1/\lambda$ is an integer.

We consider, for example, the resonance $\lambda = 2$. In this case, the Poincaré normal form is

$$\dot{z}_1 = \lambda_1 z_1 + c z_2^2, \qquad \dot{z}_2 = \lambda_2 z_2.$$

This system has only one separatrix for $c \neq 0$, and the foliation on S^3 has only one cycle. We replace λ by a nonreal value close to 2. The resulting system obtained on S^3 is, on the one hand, close to a resonant system; on the other hand, it is diffeomorphic to the system constructed from a linear focus studied earlier and has two cycles with linkage coefficient 1. It can be shown that one of these cycles, C_1, is close to the unique cycle C of the resonant system. The other cycle, C_2, lies on a thin torus with axis C_1 and closes after two trips along C_1, making one revolution on the meridian (so that the linkage coefficient of C_2 and C_1 is equal to 1). Hence, close to the resonance $\lambda = 2$, the metamorphosis of the system on S^3 consists of a bifurcation of a two-fold periodic trajectory from a periodic trajectory with eigenvalues $(-1, -1)$.

Remark. The preceding exposition follows the author's article *Remarks on singularities of finite codimension in complex dynamical systems*, Funct. Anal. Appl. **3**, 1 (1969), 1–6. The results of this paper were later generalized by J. Guckenheimer, N. Kuiper, N. N. Ladis, Ju. S. Il'jasenko, et al. Most complete are the results on the topological type of the foliation given by a linear system in the neighborhood of a singular point in a complex space.

We consider, in particular, the case where the phase space is three-dimensional and the triangle of the eigenvalues contains zero in its interior. It turns out that the topological type of the foliation in a complex space is determined by the triple of the reciprocals of the eigenvalues considered as a triple of vectors in a real plane (i.e., considered up to linear transformations of the decomplexified plane of one complex variable). In the multi-dimensional case, the real type of the collection of the reciprocals of the eigenvalues determines the topological type of the complex foliation in a one-to-one manner in the neighborhood of a singular point of a linear system, if zero belongs to the convex hull of the eigenvalues and their pairwise ratios are not real [cf., C. Camacho, N. Kuiper, J. Palis, C. R. Acad. Sci. Paris, **282**, (1976), 959–961; N. N. Ladis *Topological invariants of complex linear flows*, Differential Equations **12**, 12 (1976), 2159–2169; Ju. S. Il'jasenko, *Remarks on the topology of singular points of differential equations in a complex domain and Ladis' theorem*, Funct. Anal. Appl. **11**, 2 (1977), 28–38].

C. Versal Deformations in the Poincaré Case

We consider an analytic (smooth) vector field with singular point O. We assume that this singular point is of Poincaré type, i.e., that the convex hull of the collection of the eigenvalues does not contain the point 0.

Theorem. *At a singular point of Poincaré type, the germ of an analytic (holomorphic, smooth) vector field has a finite-parameter analytic (holomorphic, smooth) versal deformation consisting of polynomial vector fields.*

In other words:

In the neighborhood of the point O a local family of analytic (holomorphic, smooth) vector fields with singular point O of Poincaré type is analytically (holomorphically, smoothly) equivalent to a family consisting of sufficiently long segments of the Taylor series of these fields at O.

◀ By assumption, the singular point is nondegenerate and, consequently, depends smoothly on the parameter. Therefore, the singular point can be moved to the origin of coordinates by a smooth change of variables depending smoothly on the parameters. We assume that the eigenvalues are simple. Then the eigenvalues can be chosen to depend smoothly on the parameters.

In the resulting coordinate system, the family of differential equations corresponding to our family of fields has the form

$$\dot{x}_k = \lambda_k(\varepsilon)x_k + \cdots, \qquad k = 1, \ldots, n.$$

Applying the Poincaré method (Chap. 5), we shall annihilate only those terms which remain nonresonant for $\varepsilon = 0$. Then the substitutions depend smoothly on the parameter. Since the eigenvalues belong to the Poincaré domain, there are finitely many resonances, and the convergence of the whole procedure can be proved without difficulty.

We obtain a coordinate system in which the right-hand sides of all equations of the family are polynomials. ▶

The case of finitely (or infinitely) differentiable right-hand sides can be studied without difficulty, too. For details, cf., N. N. Brušlinskaja *A finiteness theorem for families of vector fields in the neighborhood of a singular point of Poincaré type*, Funct. Anal. Appl. **5**, 3 (1971), 10–15, where the case of multiple eigenvalues is also considered.

D. Materialization of Resonances

In reducing to normal form in the Siegel domain, there arise difficulties connected with small denominators. At the same time, the topological picture may be simple. For example, an ordinary saddle is built topologically the same way for both rational and irrational ratios of eigenvalues. The same phenomenon takes place in the Poincaré domain: resonances may not affect the topology of the phase portrait.

A natural question arises: why does a resonance without topological effect interfere with the analytical (even finitely smooth) reduction to normal form? To understand this, it is useful to take into account the behavior of resonances in the perturbation theory of quasiperiodic motions.

We consider the following differential equation on the n-dimensional torus T^n:

$$\dot{\theta} = \omega + \varepsilon \cdots, \qquad \theta \bmod 2\pi \in T^n, \qquad \omega \in \mathbb{R}^n, \qquad \varepsilon \ll 1. \qquad (*)$$

The following change in the topological properties of the system corresponds to the resonance $(\omega, k) = 0$ (at least in the absence of perturbation, i.e., for $\varepsilon = 0$): as opposed to the nonresonant case, the phase curves are everywhere dense not on an n-dimensional, but on an $(n - 1)$-dimensional torus. For example, for $n = 2$ at resonance, robust periodic regimes usually arise (stable and unstable limit cycles on a torus). It is clear that the existence of such cycles prevents the reduction of the equations to the normal form $\dot{\theta} = \omega$, typical for the nonresonant case.

A similar argument lies at the basis of Poincaré's proof of the nonexistence of first integrals in the three-body problem.

It can be assumed that in the local problem treated above, the effect of resonances on divergence has an analogous nature, but is connected with changes in the topology of the foliation formed by the phase curves in the complex rather than in the real domain. Such a change, even if it does not manifest itself at all on the real part of the phase space, it necessarily obstructs the analytic reduction and may interfere with the C^r-smooth reduction.

We note that the system $\dot{x}_k = \lambda_k x_k + \cdots$ can be reduced to the form (*) by the substitution $x = e^{i\theta}$ (real ω correspond to purely imaginary λ). The usual methods of investigating the limit cycles of the system (*) lead to the consideration of the first integral $\rho = e^{i(\theta, k)}$ of the unperturbed system; in the notation of the original system, one obtains $\rho = x^k$. An equation of first approximation for the invariant manifold corresponding to resonance can be obtained from the relation

$$\dot{\rho} = \rho[(k, \lambda) + (k, c)\rho + \cdots].$$

We find that formally

$$\rho \approx -\frac{(k, \lambda(\varepsilon))}{(k, c(\varepsilon))}.$$

Our series are divergent in general, and the inference needs a verification.

For $n = 2$, our arguments can be confirmed by a rigorous proof of the existence of a complex limit cycle, which has the indicated asymptotics near resonance [A. S. Pjartli, *Birth of complex invariant manifolds near a singular point of a vector field depending on parameters*, Funct. Anal. Appl. **6**, 4 (1972), 95–96; for details of the proofs, see: *Cycles of a system of two complex differential equations in a singular point neighborhood*, Trudy Mosc. Ob. **37** (1978), 95–106].

At the moment of resonance, when $(k, \lambda) = 0$, a cycle (a complex non-simply connected phase curve) approaches the complex seperatrices of the singular point. The noncontractible path which exists on this cycle disappears at resonance, merging with an equilibrium position. A particular case is the birth (or destruction) of a cycle from an equilibrium position with loss of stability (cf., § 33). In this case, $k = (1, 1, 0, \cdots)$, $\lambda_1 + \lambda_2 = 0$. The phenomenon can be observed in the set of real numbers (cf., Fig. 129). In other cases (even at the same resonance; for example, in the case of a saddle), the topology of real phase curves can persist at resonance.

The difference between the topologies of complex phase curves of an equation (or family) and of its normal form is an obstruction to the analytic reduction to the normal form. Moreover, if this difference is determined by a jet of finite order (as is usually the case), then it prevents not only the analytic but also a finitely smooth reduction to normal form. For example, in the case where the ratio of the eigenvalues can be approximated well by rational numbers, the divergence of the reducing series can be explained by the existence of complex limit cycles originating from nearby resonances of high orders in any neighborhood of a stationary point: the system in normal form does not have such cycles, and, therefore, a transformation to normal form is bound to be divergent.

The study of the question of divergence of Poincaré series is far from complete. The divergence proofs preceding Pjartli's work (Poincaré, Siegel, and Brjuno) are based on a calculation of the growth of coefficients and do not illuminate the causes of divergence in the same sense as the calculation of the coefficients of the series of arctan z proves divergence for $|z| > 1$ but does not show the reason, i.e., singularities at $z = \pm i$.

A. S. Pjartli has established the following results.

1. In the generic case, in the passage of the resonance $k_1\lambda_1 + k_2\lambda_2 = 0$ in $\mathbb{C}^2$ a branching occurs from the separatrices of the singular point of an invariant manifold, whose equation has the form $z_1^{k_1}z_2^{k_2} = \varepsilon$ in first approximation, where ε characterizes deviation from resonance and z_1, z_2 are the phase coordinates.

2. An analogous result for the same resonance is obtained in $\mathbb{C}^n$ under restrictive conditions on the remaining eigenvalues.

3. For "abnormally commensurable" λ_1 and λ_2, an infinite number of invariant manifolds exists in the generic case, corresponding to different resonances in any neighborhood of a singular point, which implies divergenece of the Poincaré series.

The work of Pjartli is based on a method of E. Hopf. Other proofs and generalizations of the first two results of Pjartli have been proposed by Brjuno [A. D. Brjuno, *Normal form of differential equations with a small parameter*, Matem. Zametki **16**, 3 (1974), 407–414; *Analytic invariant manifolds*, Dokl. Akad. Nauk USSR **216**, 2 (1974), 253–256; *Integral analytic sets*, Dokl. Akad. Nauk USSR **220**, 6 (1975), 1255–1258].

E. Resonance among Three Eigenvalues

The resonance next in complexity is

$$k_1\lambda_1 + k_2\lambda_2 + k_3\lambda_3 = 0,$$

where the triangle with vertices $\lambda_1, \lambda_2, \lambda_3$ contains 0. This resonance also has been studied by Pjartli and Brjuno. Here a branching from the separatrices of a singular point of an invariant manifold has been proved for $n = 3$. Let (z_1, z_2, z_3) be the phase coordinates and let ε be the parameter of the deformation (the resonance corresponds to $\varepsilon = 0$). Then in the space with coordinates (z, ε), the invariant manifolds fill a holomophic hypersurface whose equation has the form $z_1^{k_1}z_2^{k_2}z_3^{k} = \varepsilon$ in first approximation in an appropriate coordinate system.

Pjartli has proved that in the generic case for "abnormally commensurable" $(\lambda_1, \lambda_2, \lambda_3)$ forming a Siegel triangle in any neighborhood of the equilibrium position $0 \in \mathbb{C}^3$, an infinite number of invariant manifolds of the described type exists cooresponding to distinct resonances.

From the following it will become clear that the presence of a sufficiently large portion of a resonant invariant manifold in a neighborhood of the nonresonant point $0 \in \mathbb{C}^3$ prevents the convergence of the Poincaré–Siegel series in this neighborhood. It follows from Pjartli's result cited above that in the generic case for "abnormally commensurable" $(\lambda_1, \lambda_2, \lambda_3)$, the indicated series diverge in any neighborhood of the origin.

F. The Case of Discrete Time

We consider a local autodiffeomorphism $A \colon \mathbb{C}^n \to \mathbb{C}^n$ in the neighborhood of the fixed point 0. We denote by $(\lambda_1, \ldots, \lambda_n)$ the eigenvalues of the linearization of A at 0.

The preceding theory can be carried over to this case with the following changes.

Resonances: $\lambda_k = \lambda_1^{m_1} \cdots \lambda_n^{m_n}$ $(m_i \geqslant 0, \sum m_i \geqslant 2)$.
Poincaré domain: All $|\lambda_s| > 1$ or all $|\lambda_s| < 1$.
Siegel domain: There exist $|\lambda_i| \geqslant 1$ and $|\lambda_j| \leqslant 1$.

Remark. To every linear vector field in $\mathbb{C}^n$, there corresponds a linear transformation in $\mathbb{C}^{n-1}$ (the "Poincaré map"). Namely, let the field define the differential equation

$$\dot{z} = \alpha_1 z_1, \ldots, \dot{z}_n = \alpha_n z_n,$$

where $\alpha_n \neq 0$. We consider solutions with initial conditions for $t = 0$ in the plane $z_n = 1$. The value of a solution for $t = 2\pi i/\alpha_n$ belongs to the same plane $\mathbb{C}^{n-1}$. The mapping $A \colon \mathbb{C}^{n-1} \to \mathbb{C}^{n-1}$ thus obtained has the eigenvalues $\lambda_s = e^{2\pi i \alpha_s/\alpha_n}$ $(s = 1, \ldots, n-1)$.

An analogous construction of the Poincaré map exists (under weak additional assumptions) in the nonlinear case, as well. Therefore, results concerning invariant manifolds and bifurcations for mappings imply corresponding results for vector fields.

However, in the majority of cases, it is better to use the indicated connection between equations and mappings as a heuristic means to conjecture results in one area from existing results in the other; it is more convenient to prove the theorems independently in both situations.

G. Bifurcation of Invariant Manifolds of a Diffeomorphism

The resonance among three eigenvalues of a vector field

$$k_1 \alpha_1 + k_2 \alpha_2 + k_3 \alpha_3 = 0$$

corresponds to a resonance of the form

$$\lambda_1^{m_1} \lambda_2^{m_2} = 1.$$

In the Siegel domain, we have $m_1 > 0, m_2 > 0, |\lambda_1| \neq 1 \neq |\lambda_2|$. We assume that λ_1 is larger than 1 in absolute value and λ_2 is smaller than 1 in absolute value.

In this case, the results of Pjartli and Brjuno show the existence of invariant manifolds filling a surface in the space with coordinates (ε, z_1, z_2), the equation of which starts with terms of the form $\varepsilon = z_1^{m_1} z_2^{m_2}$. Here ε is the deviation from resonance and (z_1, z_2) are appropriate phase coordinates (depending smoothly on ε). For fixed $\varepsilon \neq 0$, the invariant manifold thus constructed is homeomorphic to a cylinder. The linkage coefficients of the directrix circle of this cylinder with the coordinate axes in $\mathbb{C}^2$ are equal to m_1 and m_2.

We show that the existence of a sufficiently large portion of such a resonant invariant manifold in a neighborhood of a nonresonant fixed point of a diffeomorphism prevents the linearization of the diffeomorphism in that neighborhood. Therefore, in the case where there are resonant manifolds in any neighborhood of a fixed point, the linearizing series diverge everywhere.

All mappings of $\mathbb{C}^2$ with $|\lambda_1| > 1 > |\lambda_2|$ are equivalent to each other topologically. In particular, all of them can be linearized and have many invariant cylinders. Nevertheless, analytic invariant cylinders are quite rare, as we will see.

H. Local Shifts

We intend to associate an elliptic curve imbedded in a holomorphic surface with a resonant invariant manifold of a mapping $\mathbb{C}^2 \to \mathbb{C}^2$. This surface is the manifold of orbits of our mapping (or the manifold of phase curves of the original differential equation in $\mathbb{C}^3$). In order to define the manifold of orbits, we introduce the following terminology.

Consider the cylinder $S^1 \times \mathbb{R}$. The standard translation of the cylinder is, by definition, the addition of 1 to the second coordinate. Let D_0 be a domain on the cylinder containing $S^1 \times [0, 1]$. The restriction to D_0 of the standard translation defines a diffeomorphism $t \colon D_0 \to D_1 = t(D_0)$. We note that the intersection $D_0 \cap D_1$ contains the circle $S^1 \times 1$. We denote by D the union $D_0 \cup D_1$.

Let M be a two-dimensional manifold, let M_0 and M_1 be its domains, and let $f: M_0 \to M_1$ be a homeomorphism.

Definition. The homeomorphism f is called a *local shift* if there exists a homeomorphism $h: M \to D$ turning M_0 into D_0, M_1 into D_1, and f into t.

Let M be a complex curve and let $f: M_0 \to M_1$ be a holomorphic local shift. Identification of every point $z \in M_0$ with its image $f(z) \in M_1$ determines a compact complex curve homeomorphic to the torus; i.e., the elliptic curve $\Gamma = M/f$.
◀ The proof is obvious. ▶

Now we consider the direct product $\Pi = (S^1 \times \mathbb{R}) \times \mathbb{R}^2$ of the cylinder with the plane. Let $T: \Pi \to \Pi$ be a translation by 1 along $\mathbb{R}$ and let E_0 be the neighborhood of $D_0 \times 0$, $E_1 = TE_0$, $E = E_0 \cup E_1$.

Let N be a smooth real four-dimensional manifold, $M \subset N$ a two-dimensional submanifold, N_1 and N_2 domains in $N = N_1 \cup N_2$ and $F: N_1 \to N_2$ a homeomorphism.

Definition. The homeomorphism F is said to be a *local shift* of N along M if a homeomorphism $H: N \to E$ exists turning N_0 into E_0, N_1 into E_1, F into T, and M into D.

Let N be a complex surface, $M \subset N$ a complex curve, $F: N_0 \to N_1$ a holomorphic local shift along M. Identification of every point $z \in N_0$ with its image $F(z) \in N_1$ determines a holomorphic complex surface $\Sigma = N/F$ which is a neighborhood of the elliptic curve $\Gamma = M/f$.
◀ The proof is obvious. ▶

I. Construction of an Elliptic Curve from a Resonant Invariant Manifold of a Linear Transformation

Let $A: \mathbb{C}^2 \to \mathbb{C}^2$ be a linear transformation with eigenvalues $\lambda_1, \lambda_2, |\lambda_1| > 1 > |\lambda_2|$. We assume that the eigenvalues satisfy the resonance relation $\lambda_1^{m_1} \lambda_2^{m_2} = 1$, where m_1 and m_2 are relatively prime. Then the cylinder with equation $z_1^{m_1} z_2^{m_2} = 1$ (where z_1 and z_2 are coordinates in an eigenbasis) is invariant with respect to A.

The restriction of A to this cylinder gives a holomorphic shift. Indeed, uniformize the curve $\lambda_1^{m_1} \lambda_2^{m_2} = 1$ by a parameter $\Lambda \neq 0$ by the formula $\lambda_1 = \Lambda^{m_2}$, $\lambda_2 = \Lambda^{-m_1}$; on the cylinder, we introduce a parameter $Z \neq 0$ by $z_1 = Z^{m_2}$, $z_2 = Z^{-m_1}$. Then the action of A on the cylinder assumes the form $Z \mapsto \Lambda Z$. This transformation is a holomorphic shift, since $|\Lambda| > 1$. The corresponding elliptic curve is $\mathbb{C}^*/\{\Lambda\} \cong \mathbb{C}/(2\pi\mathbb{Z} + \omega\mathbb{Z})$ where $\Lambda = e^{i\omega}$.

We note that in the case of resonance, a linear transformation has an entire one-parameter family of holomorphic invariant cylinders $z_1^{m_1} z_2^{m_2} = c$, $c \neq 0$. The elliptic curves constructed from these cylinders are all isomorphic.

An appropriate neighborhood of such a cylinder (or a sufficiently large finite portion of it) turns into a neighborhood of an elliptic curve on a complex surface under factorization with respect to the action of A. This surface is the direct product of the elliptic curve with $\mathbb{C}$. Indeed, the homotheties $z \mapsto kz$ determine the projection onto the elliptic curve and the mapping $z \mapsto z_1^{m_1} z_2^{m_2}$ defines the projection onto the second factor.

In particular, the index of self-intersection of the elliptic curve on the surface thus constructed is equal to zero.

J. Construction of an Elliptic Curve from a Resonant Invariant Manifold of a Nonlinear Transformation

Let $A(\varepsilon): U \to \mathbb{C}^2$ be a biholomorphic mapping of the domain $U \subset \mathbb{C}^2$ into $\mathbb{C}^2$ depending holomorphically on the parameter ε. We assume that ε varies in the neighborhood of zero in $\mathbb{C}$ and that all mappings $A(\varepsilon)$ leave the origin in $\mathbb{C}^2$ fixed.

We denote by λ_1, λ_2 the eigenvalues of the linearization of the mapping $A(0)$ at 0. We assume that $|\lambda_1| > 1 > |\lambda_2|$ and $\lambda_1^{m_1} = \lambda_2^{-m_2}$, where m_1 and m_2 are relatively prime.

For a generic family A in the passage through the resonance for $\varepsilon = 0$, an invariant holomorphic resonant cylinder branches off from the separatrices of the fixed point (cf., § 36G). We fix a sufficiently small ε and consider the restriction of $A(\varepsilon)$ to this cylinder. It can be verified that $A(\varepsilon)$ induces a local holomorphic shift on a certain part of the cylinder. [This follows from the facts that (1) in the first approximation, the cyclinder has the equation $z_1^{m_1} z_2^{m_2} = c(\varepsilon)$; (2) $A(\varepsilon)$ is close to the linearization of $A(0)$ at 0; (3) the linearization of $A(0)$ at 0 acts on the cylinder $z_1^{m_1} z_2^{m_2} = c$ as a local shift (cf., § 36I).]

Thus, for sufficiently small $|\varepsilon|$, the mapping $A(\varepsilon)$ determines an elliptic curve $\Gamma(\varepsilon)$ imbedded in a surface $\Sigma(\varepsilon)$. In the homologies of $\Gamma(\varepsilon)$, a distinguished circle exists (the image of the directrix circle of the cylinder). The curve $\Gamma(\varepsilon)$ can be represented in the form

$$\Gamma(\varepsilon) \approx \frac{\mathbb{C}}{2\pi \mathbb{Z} + \omega(\varepsilon)\mathbb{Z}},$$

where 2π corresponds to the distinguished circle. The function $\omega(\varepsilon)$ has the limit ω_0 for $\varepsilon \to 0$. From the formulas of § 36I, it follows that $\lambda_1 = e^{i\omega_0 m_2}$, $\lambda_2 = e^{-i\omega_0 m_1}$.

The index of self-intersection of the curve Γ with the surface Σ is equal to zero. Therefore, Σ is topologically the direct product of Γ with a disk. Nevertheless, Σ is not necessarily a direct product analytically.

Moreover:

(1) Σ may not be a holomorphic fibration over Γ; the neighborhood of Γ in Σ may not admit holomorphic mappings onto Γ identical on Γ. This will happen, for example, in the case where, near Γ, there is a family of elliptic curves with distinct values of the modulus of ω in Σ.

(2) Γ may not admit deformations in Σ different from translations of Γ along itself. This is, for example, the case where the normal bundle of Γ in Σ is analytically nontrivial.

Positive results on the structure of Σ are given in § 27.

K. Nonlinearizability of a Mapping in a Domain Containing a Resonant Cylinder

Theorem. *If the elliptic curve Γ is not deformable in its neighborhood Σ, then the mapping A cannot be linearized by a biholomorphic change of variables in any neighborhood of the point 0 containing a part of the holomorphic invariant cylinder large enough for the construction of the curve Γ.*

◀ Indeed, a holomorphic invariant cylinder of a linear mapping can always be deformed by means of a small homothety. ▶

Consequently, we obtain an upper estimate of the radius of covergence of the Poincaré–Siegel series in terms of a holomorphic invariant cylinder with a nontrivial normal bundle over the corresponding elliptic curve.

Theorem (Ju. S. Il'jašenko). *If a linear mapping has a holomorphic invariant cylinder whose directrix circle has linkage coefficients (m_1, m_2) with the eigenaxes and the corresponding elliptic curve is $\mathbb{C}/(2\pi\mathbb{Z} + \omega\mathbb{Z})$ (where 2π corresponds to the directrix of the cylinder), then the eigenvalues are equal to $\lambda_1 = e^{i\omega m_2}$, $\lambda_2 = e^{-i\omega m_1}$. Consequently, we have the resonance $\lambda_1^{m_1}\lambda_2^{m_2} = 1$.*

◀ Let (z_1, z_2) be the eigencoordinates. The differential forms dz_k/z_k on the cylinder are holomorphic and invariant with respect to the mapping and determine holomorphic forms on the elliptic curve.

We calculate the integrals of these forms on generators of the group of homologies of the torus. One of the generators corresponds to the directrix circle γ on the cylinder. For this, we have

$$\oint \frac{dz_1}{z_1} = 2\pi i m_2, \qquad \oint \frac{dz_2}{z_2} = -2\pi i m_1,$$

since the linkage coefficient of γ with the axis $z_1 = 0$ is equal to m_2 and m_1 with the axis $z_2 = 0$. The second generator corresponds to the segment δ connecting the point z with its image Az on the surface of the cylinder. For this, $\int dz_1/z_1 = \ln \lambda_1$, $\int dz_2/z_2 = \ln \lambda_2$. (These relations determine branches of the logarithm function.) On the other hand, all holomorphic forms on an elliptic curve are identical up to constant factors. Hence

$$\frac{\omega}{2\pi} = \frac{\ln \lambda_1}{2\pi i m_2} = -\frac{\ln \lambda_2}{2\pi i m_1}. \blacktriangleright$$

Corollary. *Let A be a local diffeomorphism with fixed point 0 and eigenvalues $\lambda_{1,2}$. Assume that a holomorphic cylinder exists in some neighborhood U of the fixed point. Let A be a holomorphic shift along this cylinder, and let ω be the period of the corresponding elliptic curve. If λ and ω are not related by the formula $\lambda_1 = e^{i\omega m_2}$, $\lambda_2 = e^{-i\omega m_1}$ (where m_1, m_2 are the linkage coefficients with the separatrices of the singular point), then the diffeomorphism A is not analytically equivalent to a linear diffeomorphism in the domain U.*

L. Divergence of the Poincaré Series

The results obtained so far imply the following theorem.

Theorem. *If in an arbitrarily small neighborhood of the nonresonant fixed point 0 of a local diffeomorphism $\mathbb{C}^2 \to \mathbb{C}^2$ there are holomorphic cylinders on which the diffeomorphism acts as a local shift, then the diffeomorphism is not analytically equivalent to a linear diffeomorphism in any neighborhood of the fixed point 0. (Consequently, the Poincaré series are divergent everywhere.)*

Pjartli established that such an accumulation of invariant cylinders at the singular point is a generic phenomenon for mappings whose eigenvalues "can be approximated abnormally well by resonant ones". Consequently, for such eigenvalues, the divergence (everywhere) of the Poincaré series is a generic phenomenon.

The results of this section can easily be carried over to vector fields in $\mathbb{C}^3$ close to a singular point of Siegel type. The materialization of the resonance $m_1\lambda_1 + m_2\lambda_2 + m_3\lambda_3 = 0$ is an elliptic curve. The points of this curve are the phase curves of the field lying on the invariant resonant surface $z_1^{m_1} z_2^{m_2} z_3^{m_3} = c\varepsilon + \cdots$.

M. Bifurcations of Elliptic Curves on Complex Surfaces

The theory (expounded above) of bifurcations of invariant manifolds of differential equations is analogous to the theory of bifurcations of elliptic curves with index of self-intersection zero on complex surfaces.

An elliptic curve and its normal bundle on a surface are given by a pair of complex numbers (ω, λ) (cf., § 27). They can be obtained from the complex φ-axis and the plane of two complex variables (r, φ) through the pastings

$$(r, \varphi) \sim (r, \varphi + 2\pi) \sim (\lambda r, \varphi + \omega).$$

A bundle is said to be *resonant* if it becomes analytically trivial upon passing to some finitely sheeted cyclic covering.

In the space of pairs (λ, ω), the resonant bundles correspond to the hypersurfaces $\lambda^n = e^{ik\omega}$. It turns out that if one continuously varies the pair (elliptic curve, surface) at the moment of passage through resonance, the elliptic curve is approached by another elliptic curve which covers the former one topologically. Therefore, the materialization of the resonance is a bifurcation of a multiple elliptic curve.

We consider a one-parameter family $\Gamma(\varepsilon) \subset \Sigma(\varepsilon)$ of pairs. We assume that resonance $\lambda^n = e^{ik\omega}$ corresponds to $\varepsilon = 0$. It turns out that the equation of the branching curve has the form $r^n e^{ik\varphi} = \varepsilon$ [after the choice of an appropriate parameter ε of the family and after an appropriate change of coordinates (r, φ) depending on ε; we assume that the resonance corresponds to $\varepsilon = 0$ and that there exist no resonances of smaller order: $\lambda^m e^{il\varphi} \neq 1$ for $0 < m < n$].

Let us deduce the equation of the branching curve by formal series. Arguing as in § 27, we can reduce the pasting to the form

$$\begin{cases} r \\ \varphi \end{cases} \mapsto \begin{cases} r\lambda(1 + \alpha\varepsilon + aw + A), \\ \varphi + \omega + \beta\varepsilon + bw + B, \end{cases}$$

where α, β, a, and b are constants, $w = r^n e^{ik\varphi}$, and A and B are power series in ε and w beginning with terms of degree 2. This substitution transforms w into $w(1 + \gamma\varepsilon + cw + C)$, where $\gamma = n\alpha + ik\beta$, $c = na + ik\beta$; C can be written in terms of second and higher degree in ε and w.

The equation $\gamma\varepsilon + cw + C = 0$ determines a branching curve. For a generic family, we have $\gamma \neq 0$, $c \neq 0$. After an appropriate change of the coordinates ε and r, this equation has the form $r^n e^{ik\varphi} = \varepsilon$.

The convergence can be studied in the same way as in the works of Pjartli and Brjuno cited previously.

Remark. It is easy to verify that the condition $\lambda^n e^{ik\omega} = 1$ means exactly that the normal bundle is analytically trivial over some finitely sheeted cyclic covering of the elliptic curve.

All fibrations considered so far are topologically trivial. In particular, the elliptic curve branching off at resonance is projected (nonholomorphically) onto the curve $\Gamma(\varepsilon)$ "along the r-direction". This projection is a topological finitely sheeted cyclic covering of the torus. This is the same covering over which the normal bundle becomes trivial at the moment of passage through resonance.

If $(n, k) = d > 1$ (but there are no resonances of smaller orders, $\lambda^m e^{il\omega} \neq 1$ for $0 < m < n$), the branching curve is not connected. In this case, it consists of d components, each of which is an (n/d)-sheeted topological covering of the original torus.

N. Divergence of the Linearization

For some nonresonant bundles (i.e., pairs λ, ω), the series reducing the pasting to normal form are divergent.

The branching off of the curves in the case of resonance allows us to "explain" the divergence of the series linearizing the pasting. Let us assume that a pair (λ, ω) is nonresonant but very close to a resonance. Then, in a small neighborhood of the initial elliptic curve, another elliptic curve will generically exist, namely the curve materializing the resonance. If the pair (λ, ω) is sufficiently close to an infinite number of resonances, then in an arbitrarily small neighborhood of the initial elliptic curve, an infinite number of curves exist, materializing distinct resonances and cyclicly covering the initial curve.

The normal bundle of the initial curve is nonresonant. Nonresonant normal bundles of degree 0 do not have sections over any cyclic finitely sheeted covering of an elliptic curve. Therefore, in the normal bundle of the initial elliptic curve, there are no elliptic curves cyclicly covering the initial curve. This means that no neighborhood of the initial curve on the surface can be mapped biholomorphically on a neighborhood of the zeroth section of a normal bundle. Therefore, the series diverge for generic pastings if the pair (λ, ω) can be approximated too well by resonant pairs. This account follows the author's article in Funct. Anal. Appl. **10** (1976), 1–12. Details of the proofs can be found in a sequence of papers by Ju. S. Il'jašenko and A. S. Pjartly *Zero-type neighborhoods of imbedded complex tori*, Trudy Sem. I. G. Petrovskogo **5** (1979), 85–95; **7** (1981), 3–49; **8** (1982), 111–127. Ju. S. Il'jašenko, *Embeddings of elliptic curves of positive type into complex surfaces*, Trudy Moskov. Mat. Obschestva **45** (1982), 37–67.

Remark. There is an analogy between compact complex submanifolds of analytic manifolds and limit cycles of differential equations: similar to the fact that a limit cycle can vanish under a small deformation of the field only if the monodromy operator has the eigenvalue 1, an elliptic curve on a surface having a vanishing index of self-intersection does not disappear under small deformations of the surface if the normal bundle is analytically nontrivial. Bogomolov suggested the following general formulation: a compact submanifold of a complex manifold does not disappear under a small deformation of the entire manifold if the one-dimensional cohomologies of the normal sheaf are trivial. (For the definition of cohomologies, see, for example, R. O. Wells, Differential Calculus on Complex Manifolds, New York, Springer-Verlag, 1980, Ch. 2.)

§ 37. Classification of Singular Points

In this section, we give up the "universal" point of view and consider individual systems (rather than families) of differential equations in the neighborhood of a singular point of the vector field and allow degeneracies of an arbitrarily large codimension. From the generic point of view, the study of such complicated singularities has very limited significance, since complicated degeneracies have a large codimension and are seldom encountered.

However, the knowledge of the general fundamental characteristics of arbitrary singularities is interesting even in those complicated cases which are not accessible to our contemporary methods.

In particular, to know what kind of pathology may occur in the case of a high codimension is useful at least to the extent that we do not waste energy on the search for nonexistent things. It turns out that among such nonexistent objects are, for example, the algebraic criteria of Lyapunov of asymptotic stability, and algebraic criteria in the center-focus problem (for vanishing roots of the characteristic equation).

In order to understand what sort of fundamental questions we want to discuss, we start by considering a very simple example which can be analyzed completely.

A. Singular Points of Functions on the Real Line

Let f be a real-valued function smooth in the neighborhood of the point $x = 0 \in \mathbb{R}$. If the point 0 is not critical, then the function is smoothly equivalent to a linear function in the neighborhood of 0 ($f(x) = x + c$).

What happens in the critical case is also well-known: if $f'(0) = 0$, then the behavior of the function is determined by the sign of $f''(0)$, etc.

To be specific, we consider the problem of conditions for a minimum of the function at 0. The answer can be given in the following way: The space J^k of k-jets of functions at 0 can be divided into three parts,

$$J^k = \text{I} \cup \text{II} \cup \text{III};$$

    I consists of the jets guaranteeing a minimum,
   II consists of the jets guaranteeing the absence of a minimum, and
 III consists of the jets from which it cannot be determined whether there
       is a minimum or not.

The jets of types I or II are said to be *sufficient* and those of type III are said to be *susceptible*.

In our case the sets I, II, and III have the following two properties.

 1. Semialgebraic Property. Each of the sets I, II, and III is a semialgabraic submanifold in the jet space J^k.

A *semialgebraic set* in $\mathbb{R}^N$ is defined as a finite union of subsets, each of which is given by a finite system of polynomial equations and inequalities.

If inequalities are not needed, then the set is said to be algebraic. A useful property of semialgebraic sets is expressed by the following theorem. [For the proof, cf., A. Seidenberg, *A new decision method for elementary algebra*, Ann. Math., Ser. 2 **60** (1954), 356–374; E. A. Gorin, *On asymptotic properties of polynomials and algebraic functions*, Russ. Math. Surveys **16**, 1 (1961), 91–118.]

Tarsky–Seidenberg Principle. *The image of a semialgebraic set under a polynomial mapping is semialgebraic.*

A weaker but equivalent formulation is the following:
The projection of a semialgebraic set onto a subspace is a semialgebraic set.

We note that the projection of an algebraic set may not be algebraic but only semialgebraic (for example, the projection of a sphere onto a plane).

2. *Almost Finite Determinacy.* As $k \to \infty$, the codimension of the set III $\subset J^k$ of susceptible jets tends to infinity.

In other words, the susceptible jets in J^k are determined by a number of conditions increasing with k. As a result, it turns out that the set of functions for which it is undecidable whether 0 is a point of local minimum from any number of terms in the Taylor series is very thin: it has infinite codimension in the function space.

B. Other Examples

The analogous problem for functions of several variables does not admit such a simple algorithm: if the second differential degenerates, then we have to appeal to higher derivatives and we arrive at the problem of the classification of algebraic curves, surfaces, etc. Nevertheless, in this case as well, the decomposition $J^k = $ I $\cup$ II $\cup$ III of the space of k-jets of functions on $\mathbb{R}^n$ are semialgebraic and almost finitely determined, although it is hopeless to explicitly write out the equations and inequalities on the Taylor coefficients for arbitrarily large n and k. The existence of these equations and inequalities can be derived from the Tarsky–Seidenberg theorem, the proof of which also contains an algorithm for obtaining these equations and inequalities (generalized Sturm's theory).

The initial segment of the classification has been calculated explicitly and turns out to be connected (quite mysteriously) with the classification of regular polyhedra, the Coxeter, Weyl, and Lie groups of the series A_k, D_k, and E_k, automorphic functions, triangles in the Lobachevsky plane, singularities of caustics, wave fronts, and oscillating integrals of the method of stationary phase [cf., V. I. Arnold, *Singular points of smooth functions and their normal forms*, Russ. Math. Surveys **30**, (1975), 1–75 and references therein; Vasil'ev, V. A., *The asymptotics of exponential integrals, Newton diagram, and classification of points of minimum*, Funct. Anal. Appl. **11**, 3 (1977), 1–11].

The next example is the problem of the topological classification of germs of smooth mappings. In 1964, Thom announced a theorem on the semi-algebraic character and almost finite determinacy in this case; the proof was given by Varčenko [A. N. Varčenko, *Local topological properties of differentiable mappings*, Izv. Akad. Nauk Ser. Matem. **38**, 5 (1974), 1037–1090; English translation: Math. USSR-Izv., **8** (1974), 1033–1082].

C. Singular Points of Vector Fields

We return to the problem of the topological classification of singular points of vector fields. At first, the problem appears to be as simple as in the case of functions. The nondegenerate singular points can be classified according to the number of eigenvalues in the left half-plane. The space of 1-jets can be divided into a finite number of parts corresponding to the number of roots in the left half-plane. Each of these parts is a semialgebraic set in the space of jets; the polynomial inequalities defining it can even be given explicitly (Routh–Hurwitz condition; cf., for example, F. R. Gantmaher, Theory of Matrices, Moscow, Nauka, 1967).

The susceptible 1-jets form a semialgebraic submanifold of codimension 1, which separates the domains corresponding to distinct numbers of roots in the left half-plane. In the preceding sections, we considered a series of examples of investigations of what happens in these degenerate cases in the passage to 2-jets, etc. Thus, this gives the impression that here, too, we may go arbitrarily far, and only the complexity of calculations and the abundance of cases do not permit us to give an algebraic classification in the cases of an arbitrarily large codimension. It turns out that this is not so [V. I. Arnold, *Algebraic unsolvability of the problem of Lyapunov stability and the problem of topological classification of singular points of an analytic system of differential equations*, Funct. Anal. Appl. **4**, 3 (1970), 1–9].

The semialgebraic property is lost even in such a simple case as in the problem of distinguishing between a center and a focus for vanishing roots of the characteristic equation (Brjuno and Il'jašenko, cf., Ju. S. Il'jašenko, *Algebraic unsolvability and almost algebraic solvability of the center-focus problem*, Funct. Anal. Appl. **6**, 3 (1972), 30–37]. Consequently, for the problems of stability and topological classification no algebraic algorithm can exist*).

There remains some hope for the existence of a nonalgebraic algorithm, nevertheless, i.e., that the property of almost finite determinacy holds: the set of germs whose topological type (or stability) is not determined by any finite segment of the Taylor series may have infinite codimension. The question of whether this is so presents serious difficulties; its formulation has to be made precise, by indicating the exact sense of the word codimension. The sets in the space of k-jets whose codimension has to be defined are not algebraic and set-theoretical difficulties may arise. Thom conjectured that the answer to this question is negative. See F. Takens, *A nonstabilisable jet of a singularity of a vector field*, Dynamical Systems (ed. M. M. Peixoto), Academic Press, New York (1973), 583–597.

We also mention the problem of algorithmical decidability of the stability of a stationary point for a vector field with polynomial components over the ring of integers.

---

* Most recently, L. Hazin and E. Šnol' established the algebraic unsolvability of the problem of stability in the case of two pairs of purely imaginary eigenvalues with resonance 3 : 1. This case corresponds to a submanifold of codimension 3 in the function space. See L. G. Hazin, E. E. Šnol', Simplest cases of algebraic unsolvability in the problems of asymptotic stability, Dokl. Akad Nauk SSR **240** (1978), 1309–1311; Stability conditions at resonance 1 : 3, Prikl. Mat. Mekh. **44** (1980), 229–244. See, also, P. M. Elisarov, Nonalgebraicity of some manifolds of differential equations, Vestn. Mosk. Univ. Mat. Mekh. **2**, (1978), 57–64.

D. Structure of Susceptible Sets

The question of almost determinacy is related to the behavior of sets of susceptible jets in the space J^k of k-jets for k going to infinity. It is easier to investigate questions of the structure of susceptible sets for fixed k. We fix a susceptible $(k-1)$-jet of a vector field at 0 and consider the space J of all k-jets with the given $(k-1)$-jet. To be specific, consider the problem of asymptotic stability. Then the space J can be divided into two (possibly empty) parts: I (stable according to the k-jet) and II (unstable according to the k-jet), and the remainder III of susceptible jets. (In the case of the problem of topological classification, there are more parts.) A reasonable formulation of the problem of a stability criterion consists of establishing what properties the parts I, II, and the boundary between them have. For instance, if the boundary is transcendental, there is no algebraic stability criterion. How complicated may this boundary be? For example, may it (or the open parts of the domains I and II) have an infinite number of connected components? or: can points of the parts I and II alternate similarly to rational and irrational numbers?

Examples of this kind are not known. However, it can be expected that this will happen in particular cases of sufficiently large codimension in a multi-dimensional space.

The local problem of the behavior of phase curves near a singular point in $\mathbb{R}^n$ is closely related to the global problem of differential equations given by a polynomial system in the projective space $\mathbb{R}P^{n-1}$ of dimension smaller by 1. In the above-mentioned work on algebraic unsolvability, this connection was used to derive the transcendental character of the boundary of stability in the space of jets of the local problem from the transcendental character of the surface of birth of limit cycles in the space of coefficients of a polynomial system on the projective plane. However, in a multi-dimensional global situation, much more complicated phenomena are possible than limit cycles, such as systems on the torus with alternating commensurability and incommensurability of the rotation numbers or domains in the function space free from structurally stable systems. All these phenomena occur in polynomial systems in a projective space, and each of them can contribute to the complexity of the boundary of stability in the space J.

Samples of Examination Problems

In the four-hour written examination, 15 interrelated problems are given. Within square brackets, we indicate the point value of each problem. These values are revealed to the students beforehand.

Variant 1

$$\ddot{x} = -\sin x + \varepsilon \cos t. \tag{1}$$

I. Let $\varepsilon = 0$.

(1) Linearize at the point $x = \pi$, $\dot{x} = 0$, [1].
(2) Is this equilibrium position stable [1]?
(3) Find the Jacobian of the mapping of the phase flow at the point $x = \pi$, $\dot{x} = 0$ at time $t = 2\pi$ [3].
(4) Find the derivative of the solution with initial condition $x = \pi$, $\dot{x} = 0$ with respect to the parameter ε at $\varepsilon = 0$ [5].
(5) Draw the graph of the solution and its derivative with respect to t under the initial condition $x = 0$, $\dot{x} = 2$ [3].
(6) Find this solution [3].

II. Let Eq. (2) be the linearized equation along the solution indicated in problem (5).

(7) Does Eq. (2) have unbounded solutions [8]?
(8) Does Eq. (2) have nonzero bounded solutions [8]?
(9) Find the Wronskian of a fundamental system of solutions of Eq. (2), given that $W(0) = 1$ [5].
(10) Write out Eq. (2) explicitly and solve it [10].
(11) Find the eigenvalues and eigenvectors of the monodromy operator for the linearized equation along the solution with initial condition $x = \pi/2$, $\dot{x} = 0$ [16].
(12) Prove that Eq. (1) has a 2π-periodic solution depending smoothly on ε and vanishing at $x = \pi$ for $\varepsilon = 0$ [6].
(13) Find the derivative of this solution with respect to ε at $\varepsilon = 0$ [6].

III. Consider the equation $u_t + uu_x = -\sin x$.

(14) Write out the equation of characteristics [2].
(15) Find the largest value of t for which the solution of the Cauchy problem with $u|_{t=0} = 0$ can be extended to $[0, t)$ [8].

Variant 2

I. Let a vector field in three-dimensions have the origin as its singular point and let one of the eigenvalues of this singular point be equal to zero and the other two purely imaginary.

(1) Reduce to normal form the terms of degree 1 of the Taylor series expansion at zero of the components of the field [1].
(2) Do the same for terms of degree 2 [3].
(3) Do the same for terms of any degree [8].
(4) Average the system with respect to the fast rotation given by the linear part of the field [12].

II. Let us have a family of fields depending on a parameter and containing the field in part I for the zero value of the parameter.

(5) Reduce a beginning segment of the Taylor series at zero of the fields of the family to the simplest possible form by a diffeomorphism depending smoothly on parameters varying in the neighborhood of zero [10].
(6) Average the same system with respect to the fast rotation given by the linear part of the initial field [20].

III. In the space of 1-jets of vector fields in three-dimensions we consider the manifold of jets with one vanishing and two purely imaginary eigenvalues at the singular points.

(7) Find the codimension of the indicated manifold [2].
(8) Write the condition of transversality of the family, given in the form found in problem (5), to the indicated manifold [8].
(9) Analyze the bifurcations of singular points in generic two-parameter families transversal to the indicated manifold [10].
(10) Analyze the bifurcations of cycles from these singular points [15].
(11) Study the existence and smoothness of a phase curve connecting these singular points [15].

IV. Let a straight line going through zero be distinguished in the plane. A diffeomorphism of the plane is said to be *distinguished* if it transforms the distinguished line into itself. A vector field is said to be distinguished if it is tangent to the distinguished line at all of its points. Let a distinguished field be given which has a singular point at zero with two vanishing eigenvalues.

(12) Reduce a segment of the Taylor series of the field at zero to the simplest possible form by means of a distinguished diffeomorphism [12].

(13) Reduce a family of distinguished fields, which is a deformation of the given field, to formal normal form by means of distinguished formal diffeomorphisms depending formally smoothly on the parameters varying in the neighborhood of zero [16].

(14) Analyze the bifurcations of singular points in generic families obtained from the normal forms of problem (13) by omitting terms of high degree [18].

(15) Apply the results of problems (12)–(14) to the study of bifurcations of the phase portrait of a field with one vanishing and two purely imaginary eigenvalues [25].

Additional Problems

(1) Let $\dot{z} = \varepsilon z + Az|z|^2 + \bar{z}^3$. Prove that the number of limit cycles is not greater than 1 if $|\operatorname{Re} A| > 1$. *Hint*: Divide the field by $z\bar{z}$ and use the formula div $P(z, \bar{z}) = 2 \operatorname{Re}(\partial P/\partial z)$.

(2) Let $A = (3 + i)/\sqrt{2}$. Then, for $\arg \varepsilon = 5\pi/4$, the singular separatrix of every saddle-node coincides with the nonsingular separatrix of the next saddle-node. *Hint*: At the moment of saddle-node coalescence, the equation can be reduced to the form $\dot{w} = e^{i\theta} [Rw(|w|^2 - 1) + i(\bar{w}^2 - w^2)]$, where $A = (R - i)e^{i\theta}$. If $R = 2$ and $\theta = \pi/4$, the separatrices are straight lines.

(3) Analyze the curves in the complex A-plane which divide domains where, when $\arg \varepsilon$ varies, singular points coalesce on the cycle, inside the cycle, and outside the cycle. *Hint*: If one varies θ, the field vectors rotate. The curves are located approximately like the four parabolas $a^2 = 2(\pm b \pm 1)$, $A = a + ib$.

(4) For small $|\operatorname{Re} A|$ and $1 < |\operatorname{Im} A| < c \approx 4.11$, the equation of problem (1) has (for appropriate ε) two limit cycles, with nine singular point inside the inner cycle. For $|\operatorname{Im} A| > c$, only one cycle exists (Neĭstadt). The boundary between domains of existence of one or two cycles looks like an ellipse with the major axis $1 \leqslant |\operatorname{Im} A| \leqslant 4.11$ and the minor axis of length 2.

(5) Prove that there are no limit cycles in the generalized system of Lotka–Volterra, $\dot{x} = x (\alpha + ax + by)$, $\dot{y} = y (\beta + cx + dy)$. *Hint*: At stability loss, the system has a first integral: a product of degrees of three linear functions (Bautin).

Index

absolute 133
action variable 165, 169
adiabatic invariant 168
  almost 171
algorithm, Euclidean 112
angular function 104
annihilation 45
Anosov
  diffeomorphism 128
  flow 132, 137
  system 128, 132, 163, 277
Anosov's theorem 124, 269
attractor, hyperbolic, strange 278
automorphism, torus 121
average
  space 100, 147
  time 99, 147
averaged
  equation 146, 149
  field 174
  motion 146
  system 165, 171
averaging in Seifert's foliation 174,
    199, 301

base manifold 233
bifurcation
  diagram 246, 282
  Hopf 274, 279
  of invariant manifold 331
  nondegenerate point 263
  of periodic solutions 265
  of the phase portrait 270, 302, 316
  regular value 263
  of singular points 262, 314
  value of a parameter 263
billiard system 138
bound states 42
bundle
  contangent 84

fiber 203
  negative, nonpositive, zero 214
  nonresonant 336
  normal 47, 50, 206, 211, 213, 334,
    336
  resonant 336
  rigid vector 210
  tangent 229
  vector 205, 210

Cartan replicas theorem 186
Cauchy problem 60, 63, 67, 77, 80, 139
center manifold 269
centralizer 242
characteristic
  direction 74, 76
    field 66
  Euler 205
  plane 73
  point 60
  vector 65
    field 65
Clairaut equation 18
codimension 1 degeneracies 224
codimension k degeneracies 224, 282
coefficient
  periodic 195
  quasiperiodic 201
  reflexion and transmission 38
condition A 158
condition $\bar{A}$ 159
configuration space 84
contact
  form, standard 70, 72
  hyperplane 71
  manifold 73
  1-form 70
  plane 15
  structure 15, 68, 71, 72
continuous fractions 112

Grundlehren der mathematischen Wissenschaften

Continued from page ii